T0212551

Stability of Functional Equations in Banach
Algebras

Yeol Je Cho • Choonkil Park
Themistocles M. Rassias • Reza Saadati

# Stability of Functional Equations in Banach Algebras

 Springer

Yeol Je Cho
Department of Mathematics Education
and the RINS
Gyeongsang National University
College of Education
Jinju, Korea, Republic of South Korea

Themistocles M. Rassias
Department of Mathematics
National Technical University of Athens
Athens, Greece

Choonkil Park
Department of Mathematics
Hanyang University
Seoul, Korea, Republic of South Korea

Reza Saadati
Department of Mathematics
Iran University of Science and Technology
Tehran, Iran

ISBN 978-3-319-38464-1          ISBN 978-3-319-18708-2    (eBook)
DOI 10.1007/978-3-319-18708-2

Mathematics Subject Classification (2010): 39-XX, 26-XX, 41-XX, 46-XX, 47-XX

Springer Cham Heidelberg New York Dordrecht London
© Springer International Publishing Switzerland 2015
Softcover reprint of the hardcover 1st edition 2015

Printed on acid-free paper

Springer International Publishing AG Switzerland is part of Springer Science+Business Media (www.
springer.com)

*To S.M. Ulam for the 75th anniversary of his stability problem for approximate homomorphisms*

# Preface

The main purpose of this book is to present some of the old and recent results on homomorphisms and derivations in Banach algebras, quasi-Banach algebras, $C^*$-algebras, $C^*$-ternary algebras, non-Archimedean Banach algebras, and multi-normed algebras.

In 1940, S. M. Ulam [321] proposed a stability problem on group homomorphisms in metric groups. In 1941, D. H. Hyers [133] proved the stability of additive mappings in Banach spaces associated with the Cauchy equation. In 1978, Th. M. Rassias [267] proved the stability of $\mathbb{R}$-linear mappings associated with the Cauchy equation, and in 2002 C. Park [220] proved the stability of $\mathbb{C}$-linear mappings in the spirit of Hyers, Ulam, and Rassias in Banach modules. Homomorphisms and derivations in Banach algebras, quasi-Banach algebras, $C^*$-algebras, $C^*$-ternary algebras, non-Archimedean Banach algebras and multi-normed algebras are additive and $\mathbb{R}$-linear or $\mathbb{C}$-linear, and so we study the stability problems for additive functional equations and additive mappings. Using the direct method and the fixed point method, the authors have studied the stability and the superstability of homomorphisms and derivations in Banach algebras, quasi-Banach algebras, $C^*$-algebras, $C^*$-ternary algebras, non-Archimedean Banach algebras, and multi-normed algebras, which are also associated with additive functional equations and additive functional inequalities.

The book provides a survey of both the latest and new results especially on the following topics:

(1) Stability theory for several new functional equations in Banach algebras and $C^*$-algebras via fixed point method and direct method.
(2) Stability theory for several new functional inequalities in Banach algebras and $C^*$-algebras via fixed point method and direct method.
(3) Stability theory of well-known new functional equations in non-Archimedean Banach algebras and non-Archimedean $C^*$-algebras.
(4) Stability theory for several new functional equations and functional inequalities in multi-Banach algebras and multi-$C^*$-algebras via fixed point method and direct method.

The book is intended to be accessible especially to graduate students who have a basic background with operator theory, functional analysis, functional equations, and analytic inequalities including an introduction to Banach algebras, quasi-Banach algebras, $C^*$-algebras, $C^*$-ternary algebras, non-Archimedean Banach algebras, and multi-normed algebras.

In Chap. 1, we provide a brief introduction to concepts with historic remarks for functional equations and their stability and the definitions of Banach algebras, quasi-Banach algebras, $C^*$-algebras, $C^*$-ternary algebras, non-Archimedean Banach algebras, and multi-normed algebras.

In Chap. 2, we study the stability of additive functional equations in Banach spaces as well as the stability and the superstability of isomorphisms, homomorphisms, derivations, and generalized derivations in Banach algebras and quasi-Banach algebras associated with additive functional equations.

In Chap. 3, we study the stability and the superstability of isomorphisms, homomorphisms, and derivations in $C^*$-algebras, Lie $C^*$-algebras, and $JC^*$-algebras, as well as the stability and the superstability of linear mappings in Banach modules over unital $C^*$-algebras. Moreover, we study Jordan $*$-derivations, quadratic Jordan $*$-derivations, $(\alpha, \beta, \gamma)$-derivations on Lie $C^*$-algebras, square root functional equations, 3rd root functional equations, and positive-additive functional equations.

In Chap. 4, we study the stability of $\mathbb{C}$-linear mappings in Banach spaces and linear mappings in normed modules over a $C^*$-algebra as well as the stability of homomorphisms and derivations in proper $CQ^*$-algebras associated with functional inequalities.

In Chap. 5, we study the stability and the superstability of $C^*$-ternary homomorphisms, $C^*$-ternary derivations, $C^*$-ternary 3-homomorphisms, and $C^*$-ternary 3-derivations in $C^*$-ternary algebras as well as investigate the stability of $JB^*$-triple homomorphisms and $JB^*$-triple derivations in $JB^*$-triples by using the direct method and the fixed point method.

In Chap. 6, we study the stability of linear mappings in multi-Banach spaces as well as the stability and the superstability of isomorphisms, homomorphisms, and derivations in multi-Banach algebras, multi-$C^*$-algebras, proper multi-$CQ^*$-algebras, and multi-$C^*$-ternary algebras. Moreover, we study the stability of ternary Jordan homomorphisms and ternary Jordan derivations in multi-$C^*$-ternary algebras.

Finally, in Chap. 7, we study the stability of additive functional equations in non-Archimedean Banach spaces as well as the stability of homomorphisms and derivations in non-Archimedean $C^*$-algebras and non-Archimedean Lie $C^*$-algebras.

Jinju, South Korea                                                                      Yeol Je Cho
Seoul, South Korea                                                                   Choonkil Park
Athens, Greece                                                            Themistocles M. Rassias
Tehran, Iran                                                                          Reza Saadati
March 2015

# Acknowledgments

We would like to express our thanks to the referees for reading the manuscript and providing valuable suggestions and comments which have helped to improve the presentation of the book.

This work was supported by the Basic Science Research Program through the National Research Foundation of Korea funded by the Ministry of Education, Science and Technology.

Last but not least, it is our pleasure to acknowledge the superb assistance provided by the staff of Springer for the publication of the book.

Yeol Je Cho
Choonkil Park
Themistocles M. Rassias
Reza Saadati

# Contents

# Chapter 1
# Introduction

In this chapter, we recall some definitions and results which will be used later on in the book.

The study of functional equations has a long history. In 1791 and 1809, Legendre [184] and Gauss [121] attempted to provide a solution of the following functional equation:

$$f(x+y) = f(x) + f(y)$$

for all $x, y \in \mathbb{R}$, which is called the *Cauchy functional equation*. A function $f : \mathbb{R} \to \mathbb{R}$ is called an *additive function* if it satisfies the Cauchy functional equation. In 1821, Cauchy [67] first found the general solution of the Cauchy functional equation, that is, if $f : \mathbb{R} \to \mathbb{R}$ is a continuous additive function, then $f$ is $\mathbb{R}$–linear, that is, $f(x) = mx$, where $m$ is a constant. Further, we can consider the biadditive function on $\mathbb{R} \times \mathbb{R}$ as follows:

A function $f : \mathbb{R} \times \mathbb{R} \to \mathbb{R}$ is called a *biadditive function* if it is additive in each variable, that is,

$$f(x+y, z) = f(x, z) + f(y, z)$$

and

$$f(x, y+z) = f(x, y) + f(x, z)$$

for all $x, y, z \in \mathbb{R}$. It is well–known that every continuous biadditive function $f : \mathbb{R} \times \mathbb{R} \to \mathbb{R}$ is of the form

$$f(x, y) = mxy$$

for all $x, y \in \mathbb{R}$, where $m$ is a constant.

© Springer International Publishing Switzerland 2015
Y.J. Cho et al., *Stability of Functional Equations in Banach Algebras*, DOI 10.1007/978-3-319-18708-2_1

Since the time of Legendre and Gauss, several mathematicians had dealt with additive functional equations in their books [4–6, 178, 313] and a number of them have studied Lagrange's mean value theorem and related functional equations, Pompeiu's mean value theorem and associated functional equations, two-dimensional mean value theorem and functional equations as well as several kinds of functional equations. We know that the mean value theorems have been motivated to study the functional equations (see the book "Mean Value Theorems and Functional Equations" by Sahoo and Riedel [305]) in 1998.

In 1940, Ulam [321] proposed the following stability problem of functional equations:

*Given a group $G_1$, a metric group $G_2$ with the metric $d(\cdot, \cdot)$ and a positive number $\varepsilon$, does there exist $\delta > 0$ such that, if a mapping $f : G_1 \to G_2$ satisfies*

$$d(f(xy), f(x)f(y)) \leq \delta$$

*for all $x, y \in G_1$, then there exists a homomorphism $h : G_1 \to G_2$ such that*

$$d(f(x), h(x)) \leq \varepsilon$$

*for all $x \in G_1$?*

Since then, several mathematicians have dealt with special cases as well as generalizations of Ulam's problem.

In fact, in 1941, Hyers [133] provided a partial solution to Ulam's problem for the case of approximately additive mappings in which $G_1$ and $G_2$ are Banach spaces with $\delta = \varepsilon$ as follows:

*Let $X$ and $Y$ be Banach spaces and let $\varepsilon > 0$. Then, for all $g : X \to Y$ with*

$$\sup_{x,y \in X} \|g(x + y) - g(x) - g(y)\| \leq \varepsilon,$$

*there exists a unique mapping $f : X \to Y$ such that*

$$\sup_{x \in X} \|g(x) - f(x)\| \leq \varepsilon$$

*and*

$$f(x + y) = f(x) + f(y)$$

*for all $x, y \in X$.*

This proof remains unchanged if $G_1$ is an Abelian semigroup. Particularly, in 1968, it was proved by Forti ([115], Proposition 1) that the following theorem can be proved:

**Theorem F (Forti).** *Let $(S, +)$ be an arbitrary semigroup and $E$ be a Banach space. Assume that $f : S \to E$ satisfies*

$$\|f(x + y) - f(x) - f(y)\| \leq \varepsilon. \tag{A}$$

*Then the limit*

$$g(x) = \lim_{n \to \infty} \frac{f(2^n x)}{2^n} \tag{B}$$

*exists for all $x \in S$ and $g : S \to E$ is the unique function satisfying*

$$\|f(x) - g(x)\| \leq \varepsilon, \quad g(2x) = 2g(x).$$

*Finally, if the semigroup $S$ is Abelian, then $G$ is additive.*

Here, the proof method which generates the solution $g$ by the formula like $(B)$ is called the *direct method*.

If $f$ is a mapping of a group or a semigroup $(S, \cdot)$ into a vector space $E$, then we call the following expression:

$$Cf(x, y) = f(x \cdot y) - f(x) - f(y)$$

the *Cauchy difference* of $f$ on $S \times S$. In the case that $E$ is a topological vector space, we call the equation of homomorphism *stable* if, whenever the Cauchy difference $Cf$ is bounded on $S \times S$, there exists a homomorphism $g : S \to E$ such that $f - g$ is bounded on $S$.

In 1980, Rätz [298] generalized Theorem F as follows: Let $(X, *)$ be a power-associative groupoid, i.e., $X$ is a nonempty set with a binary relation $x_1 * x_2 \in X$ such that the left powers satisfy $x^{m+n} = x^m * x^n$ for all $m, n \geq 1$ and $x \in X$. Let $(Y, |\cdot|)$ be a topological vector space over the field $\mathbb{Q}$ of rational numbers with $\mathbb{Q}$ topologized by its usual absolute value $|\cdot|$.

**Theorem R (Rätz).** *Let $V$ be a nonempty bounded $\mathbb{Q}$-convex subset of $Y$ containing the origin and assume that $Y$ is sequentially complete. Let $f : X \to Y$ satisfy the following conditions: for all $x_1, x_2 \in X$, there exist $k \geq 2$ such that*

$$f((x_1 * x_2)^{k^n}) = f(x_1^{k^n} * x_2^{k^n}) \tag{C}$$

*for all $n \geq 1$ and*

$$f(x_1) + f(x_2) - f(x_1 * x_2) \in V. \tag{D}$$

*Then there exists a function $g : X \to Y$ such that $g(x_1) + g(x_2) = g(x_1 * x_2)$ and $f(x) - g(x) \in \overline{V}$, where $\overline{V}$ is the sequential closure of $V$ for all $x \in X$. When $Y$ is a Hausdorff space, then $g$ is uniquely determined.*

Note that the condition $(C)$ is satisfied when $X$ is commutative and it takes the place of the commutativity in proving the additivity of $g$. However, as Rätz pointed out in his paper, the condition

$$(x_1 * x_2)^{k^n} = x_1^{k^n} * x_2^{k^n}$$

for all $x_1, x_2 \in X$, where $X$ is a semigroup, and, for all $k \geq 1$, does not imply the commutativity.

In the proofs of Theorems F and R, the completeness of the image space $E$ and the sequential completeness of $Y$, respectively, were essential in proving the existence of the limit which defined the additive function $g$. The question arises whether the completeness is necessary for the existence of an odd additive function $g$ such that $f - g$ is uniformly bounded, given that the Cauchy difference is bounded.

For this problem, in 1988, Schwaiger [306] proved the following:

**Theorem S (Schwaiger).** *Let $E$ be a normed space with the property that, for each function $f : \mathbb{Z} \to E$, whose Cauchy difference $Cf = f(x+y) - f(x) - f(y)$ is bounded for all $x, y \in \mathbb{Z}$ and there exists an additive mapping $g : \mathbb{Z} \to E$ such that $f(x) - g(x)$ is bounded for all $x \in \mathbb{Z}$. Then $E$ is complete.*

**Corollary 1.** *The statement of Theorem S remains true if $\mathbb{Z}$ is replaced by any vector space over $\mathbb{Q}$.*

In 1950, Aoki [17] generalized Hyers' theorem as follows:

**Theorem A (Aoki).** *Let $E_1$ and $E_2$ be two Banach spaces. If there exist $K > 0$ and $0 \leq p < 1$ such that*

$$\|f(x + y) - f(x) - f(y)\| \leq K(\|x\|^p + \|y\|^p)$$

*for all $x, y \in E_1$, then there exists a unique additive mapping $g : E_1 \to E_2$ such that*

$$\|f(x) - g(x)\| \leq \frac{2K}{2 - 2^p} \|x\|^p$$

*for all $x \in E_1$.*

In 1978, Th. M. Rassias [267] formulated and proved the stability theorem for the linear mapping between Banach spaces $E_1$ and $E_2$ subject to the continuity of $f(tx)$ with respect to $t \in \mathbb{R}$ for each fixed $x \in E_1$. Thus Rassias' Theorem implies Aoki's Theorem as a special case. Later, in 1990, Th. M. Rassias [274] observed that the proof of his stability theorem also holds true for $p < 0$. In 1991, Gajda [119] showed that the proof of Rassias' Theorem can be proved also for the case $p > 1$ by just replacing $n$ by $-n$ in $(B)$. These results are stated in a generalized form as follows (see Rassias and Šemrl [293]):

**Theorem RS (Th.M. Rassias and P. Semrl).** *Let $\beta(s,t)$ be nonnegative function for all nonnegative real numbers $s,t$ and positive homogeneous of degree $p$, where $p$ is real and $p \neq 1$, i.e., $\beta(\lambda s, \lambda t) = \lambda^p \beta(s,t)$ for all nonnegative $\lambda, s, t$. Given a normed space $E_1$ and a Banach space $E_2$, assume that $f : E_1 \to E_2$ satisfies the inequality*

$$\|f(x+y) - f(x) - f(y)\| \leq \beta(\|x\|, \|y\|)$$

*for all $x, y \in E_1$. Then there exists a unique additive mapping $g : E_1 \to E_2$ such that*

$$\|f(x) - g(x)\| \leq \delta \|x\|^p$$

*for all $x \in E_1$, where*

$$\delta := \begin{cases} \frac{\beta(1,1)}{2-2^p}, & p < 1, \\ \frac{\beta(1,1)}{2-2^p}, & p > 1. \end{cases}$$

The proofs for the cases $p < 1$ and $p > 1$ were provided by applying the direct methods. For $p < 1$, the additive mapping $g$ is given by $(B)$, while, in case $p > 1$, the formula is

$$g(x) = \lim_{n \to \infty} 2^n f\left(\frac{x}{2^n}\right).$$

**Corollary 2.** *Let $f : E_1 \to E_2$ be a mapping satisfying the hypotheses of Theorem RS and suppose that $f$ is continuous at a single point $y \in E_1$. Then the additive mapping $g$ is continuous.*

**Corollary 3.** *If, under the hypotheses of Theorem RS, we assume that, for each fixed $x \in E_1$, the mapping $t \to f(tx)$ from $\mathbb{R}$ to $E_2$ is continuous, then the additive mapping $g$ is $\mathbb{R}$–linear.*

*Remark 4.* (1) For $p = 0$, Theorem RS, Corollaries 2 and 3 reduce to the results of Hyers in 1941. If we put $\beta(s,t) = \varepsilon(s^p + t^p)$, then we obtain the results of Rassias [267] in 1978 and Gajda [119] in 1991.

(2) The case $p = 1$ was excluded in Theorem RS. Simple counterexamples prove that one can not extend Rassias' Theorem when $p$ takes the value one (see Z. Gajda [119], Rassias and Šemrl [293] and Hyers and Rassias [135] in 1992).

A further generalization of the Hyers-Ulam stability for a large class of mappings was obtained by Isac and Rassias [139] by introducing the following:

**Definition 5.** A mapping $f : E_1 \to E_2$ is said to be $\phi$-*additive* if there exist $\Phi \geq 0$ and a function $\phi : \mathbb{R}_+ \to \mathbb{R}_+$ satisfying

$$\lim_{t \to +\infty} \frac{\phi(t)}{t} = 0$$

such that

$$\|f(x+y) - f(x) - f(y)\| \le \Phi[\phi(\|x\|) + \phi(\|y\|)]$$

for all $x, y \in E_1$.

In [139], Isac and Rassias proved the following:

**Theorem IR (Isac and Rassias).** *Let $E_1$ be a real normed vector space and $E_2$ be a real Banach space. Let $f : E_1 \to E_2$ be a mapping such that $f(tx)$ is continuous in $t$ for each fixed $x \in E_1$. If $f$ is $\phi$-additive and $\phi$ satisfies the following conditions:*

(1) $\phi(ts) \le \phi(t)\phi(s)$ *for all $s, t \in \mathbb{R}$;*
(2) $\phi(t) < t$ *for all $t > 1$,*
    *then there exists a unique $\mathbb{R}$–linear mapping $T : E_1 \to E_2$ such that*

$$\|f(x) - T(x)\| \le \frac{2\theta}{2 - \phi(2)} \phi(\|x\|)$$

*for all $x \in E_1$.*

*Remark 6.* (1) If $\phi(t) = t^p$ with $p < 1$, then, from Theorem IR, we obtain Rassias' Theorem [267].
(2) If $p < 0$ and $\phi(t) = t^p$ with $t > 0$, then Theorem IR is implied by the result of Gajda in 1991.

Since the time the above stated results have been proven, several mathematicians (cf. [1, 3, 14, 44, 46, 49–65, 69–80, 82–85, 87–90, 95–99, 101–107, 109, 116–118, 120, 122, 124, 125, 129–134, 136, 137, 140–149, 151–153, 156–158, 160–162, 164, 173–179, 187, 189, 190, 195–201, 207, 208, 212, 214, 217, 219, 221–223, 226, 228, 230, 236, 239–241, 262, 266, 268, 269, 275–288, 296–303, 309, 311–322, 324, 330, 331] and also very recent survey papers [42, 60, 61]) have extensively studied stability theorems for several kinds of functional equations in various spaces, for example, Banach spaces, 2-Banach spaces, Banach $n$-Lie algebras, quasi-Banach spaces, Banach ternary algebras, non-Archimedean normed and Banach spaces, metric and ultra metric spaces, Menger probabilistic normed spaces, probabilistic normed space, $p$-2-normed spaces, $C^*$-algebras, $C^*$-ternary algebras, Banach ternary algebras, Banach modules, inner product spaces, Heisenberg groups and others. Further, we have to pay attention to applications of the Hyers-Ulam-Rassias stability problems, for example, (partial) differential equations, Fréchet functional equations, Riccati differential equations, Volterra integral equations, group and ring theory and some kinds of equations (see [66, 142, 150, 154, 159, 176, 185, 186, 192–194, 259–261, 327, 329]). For more details on recent development in Ulam's type stability and its applications, see the papers of Brillouët-Belluot et al. [53] and Ciepliński [86] in 2012 (see also [3–22, 47, 78, 79, 81, 110–112, 138, 148–150, 154, 157–163, 165, 166, 183, 204, 205, 213, 215, 216, 245, 246, 249, 255, 257, 263, 264, 270–295, 302, 304, 328]).

A functional equation is called *stable* if any function satisfying the functional equation "approximately" is near to a true solution of the functional equation. We say that a functional equation is *superstable* if every approximate solution is an exact solution of it (see some recent papers [40, 41, 55, 59]).

## 1.1 Fixed Point Theorems

In this section, we present some fixed point theorems which will play important roles in proving our main theorems. All stability results for functional equations were proved by applying direct method. Since the direct method sometimes does not work. In consequence, the fixed point method for studying the stability of functional equations was used for the first time by Baker in 1991 (see [43]). Next, in 2003, Radu [265] gave a lecture at Seminar on Fixed Point Theory Cluj-Napoca and proved a stability of functional equation by fixed method. Then, in 2003, Cădariu and Radu [62, 64] considered Jensen functional equation and proved a stability result via fixed point method. Jung and Chang [155] proved the stability of a cubic type functional equation with the fixed point alternative. Since then, some authors [151–153, 156, 157, 162, 164, 191, 211, 234, 251, 256] considered some important functional equations and proved the stability results via fixed point method introduced by Baker and Radu.

The Banach fixed point theorem [45] (also known as the Banach contraction principle) is an important tool in the theory of metric spaces because it guarantees the existence and uniqueness of fixed points of certain self mappings of metric spaces and provides a constructive method to find those fixed points. The theorem is named after Banach (1892–1945) and was first stated by him in 1922.

**Theorem 1.1 (Banach [45]).** *Let $(X, d)$ be a complete metric space and $T : X \longrightarrow X$ be a contraction, i.e., there exists $\alpha \in [0, 1)$ such that*

$$d(Tx, Ty) \leq \alpha d(x, y)$$

*for all $x, y \in X$. Then there exists a unique $a \in X$ such that $Ta = a$. Moreover, for all $x \in X$,*

$$\lim_{n \to \infty} T^n x = a$$

*and, in fact, for all $x \in X$,*

$$d(x, a) \leq \frac{1}{1 - \alpha} d(x, Tx).$$

**Theorem 1.2 ([62, 265]).** *Let $(X, d)$ be a complete metric space and $J : X \to X$ be a strictly contractive mapping, i.e., there exists a Lipschitz constant $L < 1$ such that*

$$d(Jx, Jy) \leq L d(x, y)$$

*for all $x, y \in X$. Then we have*

(1) *The mapping $J$ has a unique fixed point $x^* \in X$;*
(2) *The fixed point $x^*$ is globally attractive, i.e.,*

$$\lim_{n \to \infty} J^n x = x^*$$

*for all $x \in X$;*
(3) *The following inequalities hold:*

$$d(J^n x, x^*) \leq L^n d(x, x^*),$$

$$d(J^n x, x^*) \leq \frac{1}{1 - L} d(J^n x, J^{n+1} x),$$

$$d(x, x^*) \leq \frac{1}{1 - L} d(x, Jx)$$

*for all $x \in X$ and $n \geq 1$.*

Following Luxemburg [188], the concept of a *generalized complete metric space* may be introduced as in this quotation:

Let $X$ be a nonempty set. A function $d : X \times X \to [0, \infty]$ is called a *generalized metric* on $X$ if, for any $x, y, z \in X$,

(1) $d(x, y) = 0$ if and only if $x = y$;
(2) $d(x, y) = d(y, x)$;
(3) $d(x, z) \leq d(x, y) + d(y, z)$.

This concept differs from the usual concept of a complete metric space by the fact that not every two points in $X$ have necessarily a finite distance. One might call such a space a generalized complete metric space.

Next, Diaz and Margolis [95] proved a *theorem of the alternative* for any contraction mapping $T$ on a generalized complete metric space $X$. The conclusion of the theorem, speaking in general terms, asserts that: either all consecutive pairs of the sequence of successive approximations (starting from an element $x_0$ of $X$) are infinitely far apart or the sequence of successive approximations, with initial element $x_0$ converges to a fixed point of $T$ (what particular fixed point depends, in general, on the initial element $x_0$).

**Theorem 1.3 ([62, 95]).** *Let $(X, d)$ be a complete generalized metric space and $J : X \to X$ be a strictly contractive mapping with a Lipschitz constant $L < 1$. Then, for each $x \in X$, either*

$$d(J^n x, J^{n+1} x) = \infty$$

*for all $n \geq 0$ or there exists a positive integer $n_0$ such that*

(1) $d(J^n x, J^{n+1} x) < \infty$ *for all $n \geq n_0$;*
(2) *The sequence $(J^n x)$ converges to a fixed point $y^*$ of $J$;*
(3) $y^*$ *is the unique fixed point of $J$ in the set $Y = \{y \in X : d(J^{n_0} x, y) < \infty\}$;*
(4) $d(y, y^*) \leq \frac{1}{1-L} d(y, Jy)$ *for all $y \in Y$.*

## 1.2 Quasi-Banach Algebras

Let $X$ be a vector space on field $\mathbb{C}$. A normed space $X$ in which, for all $x, y \in X$, $xy \in X$ and $\|xy\| \leq \|x\| \|y\|$ is called a *complex normed algebra*. A complete normed algebra is called a *Banach algebra*. Moreover, if there exists a unit element $e$ such that $ex = xe = x$ for all $x \in X$, then $\|e\| = 1$ and $X$ is called a *unital Banach algebra*.

Let $X, Y$ be Banach algebras. A $\mathbb{C}$-linear mapping $H : X \to Y$ is called a *homomorphism* in Banach algebras if $H$ satisfies

$$H(xy) = H(x)H(y)$$

for all $x, y \in X$. A $\mathbb{C}$-linear mapping $\delta : X \to X$ is called a *derivation* on $X$ if $\delta$ satisfies

$$\delta(xy) = \delta(x)y + x\delta(y)$$

for all $x, y \in X$.

We recall some basic facts concerning quasi-Banach spaces and some preliminary results.

**Definition 1.4 ([48, 300]).** Let $X$ be a real linear space. A *quasi-norm* is a real-valued function on $X$ satisfying the following:

(1) $\|x\| \geq 0$ for all $x \in X$ and $\|x\| = 0$ if and only if $x = 0$;
(2) $\|\lambda x\| = |\lambda| \cdot \|x\|$ for all $\lambda \in \mathbb{R}$ and $x \in X$;
(3) There is a constant $K \geq 1$ such that $\|x + y\| \leq K(\|x\| + \|y\|)$ for all $x, y \in X$.

The pair $(X, \| \cdot \|)$ is called a *quasi-normed space* if $\| \cdot \|$ is a quasi-norm on $X$. The smallest possible $K$ is called the *modulus of concavity* of $\| \cdot \|$. Obviously, the balls with respect to $\| \cdot \|$ define a linear topology on $X$. By a *quasi-Banach space* we mean a complete quasi-normed space, i.e., a quasi-normed space in which every $\| \cdot \|$-Cauchy sequence in $X$ converges. This class includes Banach spaces and the most significant class of quasi-Banach spaces which are not Banach spaces are the $L_p$ spaces for $0 < p < 1$ with the quasi-norm $\| \cdot \|_p$.

A quasi-norm $\| \cdot \|$ is called a *p-norm* $(0 < p \leq 1)$ if

$$\|x + y\|^p \leq \|x\|^p + \|y\|^p$$

for all $x, y \in X$. In this case, a quasi-Banach space is called a *p-Banach space*.

For any *p*-norm, the formula $d(x, y) := \|x - y\|^p$ gives us a translation invariant metric on $X$. By the *Aoki–Rolewicz theorem* [300] (see also [48]), each quasi-norm is equivalent to some *p*-norm. Since it is much easier to work with *p*-norms than quasi-norms, henceforth we restrict our attention mainly to *p*-norms.

**Definition 1.5 ([10]).** Let $(A, \| \cdot \|)$ be a quasi-normed space. The quasi-normed space $(A, \| \cdot \|)$ is called a *quasi-normed algebra* if $A$ is an algebra and there exists a constant $C > 0$ such that

$$\|xy\| \leq C\|x\| \cdot \|y\|$$

for all $x, y \in A$. A *quasi-Banach algebra* is a complete quasi-normed algebra.

If the quasi-norm $\| \cdot \|$ is a *p-norm*, then the quasi-Banach algebra is called a *p-Banach algebra*.

## 1.3  *C\**-Algebras

Let $U$ be a Banach algebra. Then an *involution* on $U$ is a mapping $u \rightarrow u^*$ from $U$ into $U$ which satisfies the following conditions:

(1)  $u^{**} = u$ for all $u \in U$;
(2)  $(\alpha u + \beta v)^* = \overline{\alpha}u^* + \overline{\beta}v^*$;
(3)  $(uv)^* = v^*u^*$ for all $u, v \in U$.

If, in addition, $\|u^*u\| = \|u\|^2$ for all $u \in U$, then $U$ is a *C\*-algebra*.

Let $U, V$ be $C^*$-algebras. A $\mathbb{C}$-linear mapping $H : U \rightarrow V$ is called a *homomorphism* in $C^*$-algebras if $H$ satisfies

$$H(xy) = H(x)H(y), \quad H(x^*) = H(x)^*$$

for all $x, y \in U$. A $\mathbb{C}$-linear mapping $\delta : U \rightarrow U$ is called a *derivation* on $U$ if $\delta$ satisfies

$$\delta(xy) = \delta(x)y + x\delta(y)$$

for all $x, y \in U$.

Suppose that $\mathcal{A}$ is a complex Banach $*$-algebra. Let $\mathbb{C}$-linear mapping $\delta : D(\delta) \to \mathcal{A}$ be a *derivation* on $\mathcal{A}$, where $D(\delta)$ is the domain of $\delta$ and $D(\delta)$ is dense in $\mathcal{A}$. If $\delta$ satisfies the additional condition

$$\delta(a^*) = \delta(a)^*$$

for all $a \in \mathcal{A}$, then $\delta$ is called a $*$-*derivation* on $\mathcal{A}$.

It is well-known that, if $\mathcal{A}$ is a $C^*$-algebra and $D(\delta)$ is $\mathcal{A}$, then the derivation $\delta$ is bounded.

Now, we consider *proper CQ\*-algebras*, which arise as completions of $C^*$-algebras (see [15–39]) as follows:

Let $A$ be a Banach module over the $C^*$-algebra $A_0$ with an involution $*$ and $C^*$-norm $\| \cdot \|_0$ such that $A_0 \subset A$. We say that $(A, A_0)$ is a *proper CQ\*-algebra* if

(1) $A_0$ is dense in $A$ with respect to its norm $\| \cdot \|$;
(2) An involution $*$, which extends the involution of $A_0$, is defined in $A$ with the property $(xy)^* = y^*x^*$ for all $x, y \in A$ whenever the multiplication is defined;
(3) $\|y\|_0 = \sup_{x \in A, \|x\| \leq 1} \|xy\|$ for all $y \in A_0$.

**Definition 1.6.** Let $(A, A_0)$ and $(B, B_0)$ be proper $CQ^*$-algebras.

(1) A $\mathbb{C}$-linear mapping $h : A \to B$ is called a *proper CQ\*-algebra homomorphism* if

$$h(xy) = h(x)h(y)$$

for all $x, y \in A$ whenever the multiplication is defined;
(2) A $\mathbb{C}$-linear mapping $\delta : A \to A$ is called a derivation on $A$ if

$$\delta(xy) = \delta(x)y + x\delta(y)$$

for all $x, y \in A$ whenever the multiplication is defined.

A $C^*$-algebra $C$ endowed with the Lie product $[x, y] := \frac{xy - yx}{2}$ on $C$ is called a *Lie C\*-algebra* (see [224, 225, 227]).

**Definition 1.7.** Let $A$ and $B$ be Lie $C^*$-algebras. A $\mathbb{C}$-linear mapping $H : A \to B$ is called a *Lie C\*-algebra homomorphism* if

$$H([x, y]) = [H(x), H(y)]$$

for all $x, y \in A$.

**Definition 1.8.** Let $A$ be a Lie $C^*$-algebra. A $\mathbb{C}$-linear mapping $\delta : A \to A$ is called a *Lie derivation* if

$$\delta([x, y]) = [\delta(x), y] + [x, \delta(y)]$$

for all $x, y \in A$.

## 1.4  $C^*$-Ternary Algebras

Ternary algebraic structures appear more or less naturally in various domains of theoretical and mathematical physics, for example, the quark model inspired a particular brand of ternary algebraic system. One such attempt has been proposed by Nambu in 1973 and is now known under the name of "Nambu mechanics" [316] (see also [332]).

A $C^*$-*ternary algebra* is a complex Banach space $A$, equipped with a ternary product $(x, y, z) \mapsto [x, y, z]$ of $A^3$ into $A$, which is $\mathbb{C}$-linear in the outer variables, conjugate $\mathbb{C}$-linear in the middle variable and associative in the sense that

$$[x, y, [z, w, v]] = [x, [w, z, y], v] = [[x, y, z], w, v]$$

and satisfies

$$\|[x, y, z]\| \leq \|x\| \cdot \|y\| \cdot \|z\|, \quad \|[x, x, x]\| = \|x\|^3$$

(see [332]).

If a $C^*$-ternary algebra $(A, [\cdot, \cdot, \cdot])$ has the identity, i.e., an element $e \in A$ such that $x = [x, e, e] = [e, e, x]$ for all $x \in A$, then it is routine to verify that $A$, endowed with $x \circ y := [x, e, y]$ and $x^* := [e, x, e]$, is a *unital $C^*$-algebra*. Conversely, if $(A, \circ)$ is a unital $C^*$-algebra, then $[x, y, z] := x \circ y^* \circ z$ makes $A$ into a $C^*$-ternary algebra.

A $\mathbb{C}$-linear mapping $H : A \rightarrow B$ is called a $C^*$-*ternary algebra homomorphism* if

$$H([x, y, z]) = [H(x), H(y), H(z)]$$

for all $x, y, z \in A$. A $\mathbb{C}$-linear mapping $\delta : A \rightarrow A$ is called a $C^*$-*ternary derivation* if

$$\delta([x, y, z]) = [\delta(x), y, z] + [x, \delta(y), z] + [x, y, \delta(z)]$$

for all $x, y, z \in A$ (see [231]).

Ternary structures and their generalization, the so-called $n$-ary structures, are important in view of their applications in physics (see [171]).

Suppose that $\mathcal{J}$ is a complex vector space endowed with a real trilinear composition $\mathcal{J} \times \mathcal{J} \times \mathcal{J} \ni (x, y, z) \mapsto \{xy^*z\} \in \mathcal{J}$ which is complex bilinear in $(x, z)$ and conjugate linear in $y$. Then $\mathcal{J}$ is called a *Jordan triple system* if

$$\{xy^*z\} = \{zy^*x\}$$

and

$$\{\{xy^*z\}u^*v\} + \{\{xy^*v\}u^*z\} - \{xy^*\{zu^*v\}\} = \{z\{yx^*u\}^*v\}.$$

We are interested in Jordan triple systems having a Banach space structure. A complex Jordan triple system $\mathcal{J}$ with a Banach space norm $\|\cdot\|$ is called a $J^*$-*triple* if, for all $x \in \mathcal{J}$, the operator $x\square x^*$ is hermitian in the sense of Banach algebra theory. Here the operator $x\square x^*$ on $\mathcal{J}$ is defined by $(x\square x^*)y := \{xx^*y\}$. This implies that $x\square x^*$ has the real spectrum $\sigma(x\square x^*) \subset \mathbb{R}$. A $J^*$-triple $\mathcal{J}$ is called a $JB^*$-*triple* if every $x \in \mathcal{J}$ satisfies $\sigma(x\square x^*) \geq 0$ and $\|x\square x^*\| = \|x\|^2$.

A $\mathbb{C}$-linear mapping $H : \mathcal{J} \to \mathcal{L}$ is called a $JB^*$-*triple homomorphism* if

$$H(\{xyz\}) = \{H(x)H(y)H(z)\}$$

for all $x, y, z \in \mathcal{J}$. A $\mathbb{C}$-linear mapping $\delta : \mathcal{J} \to \mathcal{J}$ is called a $JB^*$-*triple derivation* if

$$\delta(\{xyz\}) = \{\delta(x)yz\} + \{x\delta(y)z\} + \{xy\delta(z)\}$$

for all $x, y, z \in \mathcal{J}$ (see [225]).


## 1.5  Non-Archimedean Normed Algebras


By a *non-Archimedean field* we mean a field $K$ equipped with a function (valuation) $|\cdot|$ from $K$ into $[0, \infty)$ such that

(1)  $|r| = 0$ if and only if $r = 0$;
(2)  $|rs| = |r|\,|s|$;
(3)  $|r + s| \leq \max\{|r|, |s|\}$ for all $r, s \in K$.

Clearly, $|1| = |-1| = 1$ and $|n| \leq 1$ for all $n \geq 1$. By the *trivial valuation* we mean the mapping $|\cdot|$ taking everything but 0 into 1 and $|0| = 0$.

Let $X$ be a vector space over a field $K$ with a non-Archimedean non-trivial valuation $|\cdot|$. A function $\|\cdot\| : X \to [0, \infty)$ is called a *non-Archimedean norm* if it satisfies the following conditions:

(1)  $\|x\| = 0$ for all $x \in X$ if and only if $x = 0$;
(2)  For all $r \in K, x \in X, \|rx\| = |r|\|x\|$;
(3)  The strong triangle inequality (*ultrametric*) holds, i.e.,

$$\|x + y\| \leq \max\{\|x\|, \|y\|\}$$

for all $x, y \in X$.

Then $(X, \|\cdot\|)$ is called a *non-Archimedean normed space*. From the fact that

$$\|x_n - x_m\| \leq \max\{\|x_{j+1} - x_j\| : m \leq j \leq n-1\}$$

for all $n \geq 1$ with $n > m$, a sequence $\{x_n\}$ is a Cauchy sequence if and only if $\{x_{n+1} - x_n\}$ converges to zero in a non-Archimedean normed space. By a *complete non-Archimedean normed space* we mean one in which every Cauchy sequence is convergent.

For any nonzero rational number $x$, there exists a unique integer $n_x \in \mathbb{Z}$ such that

$$x = \frac{a}{b} p^{n_x},$$

where $a$ and $b$ are integers not divisible by $p$. Then $|x|_p := p^{-n_x}$ defines a non-Archimedean norm on $\mathbb{Q}$. The completion of $\mathbb{Q}$ with respect to the metric $d(x, y) = |x - y|_p$ is denoted by $\mathbb{Q}_p$, which is called the *p-adic* number field.

A *non-Archimedean Banach algebra* is a complete non-Archimedaen algebra $\mathcal{A}$ which satisfies $\|ab\| \leq \|a\| \cdot \|b\|$ for all $a, b \in \mathcal{A}$. For more detailed definitions of non-Archimedean Banach algebras, the readers refer to [310].

If $\mathcal{U}$ is a non-Archimedean Banach algebra, then an involution on $\mathcal{U}$ is a mapping $t \to t^*$ from $\mathcal{U}$ into $\mathcal{U}$ satisfying the following:

(1) $t^{**} = t$ for all $t \in \mathcal{U}$;
(2) $(\alpha s + \beta t)^* = \overline{\alpha} s^* + \overline{\beta} t^*$;
(3) $(st)^* = t^* s^*$ for all $s, t \in \mathcal{U}$.

If, in addition $\|t^* t\| = \|t\|^2$ for all $t \in \mathcal{U}$, then $\mathcal{U}$ is a non-Archimedean $C^*$–algebra.

## 1.6   Multi–normed Algebras

The notion of multi-normed space was introduced by Dales and Polyakov in [92]. This concept is somewhat similar to the operator sequence space and has some connections with the operator spaces and Banach lattices. Motivations for the study of multi-normed spaces and many examples are given in [91, 92, 206].

Let $(\mathcal{E}, \| \cdot \|)$ be a complex normed space and let $k \in \mathbb{N}$. We denote by $\mathcal{E}^k$ the linear space $\mathcal{E} \oplus \cdots \oplus \mathcal{E}$ consisting of $k$-tuples $(x_1, \cdots, x_k)$, where $x_1, \cdots, x_k \in \mathcal{E}$. The linear operations on $\mathcal{E}^k$ are defined coordinate-wise. The zero element of either $\mathcal{E}$ or $\mathcal{E}^k$ is denoted by $0$. We denote by $\mathbb{N}_k$ the set $\{1, 2, \cdots, k\}$ and by $\Sigma_k$ the group of permutations on $k$ symbols.

**Definition 1.9.** A *multi-norm* on $\{\mathcal{E}^k : k \in \mathbb{N}\}$ is a sequence

$$(\| \cdot \|_k) = (\| \cdot \|_k : k \in \mathbb{N})$$

such that $\| \cdot \|_k$ is a norm on $\mathcal{E}^k$ for each $k \in \mathbb{N}$:

(A1) $\|(x_{\sigma(1)}, \cdots, x_{\sigma(k)})\|_k = \|(x_1, \cdots, x_k)\|_k$ for all $\sigma \in \Sigma_k$ and $x_1, \cdots, x_k \in \mathcal{E}$;
(A2) $\|(\alpha_1 x_1, \cdots, \alpha_k x_k)\|_k \leq (\max_{i \in \mathbb{N}_k} |\alpha_i|) \|x_1, \cdots, x_k\|_k$ for all $\alpha_1, \cdots, \alpha_k \in \mathbb{C}$ and $x_1, \cdots, x_k \in \mathcal{E}$;

(A3) $\|(x_1, \cdots, x_{k-1}, 0)\|_k = \|(x_1, \cdots, x_{k-1})\|_{k-1}$ for all $x_1, \cdots, x_{k-1} \in \mathcal{E}$;

(A4) $\|(x_1, \cdots, x_{k-1}, x_{k-1})\|_k = \|(x_1, \cdots, x_{k-1})\|_{k-1}$ for all $x_1, \cdots, x_{k-1} \in \mathcal{E}$.

In this case, we say that $((\mathcal{E}^k, \| \cdot \|_k) : k \in \mathbb{N})$ is a *multi-normed space*.

**Lemma 1.10 ([206]).** *Suppose that* $((\mathcal{E}^k, \| \cdot \|_k) : k \in \mathbb{N})$ *is a multi-normed space and let* $k \in \mathbb{N}$. *Then*

(1) $\|(x, \cdots, x)\|_k = \|x\|$ *for all* $x \in \mathcal{E}$;

(2) $\max_{i \in \mathbb{N}_k} \|x_i\| \leq \|(x_1, \cdots, x_k)\|_k \leq \sum_{i=1}^k \|x_i\| \leq k \max_{i \in \mathbb{N}_k} \|x_i\|$ *for all* $x_1, \cdots, x_k \in \mathcal{E}$.

It follows from (2) that, if $(\mathcal{E}, \| \cdot \|)$ is a Banach space, then $(\mathcal{E}^k, \| \cdot \|_k)$ is a Banach space for each $k \in \mathbb{N}$. In this case, $((\mathcal{E}^k, \| \cdot \|_k) : k \in \mathbb{N})$ is a *multi-Banach space*.

Now, we state two important examples of multi-norms for an arbitrary normed space $\mathcal{E}$ (see [92]).

*Example 1.11.* The sequence $(\| \cdot \|_k : k \in \mathbb{N})$ on $\{\mathcal{E}^k : k \in \mathbb{N}\}$ defined by

$$\|x_1, \cdots, x_k\|_k := \max_{i \in \mathbb{N}_k} \|x_i\|$$

for all $x_1, \cdots, x_k \in \mathcal{E}$ is a multi-norm called the *minimum multi-norm*. The terminology "minimum" is justified by the property (2).

*Example 1.12.* Let $\{(\| \cdot \|_k^\alpha : k \in \mathbb{N}) : \alpha \in A\}$ be the (non-empty) family of all multi-norms on $\{\mathcal{E}^k : k \in \mathbb{N}\}$. For $k \in \mathbb{N}$, set

$$\||x_1, \cdots, x_k\||_k := \sup_{\alpha \in A} \|(x_1, \cdots, x_k)\|_k^\alpha$$

for all $x_1, \cdots, x_k \in \mathcal{E}$. Then $(\|| \cdot \||_k : k \in \mathbb{N})$ is a multi-norm on $\{\mathcal{E}^k : k \in \mathbb{N}\}$, which is called the *maximum multi-norm*.

We need the following observation which can be easily deduced from the triangle inequality for the norm $\| \cdot \|_k$ and the property (2) of multi-norms.

**Lemma 1.13.** *Suppose that* $k \in \mathbb{N}$ *and* $(x_1, \cdots, x_k) \in \mathcal{E}^k$. *For each* $j \in \{1, \cdots, k\}$, *let* $(x_n^j)_{n \geq 1}$ *be a sequence in* $\mathcal{E}$ *such that* $\lim_{n \to \infty} x_n^j = x_j$. *Then, for each* $(y_1, \cdots, y_k) \in \mathcal{E}^k$, *we have*

$$\lim_{n \to \infty} (x_n^1 - y_1, \cdots, x_n^k - y_k) = (x_1 - y_1, \cdots, x_k - y_k).$$

**Definition 1.14.** Let $((\mathcal{E}^k, \| \cdot \|_k) : k \in \mathbb{N})$ be a multi-normed space. A sequence $(x_n)$ in $\mathcal{E}$ is a *multi-null* sequence if, for any $\varepsilon > 0$, there exists $n_0 \in \mathbb{N}$ such that

$$\sup_{k \in \mathbb{N}} \|(x_n, \cdots, x_{n+k-1})\|_k < \varepsilon$$

for each $n \geq n_0$. We say that the sequence $(x_n)$ is *multi-convergent* to a point $x \in \mathcal{E}$ (write $\lim_{n \to \infty} x_n = x$) if $(x_n - x)$ is a multi-null sequence.

**Definition 1.15 ([92, 170, 202]).** Let $(\mathcal{A}, \|\cdot\|)$ be a normed algebra such that $((\mathcal{A}^k, \|\cdot\|_k) : k \in \mathbb{N})$ is a multi-normed space. Then $((\mathcal{A}^k, \|\cdot\|_k) : k \in \mathbb{N})$ is called a *multi-normed algebra* if

$$\|(a_1 b_1, \cdots, a_k b_k)\|_k \leq \|(a_1, \cdots, a_k)\|_k \cdot \|(b_1, \cdots, b_k)\|_k$$

for all $k \in \mathbb{N}$ and $a_1, \cdots, a_k, b_1, \cdots, b_k \in \mathcal{A}$. Further, the multi-normed algebra $((\mathcal{A}^k, \|\cdot\|_k) : k \in \mathbb{N})$ is called a *multi-Banach algebra* if $((\mathcal{A}^k, \|\cdot\|_k) : k \in \mathbb{N})$ is a multi-Banach space.

*Example 1.16 ([92, 202, 252]).* Let $p, q$ with $1 \leq p \leq q < \infty$ and $\mathcal{A} = \ell^p$. The algebra $\mathcal{A}$ is a Banach sequence algebra with respect to coordinatewise multiplication of sequences. Let $(\|\cdot\|_k : k \in \mathbb{N})$ be the standard $(p, q)$-multi-norm on $\{\mathcal{A}^k : k \in \mathbb{N}\}$. Then $((\mathcal{A}^k, \|\cdot\|_k) : k \in \mathbb{N})$ is a multi-Banach algebra.

**Definition 1.17.** Let $(\mathcal{A}, \|\cdot\|)$ be a Banach $*$-algebra with the involution $*$. A *multi-$C^*$-algebra* is a multi-Banach algebra such that

$$\|(a_1 a_1^*, \cdots, a_k a_k^*)\| = \|(a_1, \cdots, a_k)\|^2.$$

In a series of the papers [15–33, 36–38] and [318–320], many authors have considered a special class of quasi $*$-algebras, called *proper $CQ^*$-algebras*, which arise as completions of $C^*$-algebras. They can be introduced in the following way:

Let $\mathfrak{A}$ be a linear space and $\mathcal{A}$ be a $*$-algebra contained in $\mathfrak{A}$. We say that $\mathfrak{A}$ is a *quasi-$*$-algebra* over $\mathcal{A}$ if the right and left multiplications of an element of $\mathfrak{A}$ and an element of $\mathcal{A}$ are always defined and linear. An involution $*$ which extends the involution of $\mathcal{A}$ is defined in $\mathfrak{A}$ with the property $(ab)^* = b^* a^*$ whenever the multiplication is defined.

A quasi-$*$-algebra $(\mathfrak{A}, \mathcal{A})$ is said to be *topological* if there exists a locally convex topology $\tau$ on $\mathfrak{A}$ such that

(Q1)  The involution $a \mapsto a^*$ is continuous;
(Q2)  The mappings $a \mapsto ab$ and $a \mapsto ba$ are continuous for each $b \in \mathcal{A}$;
(Q3)  $\mathcal{A}$ is dense in $\mathfrak{A}$ with topology $\tau$.

In a topological quasi-$*$-algebra, the associative law holds in the following two formulations:

$$a(bc) = (ab)c, \quad b(ac) = (ba)c$$

for all $b, c \in \mathcal{A}$ and $a \in \mathfrak{A}$.

A *$CQ^*$-algebra* is a topological quasi-$*$-algebra $(\mathfrak{A}, \mathcal{A})$ with the following properties:

(CQ1)  $(\mathcal{A}, \|\cdot\|_*)$ is a $C^*$-algebra with respect to the norm $\|\cdot\|_*$ and the involution $*$;
(CQ2)  $(\mathfrak{A}, \|\cdot\|)$ is a Banach space and $\|a^*\| = \|a\|$ for all $a \in \mathfrak{A}$;

(CQ3)  For all $b \in \mathcal{A}$, we have

$$\|b\|_* = \max \left\{ \sup_{\|a\| \leq 1} \|ab\|, \ \sup_{\|a\| \leq 1} \|ba\| \right\}.$$

Bagarello and Trapani [34] showed that both $(L^p(X, \mu), C_0(X))$ and $(L^p(X, \mu), L^\infty(X))$ are $CQ^*$-algebras.

Now, we define the *multi-$CQ^*$-algebra*.

Let $(\mathfrak{A}, \mathcal{A})$ be a $CQ^*$-algebra. We say that $\{(\mathfrak{A}^k, \mathcal{A}^k) : k \in \mathbb{N}\}$ is a *multi-$CQ^*$-algebra* if, for each $k \in \mathbb{N}$, the couple $(\mathfrak{A}^k, \mathcal{A}^k)$ is a $CQ^*$-algebra, where $\{\mathfrak{A}^k : k \in \mathbb{N}\}$ and $\{\mathcal{A}^k : k \in \mathbb{N}\}$ are a multi-Banach algebra and a multi-$C^*$-algebra, respectively.

*Example 1.18.* In [34], the authors showed that the couple $(\mathfrak{A}, \mathcal{A})$ is a $CQ^*$-algebra, where $\mathfrak{A} = \ell^p$ and $\mathcal{A} = c_0$. Now, consider Example 1.16. Then $\{(\mathfrak{A}^k, \mathcal{A}^k) : k \in \mathbb{N}\}$ is a multi-$CQ^*$-algebra.

**Definition 1.19.** Let $((A^k, \|\cdot\|_k) : k \in \mathbb{N})$ be a multi-Banach space. A *multi-$C^*$-ternary algebra* is a complex multi-Banach space $((A^k, \|\cdot\|_k) : k \in \mathbb{N})$ equipped with a ternary product.

# Chapter 2
# Stability of Functional Equations in Banach Algebras

Beginning around the year 1980, the topic of approximate homomorphisms and derivations and their stability theory in the field of functional equations and inequalities was taken up by several mathematicians (see Hyers and Rassias [135], Rassias [285] and the references therein).

In this chapter, in the first section, we show that, if $X$ and $Y$ are normed spaces with the norms $\|\cdot\|_X$ and $\|\cdot\|_Y$, respectively, and $f : X \to Y$ is a mapping such that

$$\|f(x) + f(y) + f(z)\|_Y \leq \left\|\frac{1}{q}f(qx + qy + qz)\right\|_Y$$

for all $x, y, z \in X$ and for a fixed nonzero rational number $q$, then $f$ is Cauchy additive. Next, we approximate isomorphisms and derivations in Banach algebras by the direct method.

In Sect. 2.2, we consider the $m$-variable additive functional equation:

$$\sum_{i=1}^{m} f\left(mx_i + \sum_{j=1, j\neq i}^{m} x_j\right) + f\left(\sum_{i=1}^{m} x_i\right) = 2f\left(\sum_{i=1}^{m} mx_i\right)$$

for all $m \in \mathbb{N}$ and $m \geq 2$ and, by the fixed point method, we approximate homomorphisms and derivations in Banach algebras.

In Sect. 2.3, we prove the *Hyers-Ulam stability* of homomorphisms in *quasi-Banach algebras* and generalized derivations on quasi-Banach algebras for the following functional equation:

$$\sum_{i=1}^{n} f\left(\sum_{j=1}^{n} q(x_i - x_j)\right) + nf\left(\sum_{i=1}^{n} qx_i\right) = nq\sum_{i=1}^{n} f(x_i).$$

© Springer International Publishing Switzerland 2015
Y.J. Cho et al., *Stability of Functional Equations in Banach Algebras*, DOI 10.1007/978-3-319-18708-2_2

In Sect. 2.4, we approximate homomorphisms in real Banach algebras and generalized derivations on real Banach algebras for the following *Cauchy-Jensen functional equations*

$$f\left(\frac{x+y}{2}+z\right)+f\left(\frac{x-y}{2}+z\right)=f(x)+2f(z)$$

and

$$2f\left(\frac{x+y}{2}+z\right)=f(x)+f(y)+2f(z).$$

## 2.1  Stability of $\frac{1}{q}f(qx+qy+qz)=f(x)+f(y)+f(z)$

Using the *direct method*, we investigate isomorphisms in Banach algebras and derivations on Banach algebras associated with the following functional equation

$$\frac{1}{q}f(qx+qy+qz)=f(x)+f(y)+f(z) \qquad (2.1)$$

for a fixed nonzero rational number $q$.

### 2.1.1  Isomorphisms in Banach Algebras

Here we consider *isomorphisms* in Banach algebras associated with the functional equation (2.1).

**Lemma 2.1.** *Let $X$ and $Y$ be normed spaces with norms $\|\cdot\|_X$ and $\|\cdot\|_Y$, respectively. Let $f:X\to Y$ be a mapping with $f(0)=0$ such that*

$$\|f(x)+f(y)+f(z)\|_Y \le \left\|\frac{1}{q}f(qx+qy+qz)\right\|_Y \qquad (2.2)$$

*for all $x,y,z \in X$, then $f$ is Cauchy additive, i.e.,*

$$f(x+y)=f(x)+f(y).$$

*Proof.* Letting $z=0$ and $y=-x$ in (2.2), we get

$$\|f(x)+f(-x)\|_Y \le \left\|\frac{1}{q}f(0)\right\|_Y = 0$$

for all $x \in X$. Hence $f(-x) = -f(x)$ for all $x \in X$. Letting $z = -x - y$ in (2.2), we get

$$\|f(x) + f(y) - f(x + y)\|_Y = \|f(x) + f(y) + f(-x - y)\|_Y$$
$$\leq \left\| \frac{1}{q}f(0) \right\|_Y$$
$$= 0$$

for all $x, y \in X$. Thus

$$f(x + y) = f(x) + f(y)$$

for all $x, y \in X$. This completes the proof.                                    □

Here, we assume that $A$ is a Banach algebra with the norm $\|\cdot\|_A$ and $B$ is a Banach algebra with the norm $\|\cdot\|_B$.

**Theorem 2.2.** *Let $r \neq 1$, $\theta$ be nonnegative real numbers and $f : A \to B$ be a bijective mapping with $f(0) = 0$ such that*

$$\|\mu f(x) + f(y) + f(z)\|_B \leq \left\| \frac{1}{q}f(q\mu x + qy + qz) \right\|_B \tag{2.3}$$

*and*

$$\|f(xy) - f(x)f(y)\|_B \leq \theta(\|x\|_A^{2r} + \|y\|_A^{2r}) \tag{2.4}$$

*for all $\mu \in \mathbb{T}^1 := \{\lambda \in \mathbb{C} : |\lambda| = 1\}$ and $x, y, z \in A$, then the bijective mapping $f : A \to B$ is an isomorphism.*

*Proof.* Let $\mu = 1$ in (2.3). By Lemma 2.1, the mapping $f : A \to B$ is Cauchy additive. Letting $z = 0$ and $y = -\mu x$ in (2.3), we get

$$\mu f(x) - f(\mu x) = \mu f(x) + f(-\mu x) = 0$$

for all $x \in A$ and so $f(\mu x) = \mu f(x)$ for all $x \in A$. By (2.3), the mapping $f : A \to B$ is $\mathbb{C}$-linear.

(i)  Assume that $r < 1$. By (2.4), we have

$$\|f(xy) - f(x)f(y)\|_B = \lim_{n \to \infty} \frac{1}{4^n} \|f(4^n xy) - f(2^n x)f(2^n y)\|_B$$
$$\leq \lim_{n \to \infty} \frac{4^{nr}}{4^n} \theta(\|x\|_A^{2r} + \|y\|_A^{2r})$$
$$= 0$$

for all $x, y \in A$ and so

$$f(xy) = f(x)f(y)$$

for all $x, y \in A$.

(ii) Assume that $r > 1$. By a similar method to the proof of the case (i), one can prove that the mapping $f : A \to B$ satisfies

$$f(xy) = f(x)f(y)$$

for all $x, y \in A$. Therefore, the bijective mapping $f : A \to B$ is an isomorphism in Banach algebras. This completes the proof.                    □

**Theorem 2.3.** *Let* $r \neq 1$, $\theta$ *be nonnegative real numbers and* $f : A \to B$ *be a bijective mapping satisfying* $f(0) = 0$ *and* (2.3) *such that*

$$\|f(xy) - f(x)f(y)\|_B \leq \theta \cdot \|w\|_A^r \cdot \|x\|_A^r \qquad (2.5)$$

*for all* $x, y \in A$, *then the bijective mapping* $f : A \to B$ *is an isomorphism.*

*Proof.* By (2.3), the mapping $f : A \to B$ is $\mathbb{C}$-linear.

(i) Assume that $r < 1$. By (2.5), we have

$$
\begin{aligned}
\|f(xy) - f(x)f(y)\|_B &= \lim_{n \to \infty} \frac{1}{4^n} \|f(4^n xy) - f(2^n x)f(2^n y)\|_B \\
&\leq \lim_{n \to \infty} \frac{4^{nr}}{4^n} \theta \cdot \|w\|_A^r \cdot \|x\|_A^r \\
&= 0
\end{aligned}
$$

for all $x, y \in A$ and so

$$f(xy) = f(x)f(y)$$

for all $x, y \in A$.

(ii) Assume that $r > 1$. By a similar method to the proof of the case (i), one can prove that the mapping $f : A \to B$ satisfies

$$f(xy) = f(x)f(y)$$

for all $x, y \in A$. Therefore, the bijective mapping $f : A \to B$ is an isomorphism. This completes the proof.                    □

### 2.1.2 Derivations in Banach Algebras

We consider *derivations on Banach algebras* associated with the functional equation (2.1).

**Theorem 2.4.** *Let* $r \neq 1$, $\theta$ *be nonnegative real numbers and* $f : A \to A$ *be a mapping with* $f(0) = 0$ *such that*

$$\|\mu f(x) + f(y) + f(z)\|_A \leq \left\| \frac{1}{q} f(q\mu x + qy + qz) \right\|_A \tag{2.6}$$

*and*

$$\|f(xy) - f(x)y - xf(y)\|_A \leq \theta(\|x\|_A^{2r} + \|y\|_A^{2r}) \tag{2.7}$$

*for all* $\mu \in \mathbb{T}^1$ *and* $x, y, z \in A$, *then the mapping* $f : A \to A$ *is a derivation on* $A$.

*Proof.* By (2.7), the mapping $f : A \to A$ is $\mathbb{C}$-linear.

(i) Assume that $r < 1$. By (2.7),

$$\|f(xy) - f(x)y - xf(y)\|_A$$
$$= \lim_{n \to \infty} \frac{1}{4^n} \|f(4^n xy) - f(2^n x) \cdot 2^n y - 2^n xf(2^n y)\|_A$$
$$\leq \lim_{n \to \infty} \frac{4^{nr}}{4^n} \theta(\|x\|_A^{2r} + \|y\|_A^{2r})$$
$$= 0$$

for all $x, y \in A$ and so

$$f(xy) = f(x)y + xf(y)$$

for all $x, y \in A$.

(ii) Assume that $r > 1$. By a similar method to the proof of the case (i), one can prove that the mapping $f : A \to A$ satisfies

$$f(xy) = f(x)y + xf(y)$$

for all $x, y \in A$. Therefore, the mapping $f : A \to A$ is a derivation on $A$. This completes the proof. $\qquad\square$

**Theorem 2.5.** *Let* $r \neq 1$, $\theta$ *be nonnegative real numbers and* $f : A \to A$ *be a mapping satisfying* $f(0) = 0$ *and* (2.6) *such that*

$$\|f(xy) - f(x)y - xf(y)\|_A \leq \theta \cdot \|x\|_A^r \cdot \|y\|_A^r$$

*for all* $x, y \in A$, *then the mapping* $f : A \to A$ *is a derivation on* $A$.

*Proof.* The proof is similar to the proofs of Theorems 2.2 and 2.4.                    □

## 2.2   Stability of $m$-Variable Functional Equations

In this section, using the fixed point method, we prove the *Hyers-Ulam stability* of homomorphisms and of derivations on Banach algebras for the following additive functional equation (see [108]):

$$\sum_{i=1}^{m} f\left(mx_i + \sum_{j=1,j\neq i}^{m} x_j\right) + f\left(\sum_{i=1}^{m} x_i\right) = 2f\left(\sum_{i=1}^{m} mx_i\right)$$

for all $m \in \mathbb{N}$ with $m \geq 2$.

### 2.2.1   Stability of Homomorphisms in Banach Algebras

For any mapping $f : A \to B$, we define

$$D_\mu f(x_1, \cdots, x_m)$$

$$:= \sum_{i=1}^{m} \mu f\left(mx_i + \sum_{j=1,j\neq i}^{m} x_j\right) + \mu f\left(\sum_{i=1}^{m} x_i\right) - 2f\left(\mu \sum_{i=1}^{m} mx_i\right)$$

for all $\mu \in \mathbb{T}^1 := \{v \in \mathbb{C} : |v| = 1\}$ and $x_1, \cdots, x_m \in A$.

Now, we prove the *Hyers-Ulam stability of homomorphisms* in Banach algebras for the functional equation $D_\mu f(x_1, \cdots, x_m) = 0$. Assume that $A$ is a complex Banach algebra with norm $\| \cdot \|_A$ and $B$ is a complex Banach algebra with norm $\| \cdot \|_B$.

**Theorem 2.6.** *Let* $f : A \to B$ *be a mapping for which there are functions* $\varphi : A^m \to [0, \infty)$ *and* $\psi : A^2 \to [0, \infty)$ *such that*

$$\lim_{j\to\infty} m^{-j}\varphi(m^j x_1, \cdots, m^j x_m) = 0, \tag{2.8}$$

$$\|D_\mu f(x_1, \cdots, x_m)\|_B \leq \varphi(x_1, \cdots, x_m), \tag{2.9}$$

$$\|f(xy) - f(x)f(y)\|_B \leq \psi(x, y), \tag{2.10}$$

$$\lim_{j\to\infty} m^{-2j}\psi(m^j x, m^j y) = 0 \tag{2.11}$$

*for all $\mu \in \mathbb{T}^1$ and $x_1, \cdots, x_m, x, y \in A$. If there exists $L < 1$ such that*

$$\varphi(mx, 0, \cdots, 0) \le mL\varphi(x, 0, \cdots, 0)$$

*for all $x \in A$, then there exists a unique homomorphism $H : A \to B$ such that*

$$\|f(x) - H(x)\|_B \le \frac{1}{m - mL}\varphi(x, 0, \cdots, 0) \tag{2.12}$$

*for all $x \in A$.*

*Proof.* Consider the set $X := \{g : A \to B\}$ and introduce the *generalized metric* on $X$:

$$d(g, h) = \inf\{C \in \mathbb{R}_+ : \|g(x) - h(x)\|_B \le C\varphi(x, 0, \cdots, 0), \ \forall x \in A\},$$

which $(X, d)$ is complete.

Now, we consider the linear mapping $J : X \to X$ such that $Jg(x) := \frac{1}{m}g(mx)$ for all $x \in A$. Now, we have $d(Jg, Jh) \le Ld(g, h)$ for all $g, h \in X$. Letting $\mu = 1$, $x = x_1$ and $x_2 = \cdots = x_m = 0$ in (2.9), we get

$$\|f(mx) - mf(x)\|_B \le \varphi(x, 0, \cdots, 0) \tag{2.13}$$

for all $x \in A$ and so

$$\left\|f(x) - \frac{1}{m}f(mx)\right\|_B \le \frac{1}{m}\varphi(x, 0, \cdots, 0)$$

for all $x \in A$. Hence $d(f, Jf) \le \frac{1}{m}$.

By Theorem 1.3, there exists a mapping $H : A \to B$ such that

(1) $H$ is a fixed point of $J$, i.e.,

$$H(mx) = mH(x) \tag{2.14}$$

for all $x \in A$. The mapping $H$ is a unique fixed point of $J$ in the set

$$Y = \{g \in X : d(f, g) < \infty\}.$$

This implies that $H$ is a unique mapping satisfying (2.13) such that there exists $C \in (0, \infty)$ satisfying

$$\|H(x) - f(x)\|_B \le C\varphi(x, 0, \cdots, 0)$$

for all $x \in A$;

(2) $d(J^n f, H) \to 0$ as $n \to \infty$. This implies the equality

$$\lim_{n\to\infty} \frac{f(m^n x)}{m^n} = H(x) \tag{2.15}$$

for all $x \in A$;

(3) $d(f, H) \leq \frac{1}{1-L} d(f, Jf)$, which implies the inequality $d(f, H) \leq \frac{1}{m-mL}$. This implies that the inequality (2.12) holds.

Thus it follows from (2.8), (2.9) and (2.14) that

$$\left\| \sum_{i=1}^{m} H\left(mx_i + \sum_{j=1, j\neq i}^{m} x_j\right) + H\left(\sum_{i=1}^{m} x_i\right) - 2H\left(\sum_{i=1}^{m} mx_i\right) \right\|_B$$

$$= \lim_{n\to\infty} \frac{1}{m^n} \left\| \sum_{i=1}^{m} f\left(m^{n+1} x_i + \sum_{j=1, j\neq i}^{m} m^n x_j\right) \right.$$

$$\left. + f\left(\sum_{i=1}^{m} m^n x_i\right) - 2f\left(\sum_{i=1}^{m} m^{n+1} x_i\right) \right\|_B$$

$$\leq \lim_{n\to\infty} \frac{1}{m^n} \varphi(m^n x_1, \cdots, m^n x_m) = 0$$

for all $x_1, \cdots, x_m \in A$ and so

$$\sum_{i=1}^{m} H\left(mx_i + \sum_{j=1, j\neq i}^{m} x_j\right) + H\left(\sum_{i=1}^{m} x_i\right) = 2H\left(\sum_{i=1}^{m} mx_i\right) \tag{2.16}$$

for all $x_1, \cdots, x_m \in A$. By a similar method given in above, we get $\mu H(mx) = H(m\mu x)$ for all $\mu \in \mathbb{T}^1$ and $x \in A$. Thus one can show that the mapping $H : A \to B$ is $\mathbb{C}$-linear. It follows from (2.10) that

$$\|H(xy) - H(x)H(y)\|_B = \lim_{n\to\infty} \frac{1}{m^n} \|f(m^n xy) - f(m^n x)f(m^n y)\|_B$$

$$\leq \lim_{n\to\infty} \frac{1}{m^n} \psi(m^n x, m^n y)$$

$$= 0$$

for all $x, y \in A$ and so $H(xy) = H(x)H(y)$ for all $x, y \in A$. Thus $H : A \to B$ is a homomorphism satisfying (2.12). This completes the proof. $\qquad\square$

**Corollary 2.7.** *Let $r < 1$, $\theta$ be nonnegative real numbers and $f : A \to B$ be a mapping such that*

$$\|D_\mu f(x_1, \cdots, x_m)\|_B \leq \theta \cdot (\|x_1\|_A^r + \|x_2\|_A^r + \cdots + \|x_m\|_A^r) \tag{2.17}$$

*and*

$$\|f(xy) - f(x)f(y)\|_B \le \theta \cdot (\|x\|_A^r \cdot \|y\|_A^r) \tag{2.18}$$

*for all* $\mu \in \mathbb{T}^1$ *and* $x_1, \cdots, x_m, x, y \in A$. *Then there exists a unique homomorphism* $H : A \to B$ *such that*

$$\|f(x) - H(x)\|_B \le \frac{\theta}{m - m^r} \|x\|_A^r$$

*for all* $x \in A$.

*Proof.* The proof follows from Theorem 2.6 by taking

$$\varphi(x_1, \cdots, x_m) = \theta \cdot (\|x_1\|_A^r + \|x_2\|_A^r + \cdots + \|x_m\|_A^r),$$
$$\psi(x, y) := \theta \cdot (\|x\|_A^r \cdot \|y\|_A^r)$$

for all $x_1, \cdots, x_m, x, y \in A$ and $L = m^{r-1}$.  $\square$

**Theorem 2.8.** *Let* $f : A \to B$ *be a mapping for which there exist the functions* $\varphi : A^m \to [0, \infty)$ *and* $\psi : A^2 \to [0, \infty)$ *such that*

$$\lim_{j \to \infty} m^j \varphi(m^{-j} x_1, \cdots, m^{-j} x_m) = 0, \tag{2.19}$$

$$\|D_\mu f(x_1, \cdots, x_m)\|_B \le \varphi(x_1, \cdots, x_m), \tag{2.20}$$

$$\|f(xy) - f(x)f(y)\|_B \le \psi(x, y), \tag{2.21}$$

$$\lim_{j \to \infty} m^{2j} \psi(m^{-j} x, m^{-j} y) = 0 \tag{2.22}$$

*for all* $\mu \in \mathbb{T}^1$ *and* $x_1, \cdots, x_m, x, y \in A$. *If there exists* $L < 1$ *such that* $\varphi(x, 0, \cdots, 0) \le \frac{L}{m} \varphi(mx, 0, \cdots, 0)$ *for all* $x \in A$, *then there exists a unique homomorphism* $H : A \to B$ *such that*

$$\|f(x) - H(x)\|_B \le \frac{L}{m - mL} \varphi(x, 0, \cdots, 0) \tag{2.23}$$

*for all* $x \in A$.

*Proof.* We consider the linear mapping $J : X \to X$ such that

$$Jg(x) := mg\left(\frac{x}{m}\right)$$

for all $x \in A$. It follows from (2.13) that

$$\left\| f(x) - mf\left(\frac{x}{m}\right) \right\|_B \leq \varphi\left(\frac{x}{m}, 0, \cdots, 0\right) \leq \frac{L}{m}\varphi(x, 0, \cdots, 0)$$

for all $x \in A$ and so $d(f, Jf) \leq \frac{L}{m}$. By Theorem 1.3, there exists a mapping $H : A \to B$ such that

(1) $H$ is a fixed point of $J$, i.e.,

$$H(mx) = mH(x) \tag{2.24}$$

for all $x \in A$. The mapping $H$ is a unique fixed point of $J$ in the set

$$Y = \{g \in X : d(f, g) < \infty\}.$$

This implies that $H$ is a unique mapping satisfying (2.20) such that there exists $C \in (0, \infty)$ satisfying

$$\|H(x) - f(x)\|_B \leq C\varphi(x, 0, \cdots, 0)$$

for all $x \in A$;

(2) $d(J^n f, H) \to 0$ as $n \to \infty$. This implies the equality

$$\lim_{n \to \infty} m^n f\left(\frac{x}{m^n}\right) = H(x)$$

for all $x \in A$;

(3) $d(f, H) \leq \frac{1}{1-L}d(f, Jf)$, which implies the inequality

$$d(f, H) \leq \frac{L}{m - mL},$$

which implies that the inequality (2.23) holds.

The rest of the proof is similar to the proof of Theorem 2.6.      $\square$

**Corollary 2.9.** *Let $r > 1$, $\theta$ be nonnegative real numbers and let $f : A \to B$ be a mapping such that*

$$\|D_\mu f(x_1, \cdots, x_m)\|_B \leq \theta \cdot (\|x_1\|_A^r + \|x_2\|_A^r + \cdots + \|x_m\|_A^r) \tag{2.25}$$

*and*

$$\|f(xy) - f(x)f(y)\|_B \leq \theta \cdot (\|x\|_A^r \cdot \|y\|_A^r) \tag{2.26}$$

for all $\mu \in \mathbb{T}^1$ and $x_1, \cdots, x_m, x, y \in A$. *Then there exists a unique homomorphism*
$H : A \to B$ *such that*

$$\|f(x) - H(x)\|_B \leq \frac{\theta}{m^r - m} \|x\|_A^r$$

*for all* $x \in A$.

*Proof.* The proof follows from Theorem 2.6 by taking

$$\varphi(x_1, \cdots, x_m) = \theta \cdot (\|x_1\|_A^r + \|x_2\|_A^r + \cdots + \|x_m\|_A^r),$$
$$\psi(x, y) := \theta \cdot (\|x\|_A^r \cdot \|y\|_A^r)$$

for all $x_1, \cdots, x_m, x, y \in A$ and $L = m^{1-r}$.                                    □


### 2.2.2   Stability of Derivations in Banach Algebras

Now, we prove the Hyers-Ulam stability of derivations on Banach algebras for the
functional equation $D_\mu f(x_1, \cdots, x_m) = 0$.

**Theorem 2.10.** *Let* $f : A \to A$ *be a mapping for which there exist the functions*
$\varphi : A^m \to [0, \infty)$ *and* $\psi : A^2 \to [0, \infty)$ *such that*

$$\lim_{j \to \infty} m^{-j} \varphi(m^j x_1, \cdots, m^j x_m) = 0, \tag{2.27}$$

$$\|D_\mu f(x_1, \cdots, x_m)\|_A \leq \varphi(x_1, \cdots, x_m), \tag{2.28}$$

$$\|f(xy) - f(x)y - xf(y)\|_A \leq \psi(x, y), \tag{2.29}$$

$$\lim_{j \to \infty} m^{-2j} \psi(m^j x, m^j y) = 0 \tag{2.30}$$

*for all* $\mu \in \mathbb{T}^1$ *and* $x_1, \cdots, x_m, x, y \in A$. *If there exists* $L < 1$ *such that*
$\varphi(mx, 0, \cdots, 0) \leq mL\varphi(x, 0, \cdots, 0)$ *for all* $x \in A$. *Then there exists a unique*
*derivation* $\delta : A \to A$ *such that*

$$\|f(x) - \delta(x)\|_A \leq \frac{1}{m - mL} \varphi(x, 0, \cdots, 0) \tag{2.31}$$

*for all* $x \in A$.

*Proof.* By the same reasoning as in the proof of Theorem 2.6, there exists a unique
$\mathbb{C}$-linear mapping $\delta : A \to A$ satisfying (2.31). The mapping $\delta : A \to A$ is given by

$$\delta(x) = \lim_{n \to \infty} \frac{f(m^n x)}{m^n} \tag{2.32}$$

for all $x \in A$. It follows from (2.29), (2.30) and (2.32) that

$$\|\delta(xy) - \delta(x)y - x\delta(y)\|_A$$

$$= \lim_{n \to \infty} \frac{1}{m^{2n}} \|f(m^{2n}xy) - f(m^n x) \cdot m^n y - m^n x f(m^n y)\|_A$$

$$\leq \lim_{n \to \infty} \frac{1}{m^{2n}} \psi(m^n x, m^n y)$$

$$= 0$$

for all $x, y \in A$ and so

$$\delta(xy) = \delta(x)y + x\delta(y)$$

for all $x, y \in A$. Thus $\delta : A \to A$ is a derivation satisfying (2.31). This completes the proof. □

**Corollary 2.11.** *Let $r < 1$, $\theta$ be nonnegative real numbers and $f : A \to A$ be a mapping such that*

$$\|D_\mu f(x_1, \cdots, x_m)\|_A \leq \theta \cdot (\|x_1\|_A^r + \cdots + \|x_m\|_A^r) \tag{2.33}$$

*and*

$$\|f(xy) - f(x)y - xf(y)\|_A \leq \theta \cdot (\|x\|_A^r \cdot \|y\|_A^r) \tag{2.34}$$

*for all $\mu \in \mathbb{T}^1$ and $x_1, \cdots, x_m, x, y \in A$. Then there exists a unique derivation $\delta : A \to A$ such that*

$$\|f(x) - \delta(x)\|_A \leq \frac{\theta}{m - m^r} \|x\|_A^r$$

*for all $x \in A$.*

*Proof.* The proof follows from Theorem 2.10 by taking

$$\varphi(x_1, \cdots, x_m) := \theta \cdot (\|x_1\|_A^r + \cdots \|x_m\|_A^r),$$

$$\psi(x, y) := \theta \cdot (\|x\|_A^r \cdot \|y\|_A^r)$$

for all $x_1, \cdots, x_m, x, y \in A$ and $L = m^{r-1}$. □

*Remark 2.12.* Let $f : A \to B$ be a mapping for which there exist the functions $\varphi : A^m \to [0, \infty)$ and $\psi : A^2 \to [0, \infty)$ such that

$$\lim_{j \to \infty} m^j \varphi(m^{-j} x_1, \cdots, m^{-j} x_m) = 0, \tag{2.35}$$

$$\|D_\mu f(x_1, \cdots, x_m)\|_A \le \varphi(x_1, \cdots, x_m), \tag{2.36}$$

$$\|f(xy) - f(x)y - xf(y)\|_A \le \psi(x, y), \tag{2.37}$$

$$\lim_{j \to \infty} m^{2j} \psi(m^{-j}x, m^{-j}y) = 0 \tag{2.38}$$

for all $\mu \in \mathbb{T}^1$ and $x_1, \cdots, x_m, x, y \in A$. If there exists $L < 1$ such that $\varphi(mx, 0, \cdots, 0) \le \frac{L}{m}\varphi(x, 0, \cdots, 0)$ for all $x \in A$. Then there exists a unique derivation $\delta : A \to A$ such that

$$\|f(x) - \delta(x)\|_A \le \frac{L}{m - mL}\varphi(x, 0, \cdots, 0) \tag{2.39}$$

for all $x \in A$.

**Corollary 2.13.** *Let $r > 1$, $\theta$ be nonnegative real numbers and $f : A \to A$ be a mapping such that*

$$\|D_\mu f(x_1, \cdots, x_m)\|_A \le \theta \cdot (\|x_1\|_A^r + \cdots \|x_m\|_A^r) \tag{2.40}$$

*and*

$$\|f(xy) - f(x)y - xf(y)\|_A \le \theta \cdot (\|x\|_A^r \cdot \|y\|_A^r) \tag{2.41}$$

*for all $\mu \in \mathbb{T}^1$ and $x_1, \cdots, x_m, x, y \in A$. Then there exists a unique derivation $\delta : A \to A$ such that*

$$\|f(x) - \delta(x)\|_A \le \frac{\theta}{m^r - m}\|x\|_A^r$$

*for all $x \in A$.*

*Proof.* Consider Remark 2.12 and take

$$\varphi(x_1, \cdots, x_m) := \theta \cdot (\|x_1\|_A^r + \cdots \|x_m\|_A^r),$$

$$\psi(x, y) := \theta \cdot (\|x\|_A^r \cdot \|y\|_A^r)$$

for all $x_1, \cdots, x_m, x, y \in A$ and $L = m^{1-r}$. □

## 2.3 Stability in Quasi-Banach Algebras

Let $q$ be a positive rational number and $n$ be a nonnegative integer. We consider the *Hyers-Ulam stability* of homomorphisms in *quasi-Banach algebras* and generalized derivations on quasi-Banach algebras for the following functional equation:

$$\sum_{i=1}^{n} f\left(\sum_{j=1}^{n} q(x_i - x_j)\right) + nf\left(\sum_{i=1}^{n} qx_i\right) = nq \sum_{i=1}^{n} f(x_i). \qquad (2.42)$$

This is applied to investigate some isomorphisms in quasi-Banach algebras (see [180, 233, 238]).

### 2.3.1 Stability of Homomorphisms in Quasi-Banach Algebras

Let $q$ be a positive rational number. For any mapping $f : A \to B$, we define $Df : A^n \to B$ by

$$Df(x_1, \cdots, x_n)$$
$$:= \sum_{i=1}^{n} f\left(\sum_{j=1}^{n} q(x_i - x_j)\right) + nf\left(\sum_{i=1}^{n} qx_i\right) - nq \sum_{i=1}^{n} f(x_i)$$

for all $x_1, \cdots, x_n \in X$.

**Lemma 2.14.** *Let $f : A \to B$ be a mapping satisfies the functional equation (2.42). Then the mapping $f$ is Cauchy additive and $\mathbb{R}$-linear.*

*Proof.* The proof is easy (see also [229, 267]).                                          $\square$

Now, we prove the *Hyers-Ulam stability* of homomorphisms in quasi-Banach algebras.

**Theorem 2.15.** *Assume that $r > 2$ if $nq > 1$ and $0 < r < 1$ if $nq < 1$. Let $\theta$ be a positive real number and $f : A \to B$ be an odd mapping such that*

$$\|Df(x_1, \cdots, x_n)\|_B \leq \theta \sum_{j=1}^{n} \|x_j\|_A^r \qquad (2.43)$$

*and*

$$\|f(xy) - f(x)f(y)\|_B \leq \theta(\|x\|_A^r + \|y\|_A^r) \qquad (2.44)$$

*for all $x, y, x_1, \cdots, x_n \in A$. If $f(tx)$ is continuous in $t \in \mathbb{R}$ for each fixed $x \in A$, then there exists a unique homomorphism $H : A \to B$ such that*

$$\|f(x) - H(x)\|_B \leq \frac{\theta}{((nq)^{pr} - (nq)^p)^{\frac{1}{p}}} \|x\|_A^r \qquad (2.45)$$

*for all $x \in A$.*

*Proof.* Letting $x_1 = \cdots = x_n = x$ in (2.43), we get

$$\|nf(nqx) - n^2qf(x)\|_B \le n\theta\|x\|_A^r \tag{2.46}$$

for all $x \in A$ and so

$$\left\|f(x) - nqf\left(\frac{x}{nq}\right)\right\|_B \le \frac{\theta}{(nq)^r}\|x\|_A^r$$

for all $x \in A$. Since $B$ is a $p$-Banach algebra, we have

$$\left\|(nq)^l f\left(\frac{x}{(nq)^l}\right) - (nq)^m f\left(\frac{x}{(nq)^m}\right)\right\|_B^p$$

$$\le \sum_{j=l}^{m-1}\left\|(nq)^j f\left(\frac{x}{(nq)^j}\right) - (nq)^{j+1} f\left(\frac{x}{(nq)^{j+1}}\right)\right\|_B^p \tag{2.47}$$

$$\le \frac{\theta^p}{(nq)^{pr}}\sum_{j=l}^{m-1}\frac{(nq)^{pj}}{(nq)^{prj}}\|x\|_A^{pr}$$

for all $m \ge 1$, $l$ with $m > l$ and $x \in A$. It follows from (2.47) that the sequence $\{(nq)^d f(\frac{x}{(nq)^d})\}$ is a Cauchy sequence for all $x \in A$. Since $B$ is complete, the sequence $\{(nq)^d f(\frac{x}{(nq)^d})\}$ converges. So one can define a mapping $H : A \to B$ by

$$H(x) := \lim_{d\to\infty}(nq)^d f\left(\frac{x}{(nq)^d}\right)$$

for all $x \in A$. Moreover, letting $l = 0$ and $m \to \infty$ in (2.47), we get (2.46). It follows from (2.43) that

$$\|DH(x_1, \cdots, x_n)\|_B = \lim_{d\to\infty}(nq)^d\left\|Df\left(\frac{x_1}{(nq)^d}, \cdots, \frac{x_n}{(nq)^d}\right)\right\|_B$$

$$\le \lim_{d\to\infty}\frac{(nq)^d\theta}{(nq)^{dr}}\sum_{j=1}^{n}\|x_j\|_A^r$$

$$= 0$$

for all $x_1, \cdots, x_n \in A$. Thus we have

$$DH(x_1, \cdots, x_n) = 0$$

for all $x_1, \cdots, x_n \in A$. By (2.43), the mapping $H : A \to B$ is Cauchy additive and $\mathbb{R}$-linear. It follows from (2.45) that

$$\|H(xy) - H(x)H(y)\|_B$$

$$= \lim_{d \to \infty} (nq)^{2d} \left\| f\left(\frac{xy}{(nq)^d (nq)^d}\right) - f\left(\frac{x}{(nq)^d}\right) f\left(\frac{y}{(nq)^d}\right) \right\|_B$$

$$\leq \lim_{d \to \infty} \frac{(nq)^{2d\theta}}{(nq)^{dr}} (\|x\|_A^r + \|y\|_A^r)$$

$$= 0$$

for all $x, y \in A$ and so

$$H(xy) = H(x)H(y)$$

for all $x, y \in A$.

Now, let $T : A \to B$ be another mapping satisfying (2.46). Then we have

$$\|H(x) - T(x)\|_B$$

$$= (nq)^d \left\| H\left(\frac{x}{(nq)^d}\right) - T\left(\frac{x}{(nq)^d}\right) \right\|_B$$

$$\leq (nq)^d K \left( \left\| H\left(\frac{x}{(nq)^d}\right) - f\left(\frac{x}{(nq)^d}\right) \right\|_B + \left\| T\left(\frac{x}{(nq)^d}\right) - f\left(\frac{x}{(nq)^d}\right) \right\|_B \right)$$

$$\leq \frac{2 \cdot (nq)^d K\theta}{((nq)^{pr} - (nq)^p)^{\frac{1}{p}} (nq)^{dr}} \|x\|_A^r,$$

which tends to zero as $n \to \infty$ for all $x \in A$. So we can conclude that $H(x) = T(x)$ for all $x \in A$. This proves the uniqueness of $H$. Thus the mapping $H : A \to B$ is a unique homomorphism satisfying (2.46). This completes the proof.                                         □

**Theorem 2.16.** *Assume that $0 < r < 1$ if $nq > 1$ and that $r > 2$ if $nq < 1$. Let $\theta$ be a positive real number, and let $f : A \to B$ be an odd mapping satisfying* (2.43) *and* (2.45). *If $f(tx)$ is continuous in $t \in \mathbb{R}$ for each fixed $x \in A$, then there exists a unique homomorphism $H : A \to B$ such that*

$$\|f(x) - H(x)\|_B \leq \frac{\theta}{((nq)^p - (nq)^{pr})^{\frac{1}{p}}} \|x\|_A^r \qquad (2.48)$$

*for all $x \in A$.*

*Proof.* It follows from (2.46) that

$$\|f(x) - \frac{1}{nq} f(nqx)\|_B \leq \frac{\theta}{nq} \|x\|_A^r$$

for all $x \in A$. Since $B$ is a $p$-Banach algebra, we have

$$\left\|\frac{1}{(nq)^l}f((nq)^l x) - \frac{1}{(nq)^m}f((nq)^m x)\right\|_B^p$$

$$\leq \sum_{j=l}^{m-1}\left\|\frac{1}{(nq)^j}f((nq)^j x) - \frac{1}{(nq)^{j+1}}f((nq)^{j+1}x)\right\|_B^p \qquad (2.49)$$

$$\leq \frac{\theta^p}{(nq)^p}\sum_{j=l}^{m-1}\frac{(nq)^{prj}}{(nq)^{pj}}\|x\|_A^{pr}$$

for all $m \geq 1$, $l$ with $m > l$ and $x \in A$. It follows from (2.49) that the sequence $\{\frac{1}{(nq)^d}f((nq)^d x)\}$ is a Cauchy sequence for all $x \in A$. Since $B$ is complete, the sequence $\{\frac{1}{(nq)^d}f((nq)^d x)\}$ converges. So one can define a mapping $H : A \to B$ by

$$H(x) := \lim_{d\to\infty}\frac{1}{(nq)^d}f((nq)^d x)$$

for all $x \in A$. Moreover, letting $l = 0$ and $m \to \infty$ in (2.49), we get (2.48).

The rest of the proof is similar to the proof of Theorem 2.15. $\qquad\square$

### 2.3.2 Isomorphisms in Quasi-Banach Algebras

Assume that $A$ is a *quasi-Banach algebra* with the quasi-norm $\|\cdot\|_A$ and the unit $e$ and $B$ is a $p$-Banach algebra with the $p$-norm $\|\cdot\|_B$ and the unit $e'$. Let $K$ be the modulus of concavity of $\|\cdot\|_B$.

Now, we consider isomorphisms in quasi-Banach algebras.

**Theorem 2.17.** *Assume that $r > 2$ if $nq > 1$ and that $0 < r < 1$ if $nq < 1$. Let $\theta$ be a positive real number and $f : A \to B$ be an odd bijective mapping satisfying (2.43) such that*

$$f(xy) = f(x)f(y) \qquad (2.50)$$

*for all $x, y \in A$. If $\lim_{d\to\infty}(nq)^d f(\frac{e}{(nq)^d}) = e'$ and $f(tx)$ is continuous in $t \in \mathbb{R}$ for each fixed $x \in A$, then the mapping $f : A \to B$ is an isomorphism.*

*Proof.* The condition (2.50) implies that $f : A \to B$ satisfies (2.45). By the same reasoning as in the proof of Theorem 2.15, there exists a unique homomorphism $H : A \to B$, which is defined by

$$H(x) := \lim_{d\to\infty}(nq)^d f\left(\frac{x}{(nq)^d}\right)$$

for all $x \in A$. Thus we have

$$H(x) = H(ex) = \lim_{d \to \infty} (nq)^d f\left(\frac{ex}{(nq)^d}\right) = \lim_{d \to \infty} (nq)^d f\left(\frac{e}{(nq)^d} \cdot x\right)$$

$$= \lim_{d \to \infty} (nq)^d f\left(\frac{e}{(nq)^d}\right) f(x) = e'f(x) = f(x)$$

for all $x \in A$. So the bijective mapping $f : A \to B$ is an isomorphism. This completes the proof.    $\square$

*Remark 2.18.* Assume that $0 < r < 1$ if $nq > 1$ and that $r > 2$ if $nq < 1$. Let $\theta$ be a positive real number and $f : A \to B$ be an odd bijective mapping satisfying (2.43) and (2.50). If $f(tx)$ is continuous in $t \in \mathbb{R}$ for each fixed $x \in A$ and

$$\lim_{d \to \infty} \frac{1}{(nq)^d} f((nq)^d e) = e',$$

then the mapping $f : A \to B$ is an isomorphism.

### 2.3.3  Stability of Generalized Derivations in Quasi-Banach Algebras

Assume that $A$ is a *p-Banach algebra* with the *p*-norm $\| \cdot \|_A$. Let $K$ be the modulus of concavity of $\| \cdot \|_A$.

**Definition 2.19 ([18]).**  A *generalized derivation* $\delta : A \to A$ is $\mathbb{R}$-linear and fulfills the *generalized Leibniz rule*:

$$\delta(xyz) = \delta(xy)z - x\delta(y)z + x\delta(yz)$$

for all $x, y, z \in A$.

Now, we prove the *Hyers-Ulam stability* of generalized derivations on quasi-Banach algebras.

**Theorem 2.20.**  *Assume that $r > 3$ if $nq > 1$ and that $0 < r < 1$ if $nq < 1$. Let $\theta$ be a positive real number and $f : A \to A$ be an odd mapping satisfying (2.43) such that*

$$\|f(xyz) - f(xy)z + xf(y)z - xf(yz)\|_A$$
$$\leq \theta(\|x\|_A^r + \|y\|_A^r + \|z\|_A^r) \tag{2.51}$$

*for all $x, y, z \in A$. If $f(tx)$ is continuous in $t \in \mathbb{R}$ for each fixed $x \in A$, then there exists a unique generalized derivation $\delta : A \to A$ such that*

$$\|f(x) - \delta(x)\|_A \le \frac{\theta}{((nq)^{pr} - (nq)^p)^{\frac{1}{p}}} \|x\|_A^r \tag{2.52}$$

for all $x \in A$.

*Proof.* By the same reasoning as in the proof of Theorem 2.15, there exists a unique $\mathbb{R}$-linear mapping $\delta : A \to A$ satisfying (2.52). The mapping $\delta : A \to A$ is defined by

$$\delta(x) := \lim_{d \to \infty} (nq)^d f\left(\frac{x}{(nq)^d}\right)$$

for all $x \in A$. It follows from (2.51) that

$$
\begin{aligned}
&\|\delta(xyz) - \delta(xy)z + x\delta(y)z - x\delta(yz)\|_A \\
&= \lim_{d \to \infty} (nq)^{3d} \Big\| f\left(\frac{xyz}{(nq)^{3d}}\right) - f\left(\frac{xy}{(nq)^{2d}}\right)\frac{z}{(nq)^d} \\
&\quad + \frac{x}{(nq)^d} f\left(\frac{y}{(nq)^d}\right)\frac{y}{(nq)^d} - \frac{x}{(nq)^d} f\left(\frac{yz}{(nq)^{2d}}\right) \Big\|_A \\
&\le \lim_{d \to \infty} \frac{(nq)^{3d}\theta}{(nq)^{dr}}(\|x\|_A^r + \|y\|_A^r + \|z\|_A^r) \\
&= 0
\end{aligned}
$$

for all $x, y, z \in A$ and so

$$\delta(xyz) = \delta(xy)z - x\delta(y)z + x\delta(yz)$$

for all $x, y, z \in A$. Thus the mapping $\delta : A \to A$ is a unique generalized derivation satisfying (2.52). This completes the proof. $\qquad \square$

**Theorem 2.21.** *Assume that $0 < r < 1$ if $nq > 1$ and that $r > 3$ if $nq < 1$. Let $\theta$ be a positive real number and $f : A \to A$ be an odd mapping satisfying (2.43) and (2.51). If $f(tx)$ is continuous in $t \in \mathbb{R}$ for each fixed $x \in A$, then there exists a unique generalized derivation $\delta : A \to A$ such that*

$$\|f(x) - \delta(x)\|_A \le \frac{\theta}{((nq)^p - (nq)^{pr})^{\frac{1}{p}}} \|x\|_A^r \tag{2.53}$$

for all $x \in A$.

*Proof.* By the same reasoning as in the proof of Theorem 2.16, there exists a unique $\mathbb{R}$-linear mapping $\delta : A \to A$ satisfying (2.53). The mapping $\delta : A \to A$ is defined by

$$\delta(x) := \lim_{d \to \infty} \frac{1}{(nq)^d} f((nq)^d x)$$

for all $x \in A$.

The rest of the proof is similar to the proof of Theorem 2.20. $\qquad \square$

## 2.4   Stability of Cauchy–Jensen Functional Equations

In this section, we prove the *Hyers-Ulam stability* of homomorphisms in real Banach algebras and generalized derivations on real Banach algebras for the following *Cauchy-Jensen functional equations*(see also [19, 147, 232]):

$$f\left(\frac{x+y}{2} + z\right) + f\left(\frac{x-y}{2} + z\right) = f(x) + 2f(z)$$

and

$$2f\left(\frac{x+y}{2} + z\right) = f(x) + f(y) + 2f(z).$$

### 2.4.1   Stability of Homomorphisms in Real Banach Algebras

Assume that $A$ is a real Banach algebra with the norm $\|\cdot\|_A$ and $B$ is a real Banach algebra with the norm $\|\cdot\|_B$. For any mapping $f : A \rightarrow B$, we define

$$Cf(x, y, z) := f\left(\frac{x+y}{2} + z\right) + f\left(\frac{x-y}{2} + z\right) - f(x) - 2f(z)$$

for all $x, y, z \in A$.

Now, we prove the Hyers-Ulam stability of homomorphisms in real Banach algebras for the functional equation $Cf(x, y, z) = 0$.

**Lemma 2.22.** *Let $X$ and $Y$ be vector spaces. If a mapping $f : X \rightarrow Y$ satisfies*

$$f\left(\frac{x+y}{2} + z\right) + f\left(\frac{x-y}{2} + z\right) = f(x) + 2f(z),  \qquad (2.54)$$

$$f\left(\frac{x+y}{2} + z\right) - f\left(\frac{x-y}{2} + z\right) = f(y)  \qquad (2.55)$$

*or*

$$2f\left(\frac{x+y}{2} + z\right) = f(x) + f(y) + 2f(z)  \qquad (2.56)$$

*for all $x, y, z \in X$, then the mapping $f : X \rightarrow Y$ is Cauchy additive.*

*Proof.* Letting $x = y$ in (2.54), we get

$$f(x + z) + f(z) = f(x) + 2f(z)$$

for all $x, z \in X$ and so $f(x + z) = f(x) + f(z)$ for all $x, z \in X$. Hence $f : X \to Y$ is Cauchy additive. Letting $x = y$ in (2.55), we get

$$f(x + z) - f(z) = f(x)$$

for all $x, z \in X$ and so $f(x + z) = f(x) + f(z)$ for all $x, z \in X$. Hence $f : X \to Y$ is Cauchy additive. Letting $x = y$ in (2.56), we get

$$2f(x + z) = 2f(x) + 2f(z)$$

for all $x, z \in X$ and so $f(x + z) = f(x) + f(z)$ for all $x, z \in X$. Hence $f : X \to Y$ is Cauchy additive. This completes the proof. $\qquad\square$

The mappings $f : X \to Y$ given in the statement of Lemma 2.22 are called *Cauchy–Jensen type additive mappings*. Putting $z = 0$ in (2.56), we get the Jensen additive mapping $2f(\frac{x+y}{2}) = f(x) + f(y)$ and, putting $x = y$ in (2.56), we get the Cauchy additive mapping $f(x + z) = f(x) + f(z)$.

**Theorem 2.23.** *Let $f : A \to B$ be a mapping for which there exists a function $\varphi : A^3 \to [0, \infty)$ such that*

$$\sum_{j=0}^{\infty} \frac{1}{2^j} \varphi(2^j x, 2^j y, 2^j z) < \infty, \tag{2.57}$$

$$\|Cf(x, y, z)\|_B \leq \varphi(x, y, z), \tag{2.58}$$

$$\|f(xy) - f(x)f(y)\|_B \leq \varphi(x, y, 0) \tag{2.59}$$

*for all $x, y, z \in A$. If there exists $L < 1$ such that*

$$\varphi(x, x, x) \leq 2L\varphi(\frac{x}{2}, \frac{x}{2}, \frac{x}{2})$$

*for all $x \in A$ and $f(tx)$ is continuous in $t \in \mathbb{R}$ for each fixed $x \in A$, then there exists a unique homomorphism $H : A \to B$ such that*

$$\|f(x) - H(x)\|_B \leq \frac{1}{2 - 2L} \varphi(x, x, x) \tag{2.60}$$

*for all $x \in A$.*

*Proof.* Consider the set

$$X := \{g : A \to B\}$$

and introduce the generalized metric on $X$ defined by

$$d(g, h) = \inf\{C \in \mathbb{R}_+ : \|g(x) - h(x)\|_B \le C\varphi(x, x, x), \ \forall x \in A\},$$

which $(X, d)$ is complete.

Now, we consider the linear mapping $J : X \to X$ such that

$$Jg(x) := \frac{1}{2}g(2x)$$

for all $x \in A$. Note that

$$d(Jg, Jh) \le Ld(g, h)$$

for all $g, h \in X$. Letting $y = z = x$ in (2.58), we get

$$\|f(2x) - 2f(x)\|_B \le \varphi(x, x, x) \tag{2.61}$$

for all $x \in A$ and so

$$\left\| f(x) - \frac{1}{2}f(2x) \right\|_B \le \frac{1}{2}\varphi(x, x, x)$$

for all $x \in A$. Hence $d(f, Jf) \le \frac{1}{2}$. By Theorem 1.3, there exists a mapping $H : A \to B$ such that

(1) $H$ is a fixed point of $J$, i.e.,

$$H(2x) = 2H(x) \tag{2.62}$$

for all $x \in A$. The mapping $H$ is a unique fixed point of $J$ in the set

$$Y = \{g \in X : d(f, g) < \infty\}.$$

This implies that $H$ is a unique mapping satisfying (2.62) such that there exists $C \in (0, \infty)$ satisfying

$$\|H(x) - f(x)\|_B \le C\varphi(x, x, x)$$

for all $x \in A$;

(2) $d(J^n f, H) \to 0$ as $n \to \infty$. This implies the equality

$$\lim_{n \to \infty} \frac{f(2^n x)}{2^n} = H(x) \tag{2.63}$$

for all $x \in A$;

(3) $d(f, H) \le \frac{1}{1-L} d(f, Jf)$, which implies the inequality

$$d(f, H) \le \frac{1}{2 - 2L}.$$

This implies that the inequality (2.60) holds.

It follows from (2.57), (2.58) and (2.63) that

$$\left\| H\left(\frac{x+y}{2} + z\right) + H\left(\frac{x-y}{2} + z\right) - H(x) - 2H(z) \right\|_B$$

$$= \lim_{n \to \infty} \frac{1}{2^n} \| f(2^{n-1}(x+y) + 2^n z) + f(2^{n-1}(x-y) + 2^n z)$$

$$-f(2^n x) - 2f(2^n z) \|_B$$

$$\le \lim_{n \to \infty} \frac{1}{2^n} \varphi(2^n x, 2^n y, 2^n z)$$

$$= 0$$

for all $x, y, z \in A$ and so

$$H\left(\frac{x+y}{2} + z\right) + H\left(\frac{x-y}{2} + z\right) = H(x) + 2H(z)$$

for all $x, y, z \in A$. By Lemma 2.22, the mapping $H : A \to B$ is Cauchy additive. By (2.58), the mapping $H : A \to B$ is $\mathbb{R}$-linear. It follows from (2.59) that

$$\| H(xy) - H(x)H(y) \|_B = \lim_{n \to \infty} \frac{1}{4^n} \| f(4^n xy) - f(2^n x)f(2^n y) \|_B$$

$$\le \lim_{n \to \infty} \frac{1}{4^n} \varphi(2^n x, 2^n y, 0)$$

$$\le \lim_{n \to \infty} \frac{1}{2^n} \varphi(2^n x, 2^n y, 0)$$

$$= 0$$

for all $x, y \in A$ and so

$$H(xy) = H(x)H(y)$$

for all $x, y \in A$. Thus $H : A \to B$ is a homomorphism satisfying (2.60). This completes the proof. ☐

**Corollary 2.24.** *Let $r < 1$, $\theta$ be nonnegative real numbers and $f : A \to B$ be a mapping such that*

$$\| Cf(x, y, z) \|_B \le \theta(\|x\|_A^r + \|y\|_A^r + \|z\|_A^r) \tag{2.64}$$

*and*

$$\|f(xy) - f(x)f(y)\|_B \leq \theta(\|x\|_A^r + \|y\|_A^r) \tag{2.65}$$

*for all $x, y, z \in A$. If $f(tx)$ is continuous in $t \in \mathbb{R}$ for each fixed $x \in A$, then there exists a unique homomorphism $H : A \to B$ such that*

$$\|f(x) - H(x)\|_B \leq \frac{3\theta}{2 - 2^r}\|x\|_A^r$$

*for all $x \in A$.*

*Proof.* The proof follows from Theorem 2.23 by taking

$$\varphi(x, y, z) := \theta(\|x\|_A^r + \|y\|_A^r + \|z\|_A^r)$$

for all $x, y, z \in A$ and $L = 2^{r-1}$. We get the desired result. $\qquad\square$

**Theorem 2.25.** *Let $f : A \to B$ be a mapping for which there exists a function $\varphi : A^3 \to [0, \infty)$ satisfying (2.58) and (2.59) such that*

$$\sum_{j=0}^{\infty} 4^j \varphi\left(\frac{x}{2^j}, \frac{y}{2^j}, \frac{z}{2^j}\right) < \infty \tag{2.66}$$

*for all $x, y, z \in A$. If there exists $L < 1$ such that*

$$\varphi(x, x, x) \leq \frac{1}{2}L\varphi(2x, 2x, 2x)$$

*for all $x \in A$ and $f(tx)$ is continuous in $t \in \mathbb{R}$ for each fixed $x \in A$, then there exists a unique homomorphism $H : A \to B$ such that*

$$\|f(x) - H(x)\|_B \leq \frac{L}{2 - 2L}\varphi(x, x, x) \tag{2.67}$$

*for all $x \in A$.*

*Proof.* We consider the linear mapping $J : X \to X$ such that

$$Jg(x) := 2g\left(\frac{x}{2}\right)$$

for all $x \in A$. It follows from (2.61) that

$$\left\|f(x) - 2f\left(\frac{x}{2}\right)\right\|_B \leq \varphi\left(\frac{x}{2}, \frac{x}{2}, \frac{x}{2}\right) \leq \frac{L}{2}\varphi(x, x, x)$$

for all $x \in A$. Hence $d(f, Jf) \leq \frac{L}{2}$. By Theorem 1.3, there exists a mapping $H : A \to B$ such that

(1)  $H$ is a fixed point of $J$, i.e.,

$$H(2x) = 2H(x) \tag{2.68}$$

for all $x \in A$. The mapping $H$ is a unique fixed point of $J$ in the set

$$Y = \{g \in X : d(f, g) < \infty\}.$$

This implies that $H$ is a unique mapping satisfying (2.68) such that there exists $C \in (0, \infty)$ satisfying

$$\|H(x) - f(x)\|_B \leq C\varphi(x, x, x)$$

for all $x \in A$;

(2)  $d(J^n f, H) \to 0$ as $n \to \infty$. This implies the equality

$$\lim_{n \to \infty} 2^n f\left(\frac{x}{2^n}\right) = H(x) \tag{2.69}$$

for all $x \in A$;

(3)  $d(f, H) \leq \frac{1}{1-L} d(f, Jf)$, which implies the inequality

$$d(f, H) \leq \frac{L}{2 - 2L},$$

which implies that the inequality (2.67) holds.

It follows from (2.58), (2.66) and (2.69) that

$$\left\| H\left(\frac{x+y}{2} + z\right) + H\left(\frac{x-y}{2} + z\right) - H(x) - 2H(z) \right\|_B$$

$$= \lim_{n \to \infty} 2^n \left\| f\left(\frac{x+y}{2^{n+1}} + \frac{z}{2^n}\right) + f\left(\frac{x-y}{2^{n+1}} + \frac{z}{2^n}\right) - f\left(\frac{x}{2^n}\right) - 2f\left(\frac{z}{2^n}\right) \right\|_B$$

$$\leq \lim_{n \to \infty} 2^n \varphi\left(\frac{x}{2^n}, \frac{y}{2^n}, \frac{z}{2^n}\right)$$

$$\leq \lim_{n \to \infty} 4^n \varphi\left(\frac{x}{2^n}, \frac{y}{2^n}, \frac{z}{2^n}\right)$$

$$= 0$$

for all $x, y, z \in A$ and so

$$H\left(\frac{x+y}{2} + z\right) + H\left(\frac{x-y}{2} + z\right) = H(x) + 2H(z)$$

for all $x, y, z \in A$. By Lemma 2.22, the mapping $H : A \to B$ is Cauchy additive. By (2.58), the mapping $H : A \to B$ is $\mathbb{R}$-linear. It follows from (2.59) that

$$
\begin{aligned}
\|H(xy) - H(x)H(y)\|_B &= \lim_{n \to \infty} 4^n \left\| f\left(\frac{xy}{4^n}\right) - f\left(\frac{x}{2^n}\right) f\left(\frac{y}{2^n}\right) \right\|_B \\
&\leq \lim_{n \to \infty} 4^n \varphi\left(\frac{x}{2^n}, \frac{y}{2^n}, 0\right) \\
&= 0
\end{aligned}
$$

for all $x, y \in A$ and so

$$
H(xy) = H(x)H(y)
$$

for all $x, y \in A$. Thus $H : A \to B$ is a homomorphism satisfying (2.67). This completes the proof.      $\square$

**Corollary 2.26.** *Let $r > 2$, $\theta$ be nonnegative real numbers and $f : A \to B$ be a mapping satisfying (2.64) and (2.65). If $f(tx)$ is continuous in $t \in \mathbb{R}$ for each fixed $x \in A$, then there exists a unique homomorphism $H : A \to B$ such that*

$$
\|f(x) - H(x)\|_B \leq \frac{3\theta}{2^r - 2} \|x\|_A^r
$$

*for all $x \in A$.*

*Proof.* The proof follows from Theorem 2.25 by taking

$$
\varphi(x, y, z) := \theta(\|x\|_A^r + \|y\|_A^r + \|z\|_A^r)
$$

for all $x, y, z \in A$ and $L = 2^{1-r}$.      $\square$

### 2.4.2   Stability of Generalized Derivations in Real Banach Algebras

Assume that $A$ is a real Banach algebra with the norm $\| \cdot \|_A$. For any mapping $f : A \to A$, we define

$$
Df(x, y, z) := 2f\left(\frac{x + y}{2} + z\right) - f(x) - f(y) - 2f(z)
$$

for all $x, y, z \in A$.

Now, we prove the *Hyers-Ulam stability* of generalized derivations on real Banach algebras for the functional equation $Df(x, y, z) = 0$.

**Theorem 2.27.** *Let $f : A \rightarrow A$ be a mapping for which there exists a function $\varphi : A^3 \rightarrow [0, \infty)$ satisfying (2.57) such that*

$$\|Df(x, y, z)\|_A \leq \varphi(x, y, z) \tag{2.70}$$

*and*

$$\|f(xyz) - f(xy)z + xf(y)z - xf(yz)\|_A \leq \varphi(x, y, z) \tag{2.71}$$

*for all $x, y, z \in A$. If there exists $L < 1$ such that*

$$\varphi(x, x, x) \leq 2L\varphi\left(\frac{x}{2}, \frac{x}{2}, \frac{x}{2}\right)$$

*for all $x \in A$ and $f(tx)$ is continuous in $t \in \mathbb{R}$ for each fixed $x \in A$, then there exists a unique generalized derivation $\delta : A \rightarrow A$ such that*

$$\|f(x) - \delta(x)\|_A \leq \frac{1}{4 - 4L}\varphi(x, x, x) \tag{2.72}$$

*for all $x \in A$.*

*Proof.* Consider the set

$$X := \{g : A \rightarrow A\}$$

and introduce the *generalized metric* on $X$ defined by

$$d(g, h) = \inf\{C \in \mathbb{R}_+ : \|g(x) - h(x)\|_A \leq C\varphi(x, x, x), \ \forall x \in A\},$$

which $(X, d)$ is complete.

Now, we consider the linear mapping $J : X \rightarrow X$ such that

$$Jg(x) := \frac{1}{2}g(2x)$$

for all $x \in A$. Now, we have

$$d(Jg, Jh) \leq Ld(g, h)$$

for all $g, h \in X$. Letting $y = z = x$ in (2.70), we get

$$\|2f(2x) - 4f(x)\|_A \leq \varphi(x, x, x) \tag{2.73}$$

for all $x \in A$ and so

$$\left\| f(x) - \frac{1}{2} f(2x) \right\|_A \leq \frac{1}{4} \varphi(x, x, x)$$

for all $x \in A$. Hence $d(f, Jf) \leq \frac{1}{4}$. By Theorem 1.3, there exists a mapping $\delta : A \to A$ such that

(1)  $\delta$ is a fixed point of $J$, i.e.,

$$\delta(2x) = 2\delta(x) \qquad (2.74)$$

for all $x \in A$. The mapping $\delta$ is a unique fixed point of $J$ in the set

$$Y = \{ g \in X : d(f, g) < \infty \}.$$

This implies that $\delta$ is a unique mapping satisfying (2.74) such that there exists $C \in (0, \infty)$ satisfying

$$\| \delta(x) - f(x) \|_A \leq C \varphi(x, x, x)$$

for all $x \in A$;

(2)  $d(J^n f, \delta) \to 0$ as $n \to \infty$. This implies the equality

$$\lim_{n \to \infty} \frac{f(2^n x)}{2^n} = \delta(x) \qquad (2.75)$$

for all $x \in A$;

(3)  $d(f, \delta) \leq \frac{1}{1-L} d(f, Jf)$, which implies the inequality

$$d(f, \delta) \leq \frac{1}{4 - 4L}.$$

This implies that the inequality (2.72) holds.

It follows from (2.57), (2.70) and (2.75) that

$$\left\| 2\delta\left( \frac{x+y}{2} + z \right) - \delta(x) - \delta(y) - 2\delta(z) \right\|_A$$

$$= \lim_{n \to \infty} \frac{1}{2^n} \| 2f(2^{n-1}(x+y) + 2^n z) - f(2^n x) - f(2^n y) - 2f(2^n z) \|_A$$

$$\leq \lim_{n \to \infty} \frac{1}{2^n} \varphi(2^n x, 2^n y, 2^n z)$$

$$= 0$$

for all $x, y, z \in A$ and so

$$2\delta\left(\frac{x+y}{2}+z\right) = \delta(x) + \delta(y) + 2\delta(z)$$

for all $x, y, z \in A$. By Lemma 2.22, the mapping $\delta : A \to A$ is Cauchy additive. By (2.71), the mapping $\delta : A \to A$ is $\mathbb{R}$-linear. It follows from (2.71) that

$$\|\delta(xyz) - \delta(xy)z + x\delta(y)z - x\delta(yz)\|_A$$

$$= \lim_{n\to\infty} \frac{1}{8^n}\|f(8^n xyz) - f(4^n xy)\cdot 2^n z + 2^n xf(2^n y)\cdot 2^n z - 2^n xf(4^n yz)\|_A$$

$$\leq \lim_{n\to\infty} \frac{1}{8^n}\varphi(2^n x, 2^n y, 2^n z)$$

$$\leq \lim_{n\to\infty} \frac{1}{2^n}\varphi(2^n x, 2^n y, 2^n z)$$

$$= 0$$

for all $x, y, z \in A$ and so

$$\delta(xyz) = \delta(xy)z - x\delta(y)z + x\delta(yz)$$

for all $x, y, z \in A$. Thus $\delta : A \to A$ is a generalized derivation satisfying (2.72). This completes the proof. □

**Corollary 2.28.** *Let $r < 1$, $\theta$ be nonnegative real numbers and $f : A \to A$ be a mapping such that*

$$\|Df(x, y, z)\|_A \leq \theta \cdot \|x\|_A^{\frac{r}{3}}\cdot \|y\|_A^{\frac{r}{3}}\cdot \|z\|_A^{\frac{r}{3}} \tag{2.76}$$

*and*

$$\|f(xyz) - f(xy)z + xf(y)z - xf(yz)\|_A \tag{2.77}$$
$$\leq \theta \cdot \|x\|_A^{\frac{r}{3}}\cdot \|y\|_A^{\frac{r}{3}}\cdot \|z\|_A^{\frac{r}{3}}$$

*for all $x, y, z \in A$. If $f(tx)$ is continuous in $t \in \mathbb{R}$ for each fixed $x \in A$, then there exists a unique generalized derivation $\delta : A \to A$ such that*

$$\|f(x) - \delta(x)\|_A \leq \frac{\theta}{4 - 2^{r+1}}\|x\|_A^r$$

*for all $x \in A$.*

*Proof.* The proof follows from Theorem 2.27 by taking

$$\varphi(x, y, z) := \theta \cdot \|x\|_A^{\frac{r}{3}}\cdot \|y\|_A^{\frac{r}{3}}\cdot \|z\|_A^{\frac{r}{3}}$$

for all $x, y, z \in A$ and $L = 2^{r-1}$. □

**Theorem 2.29.** *Let $f : A \rightarrow A$ be a mapping for which there exists a function $\varphi : A^3 \rightarrow [0, \infty)$ satisfying (2.70) and (2.71) such that*

$$\sum_{j=0}^{\infty} 8^j \varphi\left(\frac{x}{2^j}, \frac{y}{2^j}, \frac{z}{2^j}\right) < \infty \tag{2.78}$$

*for all $x, y, z \in A$. If there exists $L < 1$ such that*

$$\varphi(x, x, x) \leq \frac{1}{2} L \varphi(2x, 2x, 2x)$$

*for all $x \in A$ and $f(tx)$ is continuous in $t \in \mathbb{R}$ for each fixed $x \in A$, then there exists a unique generalized derivation $\delta : A \rightarrow A$ such that*

$$\|f(x) - \delta(x)\|_A \leq \frac{L}{4 - 4L} \varphi(x, x, x) \tag{2.79}$$

*for all $x \in A$.*

*Proof.* We consider the linear mapping $J : X \rightarrow X$ such that

$$Jg(x) := 2g\left(\frac{x}{2}\right)$$

for all $x \in A$. It follows from (2.73) that

$$\left\|f(x) - 2f\left(\frac{x}{2}\right)\right\|_A \leq \frac{1}{2}\varphi\left(\frac{x}{2}, \frac{x}{2}, \frac{x}{2}\right) \leq \frac{L}{4}\varphi(x, x, x)$$

for all $x \in A$. Hence $d(f, Jf) \leq \frac{L}{4}$. By Theorem 1.3, there exists a mapping $\delta : A \rightarrow A$ such that

(1)  $\delta$ is a fixed point of $J$, i.e.,

$$\delta(2x) = 2\delta(x) \tag{2.80}$$

for all $x \in A$. The mapping $\delta$ is a unique fixed point of $J$ in the set

$$Y = \{g \in X : d(f, g) < \infty\}.$$

This implies that $\delta$ is a unique mapping satisfying (2.80) such that there exists $C \in (0, \infty)$ satisfying

$$\|\delta(x) - f(x)\|_A \leq C\varphi(x, x, x)$$

for all $x \in A$;

(2) $d(J^n f, \delta) \to 0$ as $n \to \infty$. This implies the equality

$$\lim_{n\to\infty} 2^n f\left(\frac{x}{2^n}\right) = \delta(x) \tag{2.81}$$

for all $x \in A$;

(3) $d(f, \delta) \leq \frac{1}{1-L} d(f, Jf)$, which implies the inequality

$$d(f, \delta) \leq \frac{L}{4 - 4L},$$

which implies that the inequality (2.79) holds.

It follows from (2.70), (2.78) and (2.81) that

$$\left\| 2\delta\left(\frac{x+y}{2} + z\right) - \delta(x) - \delta(y) - 2\delta(z) \right\|_A$$

$$= \lim_{n\to\infty} 2^n \left\| 2f\left(\frac{x+y}{2^{n+1}} + \frac{z}{2^n}\right) - f\left(\frac{x}{2^n}\right) - f\left(\frac{y}{2^n}\right) - 2f\left(\frac{z}{2^n}\right) \right\|_A$$

$$\leq \lim_{n\to\infty} 2^n \varphi\left(\frac{x}{2^n}, \frac{y}{2^n}, \frac{z}{2^n}\right)$$

$$\leq \lim_{n\to\infty} 8^n \varphi\left(\frac{x}{2^n}, \frac{y}{2^n}, \frac{z}{2^n}\right)$$

$$= 0$$

for all $x, y, z \in A$ and so

$$2\delta\left(\frac{x+y}{2} + z\right) = \delta(x) + \delta(y) + 2\delta(z)$$

for all $x, y, z \in A$. By Lemma 2.22, the mapping $\delta : A \to A$ is Cauchy additive. It is straight forward to show that the mapping $\delta : A \to A$ is $\mathbb{R}$-linear. It follows from (2.71) that

$$\|\delta(xyz) - \delta(xy)z + x\delta(y)z - x\delta(yz)\|_A$$

$$= \lim_{n\to\infty} 8^n \left\| f\left(\frac{xyz}{8^n}\right) - f\left(\frac{xy}{4^n}\right) \cdot \frac{z}{2^n} + \frac{x}{2^n} f\left(\frac{y}{2^n}\right) \cdot \frac{z}{2^n} - \frac{x}{2^n} f\left(\frac{yz}{4^n}\right) \right\|_A$$

$$\leq \lim_{n\to\infty} 8^n \varphi\left(\frac{x}{2^n}, \frac{y}{2^n}, \frac{z}{2^n}\right)$$

$$= 0$$

for all $x, y, z \in A$ and so

$$\delta(xyz) = \delta(xy)z - x\delta(y)z + x\delta(yz)$$

for all $x, y, z \in A$. Thus $\delta : A \to A$ is a generalized derivation satisfying (2.80). This completes the proof.  $\square$

**Corollary 2.30.** *Let $r > 3$, $\theta$ be nonnegative real numbers and $f : A \to A$ be a mapping satisfying (2.76) and (2.77). If $f(tx)$ is continuous in $t \in \mathbb{R}$ for each fixed $x \in A$, then there exists a unique generalized derivation $\delta : A \to A$ such that*

$$\|f(x) - \delta(x)\|_A \leq \frac{\theta}{2^{r+1} - 4} \|x\|_A^r$$

*for all $x \in A$.*

*Proof.* The proof follows from Theorem 2.29 by taking

$$\varphi(x, y, z) := \theta \cdot \|x\|_A^{\frac{r}{3}} \cdot \|y\|_A^{\frac{r}{3}} \cdot \|z\|_A^{\frac{r}{3}}$$

for all $x, y, z \in A$ and $L = 2^{1-r}$.                                    $\square$

# Chapter 3
# Stability of Functional Equations in $C^*$-Algebras

In this chapter, we study the stability of some important functional equations in $*$-algebras by using both the direct and fixed point methods.

In Sect. 3.1, we consider the linear bijection $h : A \rightarrow B$ of a unital $C^*$-algebra $A$ onto a unital $C^*$-algebra $B$ and show that it is a $C^*$-algebra isomorphism when $h(3^n uy) = h(3^n u)h(y)$ for all unitaries $u \in A$, $y \in A$ and $n \in \mathbb{Z}$ by the Ulam method.

In Sect. 3.2, we introduce a new functional equation, which is called the *Apollonius type additive functional equation*, and a solution of the functional equation is called the *Apollonius type additive mapping*:

$$L(z-x) + L(z-y) = -\frac{1}{2}L(x+y) + 2L\left(z - \frac{x+y}{4}\right).$$

Also, we investigate homomorphisms and derivations in $C^*$-algebras associated with the Apollonius type additive functional equation, homomorphisms and derivations on Lie $C^*$-algebras associated with the Apollonius type additive functional equation. Finally, we study homomorphisms and derivations on $JC^*$-algebras associated with the Apollonius type additive functional equation.

In Sect. 3.3, by using the fixed point method, we prove the *Hyers-Ulam stability* of homomorphisms in $C^*$-algebras and Lie $C^*$-algebras and derivations on $C^*$-algebras and Lie $C^*$-algebras for the following *Jensen type functional equation*

$$f\left(\frac{x+y}{2}\right) + f\left(\frac{x-y}{2}\right) = f(x).$$

In Sect. 3.4, we introduce the following *additive functional equation* :

$$\sum_{j=1}^{n} f\left(\frac{1}{2}\sum_{1\leq i\leq n, i\neq j} r_i x_i - \frac{1}{2}r_j x_j\right) + \sum_{i=1}^{n} r_i f(x_i) = nf\left(\frac{1}{2}\sum_{i=1}^{n} r_i x_i\right),$$

© Springer International Publishing Switzerland 2015
Y.J. Cho et al., *Stability of Functional Equations in Banach Algebras*, DOI 10.1007/978-3-319-18708-2_3

where $r_1, \cdots, r_n \in \mathbb{R}$. Using the fixed point method, we investigate the Hyers-Ulam stability of the above functional equation in Banach modules over a $C^*$-algebra. These results are applied to investigate $C^*$-algebra homomorphisms in unital $C^*$-algebras.

In Sect. 3.5, we show that, if an odd mapping $f : X \rightarrow Y$ satisfies the functional equation:

$$rf\left(\frac{\sum_{j=1}^{d} x_j}{r}\right) + \sum_{\substack{\iota(j) = 0, 1 \\ \sum_{j=1}^{d} \iota(j) = l}} rf\left(\frac{\sum_{j=1}^{d} (-1)^{\iota(j)} x_j}{r}\right)$$

$$= ({}_{d-1}C_l - {}_{d-1}C_{l-1} + 1) \sum_{j=1}^{d} f(x_j),$$

then the odd mapping $f : X \rightarrow Y$ is additive. Also, we prove the Hyers-Ulam stability of the above functional equation in Banach modules over a unital $C^*$-algebra. As an application, we show that every almost linear bijection $h : A \rightarrow B$ of a unital $C^*$-algebra $A$ onto a unital $C^*$-algebra $B$ is a $C^*$-algebra isomorphism when $h(\frac{2^n}{r^n}uy) = h(\frac{2^n}{r^n}u)h(y)$ for all unitaries $u \in A, y \in A$ and $n \in \mathbb{N}$.

In Sect. 3.6, we investigate the Hyers-Ulam stability of Jordan $*$-derivations and f quadratic Jordan $*$-derivations on real $C^*$-algebras and real $JC^*$-algebras. Also, we prove the superstability of Jordan $*$-derivations and quadratic Jordan $*$-derivations on real $C^*$-algebras and real $JC^*$-algebras under some conditions.

In Sect. 3.7, we investigate the Hyers-Ulam stability of $(\alpha, \beta, \gamma)$-derivations on Lie $C^*$-algebras associated with the following functional equation:

$$f\left(\frac{x_2 - x_1}{3}\right) + f\left(\frac{x_1 - 3x_3}{3}\right) + f\left(\frac{3x_1 + 3x_3 - x_2}{3}\right) = f(x_1).$$

In Sects. 3.8 and 3.9, we introduce a square root functional equation and a 3rd root functional equation. By using both the fixed point method and direct method, we prove the Hyers-Ulam stability of the square root functional equation and of the 3rd root functional equation in $C^*$-algebras.

In Sect. 3.10, we introduce the following functional equation:

$$T\left(\left(x^{\frac{1}{m}} + y^{\frac{1}{m}}\right)^m\right) = \left(T(x)^{\frac{1}{m}} + T(y)^{\frac{1}{m}}\right)^m$$

for all $x, y \in A^+$ and a fixed integer $m$ greater than 1, which is called a *positive-additive functional equation*. Using the fixed point and direct methods, we prove the stability of the positive-additive functional equation in $C^*$-algebras.

Finally, in Sect. 3.11, we show that every almost unital almost linear mapping $f : \mathcal{A} \rightarrow \mathcal{B}$ of $JC^*$-algebra $\mathcal{A}$ to a $JC^*$-algebra $\mathcal{B}$ is a homomorphism when $f(2^n u \circ y) = f(2^n u) \circ f(y)$ for all unitaries $u \in \mathcal{A}, y \in \mathcal{A}$ and $n \geq 0$ and every almost unital almost linear continuous mapping $f : \mathcal{A} \to \mathcal{B}$ of a $JC^*$-algebra $\mathcal{A}$ of real rank zero to a $JC^*$-algebra $\mathcal{B}$ is a homomorphism when $f(2^n u \circ y) = f(2^n u) \circ f(y)$ for all $u \in \{v \in \mathcal{A} : v = v^*, \|v\| = 1, v$ is invertible$\}$, $y \in \mathcal{A}$ and $n \geq 0$. Furthermore, we prove the Hyers-Ulam stability of $*$-homomorphisms in $JC^*$-algebras and $\mathbb{C}$-linear $*$-derivations on $JC^*$-algebras.

## 3.1  Isomorphisms in Unital $C^*$-Algebras

It is shown that every almost linear bijection $h : A \to B$ of a unital $C^*$-algebra $A$ onto a unital $C^*$-algebra $B$ is a $C^*$-algebra isomorphism when $h(3^n u y) = h(3^n u) h(y)$ for all unitaries $u \in A$, $y \in A$, $n \in \mathbb{Z}$ and an almost linear continuous bijection $h : A \to B$ of a unital $C^*$-algebra $A$ of real rank zero onto a unital $C^*$-algebra $B$ is a $C^*$-algebra isomorphism when $h(3^n u y) = h(3^n u) h(y)$ for all $u \in \{v \in A \mid v = v^*,$ $\|v\| = 1, v$ is invertible$\}$, $y \in A$ and $n \in \mathbb{Z}$.

Assume that $X$ and $Y$ are left normed modules over a unital $C^*$-algebra $A$. It is shown that every surjective isometry $T : X \to Y$ satisfying $T(0) = 0$ and $T(ux) = uT(x)$ for all $x \in X$ and unitaries $u \in A$ is an $A$-linear isomorphism. This is applied to investigate $C^*$-algebra isomorphisms in unital $C^*$-algebras.

Let $X$ and $Y$ be Banach spaces with the norms $\| \cdot \|$ and $\| \cdot \|$, respectively. Consider a mapping $f : X \to Y$ such that $f(tx)$ is continuous in $t \in \mathbb{R}$ for each fixed $x \in X$. Th. M. Rassias [267] introduced the following inequality, which is called the *Cauchy-Rassias inequality*. Assume that there exist constants $\theta \geq 0$ and $p \in [0, 1)$ such that

$$\|f(x + y) - f(x) - f(y)\| \leq \theta(\|x\|^p + \|y\|^p)$$

for all $x, y \in X$. Th. M. Rassias [267] showed that there exists a unique $\mathbb{R}$-linear mapping $T : X \to Y$ such that

$$\|f(x) - T(x)\| \leq \frac{2\theta}{2 - 2^p} \|x\|^p$$

for all $x \in X$. The above inequality has provided a lot of influence in the development of what we now call the *Hyers–Ulam stability* of functional equations. Beginning around the year 1980 the topic of approximate homomorphisms or the stability of the equation of homomorphism was studied by a number of mathematicians. In [145], Jun and Lee proved the following:

Denote by $\varphi : X \times X \to [0, \infty)$ a function such that

$$\tilde{\varphi}(x, y) = \sum_{j=0}^{\infty} \frac{1}{3^j} \varphi(3^j x, 3^j y) < \infty$$

for all $x, y \in X$. Suppose that $f : X \to Y$ is a mapping satisfying $f(0) = 0$ and

$$\left\| 2f(\frac{x+y}{2}) - f(x) - f(y) \right\| \leq \varphi(x, y)$$

for all $x, y \in X$. Then there exists a unique additive mapping $T : X \to Y$ such that

$$\|f(x) - T(x)\| \leq \frac{1}{3}(\tilde{\varphi}(x, -x) + \tilde{\varphi}(-x, 3x))$$

for all $x \in X$. In [242], Park and Park applied Jun and Lee's result to *Jensen's equation* in Banach modules over a $C^*$-algebra.

Throughout this section, let $A$ be a unital $C^*$-algebra with the norm $\| \cdot \|$ and the unit $e$ and $B$ be a unital $C^*$-algebra with the norm $\| \cdot \|$. Let $U(A)$ be the set of unitary elements in $A$, $A_{sa} = \{x \in A \mid x = x^*\}$ and $I_1(A_{sa}) = \{v \in A_{sa} \mid \|v\| = 1, v$ is invertible$\}$.

We prove that every almost linear bijection $h : A \to B$ is a $C^*$-algebra isomorphism when $h(3^n u y) = h(3^n u)h(y)$ for all $u \in U(A)$, $y \in A$, $n \in \mathbb{Z}$ and, for a unital $C^*$-algebra $A$ of real rank zero, every almost linear continuous bijection $h : A \to B$ is a $C^*$-algebra isomorphism when $h(3^n u y) = h(3^n u)h(y)$ for all $u \in I_1(A_{sa})$, $y \in A$ and $n \in \mathbb{Z}$. Also, we prove that every surjective isometry satisfying some conditions is a $C^*$-algebra isomorphism.

### 3.1.1   $C^*$-Algebra Isomorphisms in Unital $C^*$-Algebras

Now, we investigate $C^*$-algebra isomorphisms in unital $C^*$-algebras.

**Theorem 3.1.** *Let $h : A \to B$ be a bijective mapping satisfying $h(0) = 0$ and $h(3^n u y) = h(3^n u)h(y)$ for all $u \in U(A)$, $y \in A$ and $n \in \mathbb{Z}$, for which there exists a function $\varphi : A \times A \to [0, \infty)$ such that*

$$\tilde{\varphi}(x, y) := \sum_{j=0}^{\infty} \frac{1}{3^j} \varphi(3^j x, 3^j y) < \infty, \tag{3.1}$$

$$\left\| 2h\left(\frac{\mu x + \mu y}{2}\right) - \mu h(x) - \mu h(y) \right\| \leq \varphi(x, y), \tag{3.2}$$

$$\|h(3^n u^*) - h(3^n u)^*\| \leq \varphi(3^n u, 3^n u) \tag{3.3}$$

*for all $\mu \in S^1 := \{\lambda \in \mathbb{C} : |\lambda| = 1\}$, $u \in U(A)$, $n \in \mathbb{Z}$ and $x, y \in A$. Assume that*

$$\lim_{n \to \infty} \frac{h(3^n e)}{3^n} \tag{3.4}$$

*is invertible. Then the bijective mapping $h : A \to B$ is a $C^*$-algebra isomorphism.*

*Proof.* Put $\mu = 1 \in S^1$. By direct method there exists a unique additive mapping $H : A \to B$ such that

$$\|h(x) - H(x)\| \leq \frac{1}{3}\big(\tilde{\varphi}(x, -x) + \tilde{\varphi}(-x, 3x)\big) \tag{3.5}$$

for all $x \in A$. The additive mapping $H : A \to B$ is given by

$$H(x) = \lim_{n \to \infty} \frac{1}{3^n} h(3^n x)$$

for all $x \in A$. By the assumption, for each $\mu \in S^1$,

$$\frac{1}{3^n}\left\|2h\left(\frac{3^n \mu x}{2}\right) - \mu h(3^n x)\right\| \leq \frac{1}{3^n}\varphi(3^n x, 0),$$

which tends to zero as $n \to \infty$ for all $x \in A$. Hence we have

$$2H\left(\frac{\mu x}{2}\right) = \lim_{n \to \infty} \frac{2h\left(\frac{3^n \mu x}{2}\right)}{3^n} = \lim_{n \to \infty} \frac{\mu h(3^n x)}{3^n} = \mu H(x)$$

for all $\mu \in S^1$ and $x \in A$. Since $H : A \to B$ is additive,

$$H(\mu x) = 2H\left(\frac{\mu x}{2}\right) = \mu H(x) \tag{3.6}$$

for all $\mu \in S^1$ and $x \in A$.

Now, let $\lambda \in \mathbb{C}$ ($\lambda \neq 0$) and $M$ be an integer greater than $4|\lambda|$. Then we have

$$\left|\frac{\lambda}{M}\right| < \frac{1}{4} < 1 - \frac{2}{3} = \frac{1}{3}.$$

By Kadison and Pedersen [167], there exist three elements $\mu_1, \mu_2, \mu_3 \in S^1$ such that $3\frac{\lambda}{M} = \mu_1 + \mu_2 + \mu_3$. So, by (3.6), we have

$$H(\lambda x) = H\left(\frac{M}{3} \cdot 3\frac{\lambda}{M}x\right) = M \cdot H\left(\frac{1}{3} \cdot 3\frac{\lambda}{M}x\right) = \frac{M}{3}H\left(3\frac{\lambda}{M}x\right)$$

$$= \frac{M}{3}H(\mu_1 x + \mu_2 x + \mu_3 x) = \frac{M}{3}(H(\mu_1 x) + H(\mu_2 x) + H(\mu_3 x))$$

$$= \frac{M}{3}(\mu_1 + \mu_2 + \mu_3)H(x) = \frac{M}{3} \cdot 3\frac{\lambda}{M}H(x)$$

$$= \lambda H(x)$$

for all $x \in A$. Hence we have

$$H(\zeta x + \eta y) = H(\zeta x) + H(\eta y) = \zeta H(x) + \eta H(y)$$

for all $\zeta, \eta \in \mathbb{C}$ ($\zeta, \eta \neq 0$) and $x, y \in A$ and $H(0x) = 0 = 0H(x)$ for all $x \in A$. Thus the unique additive mapping $H : A \to B$ is a $\mathbb{C}$-linear mapping. By (3.1) and (3.3), we get

$$H(u^*) = \lim_{n \to \infty} \frac{h(3^n u^*)}{3^n} = \lim_{n \to \infty} \frac{h(3^n u)^*}{3^n}$$

$$= \left( \lim_{n \to \infty} \frac{h(3^n u)}{3^n} \right)^* = H(u)^*$$

for all $u \in U(A)$. Since $H$ is $\mathbb{C}$-linear and each $x \in A$ is a finite linear combination of unitary elements (see [168]), i.e., $x = \sum_{j=1}^{m} \lambda_j u_j$ for all $\lambda_j \in \mathbb{C}$ and $u_j \in U(A)$, we have

$$H(x^*) = H\left( \sum_{j=1}^{m} \overline{\lambda_j} u_j^* \right) = \sum_{j=1}^{m} \overline{\lambda_j} H(u_j^*) = \sum_{j=1}^{m} \overline{\lambda_j} H(u_j)^*$$

$$= \left( \sum_{j=1}^{m} \lambda_j H(u_j) \right)^* = H(\sum_{j=1}^{m} \lambda_j u_j)^* = H(x)^*$$

for all $x \in A$. Since $h(3^n uy) = h(3^n u)h(y)$ for all $u \in U(A)$, $y \in A$ and $n \in \mathbb{Z}$, we have

$$H(uy) = \lim_{n \to \infty} \frac{1}{3^n} h(3^n uy) = \lim_{n \to \infty} \frac{1}{3^n} h(3^n u)h(y) = H(u)h(y) \qquad (3.7)$$

for all $u \in U(A)$ and $y \in A$. By the additivity of $H$ and (3.7), we have

$$3^n H(uy) = H(3^n uy) = H(u(3^n y)) = H(u)h(3^n y)$$

for all $u \in U(A)$ and $y \in A$. Hence it follows that

$$H(uy) = \frac{1}{3^n} H(u)h(3^n y) = H(u)\frac{1}{3^n} h(3^n y) \qquad (3.8)$$

for all $u \in U(A)$ and $y \in A$. Taking $n \to \infty$ in (3.8), we obtain

$$H(uy) = H(u)H(y) \qquad (3.9)$$

for all $u \in U(A)$ and $y \in A$. Since $H$ is $\mathbb{C}$-linear and each $x \in A$ is a finite linear combination of unitary elements, i.e., $x = \sum_{j=1}^{m} \lambda_j u_j$ for all $\lambda_j \in \mathbb{C}$ and $u_j \in U(A)$, it follows from (3.9) that

$$H(xy) = H\left(\sum_{j=1}^{m} \lambda_j u_j y\right) = \sum_{j=1}^{m} \lambda_j H(u_j y) = \sum_{j=1}^{m} \lambda_j H(u_j) H(y)$$

$$= H\left(\sum_{j=1}^{m} \lambda_j u_j\right) H(y) = H(x) H(y)$$

for all $x, y \in A$. By (3.7) and (3.9), we have

$$H(e)H(y) = H(ey) = H(e)h(y)$$

for all $y \in A$. Since $\lim_{n \to \infty} \frac{h(3^n e)}{3^n} = H(e)$ is invertible,

$$H(y) = h(y)$$

for all $y \in A$. Therefore, the bijective mapping $h : A \to B$ is a $C^*$-algebra isomorphism. This completes the proof. $\qquad\square$

**Corollary 3.2.** *Let $h : A \to B$ be a bijective mapping satisfying $h(0) = 0$ and $h(3^n uy) = h(3^n u)h(y)$ for all $u \in U(A)$, $y \in A$ and $n \in \mathbb{Z}$, for which there exist constants $\theta \geq 0$ and $p \in [0, 1)$ such that*

$$\left\| 2h\left(\frac{\mu x + \mu y}{2}\right) - \mu h(x) - \mu h(y) \right\| \leq \theta(\|x\|^p + \|y\|^p)$$

*and*

$$\|h(3^n u^*) - h(3^n u)^*\| \leq 2 \cdot 3^{np}\theta$$

*for all $\mu \in S^1$, $u \in U(A)$, $n \in \mathbb{Z}$ and $x, y \in A$. Assume that $\lim_{n \to \infty} \frac{h(3^n e)}{3^n}$ is invertible. Then the bijective mapping $h : A \to B$ is a $C^*$-algebra isomorphism.*

*Proof.* Defining

$$\varphi(x, y) = \theta(\|x\|^p + \|y\|^p)$$

(Rassias upper bound in the Cauchy-Rassias inequality) and applying Theorem 3.1, we get the desired result. $\qquad\square$

**Theorem 3.3.** *Let $h : A \to B$ be a bijective mapping satisfying $h(0) = 0$ and $h(3^n uy) = h(3^n u)h(y)$ for all $u \in U(A)$, $y \in A$ and $n \in \mathbb{Z}$, for which there exists a function $\varphi : A \times A \to [0, \infty)$ satisfying (3.1), (3.3) and (3.4) such that*

$$\left\| 2h\left(\frac{\mu x + \mu y}{2}\right) - \mu h(x) - \mu h(y) \right\| \leq \varphi(x, y) \qquad (3.10)$$

*for $\mu = 1, i$ and $x, y \in A$. If $h(tx)$ is continuous in $t \in \mathbb{R}$ for each fixed $x \in A$, then the bijective mapping $h : A \to B$ is a $C^*$-algebra isomorphism.*

*Proof.* Put $\mu = 1$ in (3.10). Then, there exists a unique additive mapping $H : A \rightarrow B$ satisfying (3.5). Also, by (3.10), the additive mapping $H : A \rightarrow B$ is $\mathbb{R}$-linear.

Put $\mu = i$ and $y = 0$ in (3.10). By the same method as in the proof of Theorem 3.1, one can obtain that

$$H(ix) = 2H\left(\frac{ix}{2}\right) = \lim_{n\to\infty} \frac{2h(\frac{3^n ix}{2})}{3^n} = \lim_{n\to\infty} \frac{ih(3^n x)}{3^n} = iH(x)$$

for all $x \in A$. For each $\lambda \in \mathbb{C}$, let $\lambda = s + it$ for all $s, t \in \mathbb{R}$. So, we have

$$H(\lambda x) = H(sx + itx) = sH(x) + tH(ix) = sH(x) + itH(x)$$
$$= (s + it)H(x) = \lambda H(x)$$

for all $\lambda \in \mathbb{C}$ and $x \in A$. Thus we have

$$H(\zeta x + \eta y) = H(\zeta x) + H(\eta y) = \zeta H(x) + \eta H(y)$$

for all $\zeta, \eta \in \mathbb{C}$ and $x, y \in A$. Hence the additive mapping $H : A \rightarrow B$ is $\mathbb{C}$-linear.

The rest of the proof is the same as in the proof of Theorem 3.1. This completes the proof.                                                                                                    □

From now on, assume that $A$ is a unital $C^*$-algebra of real rank zero, where "real rank zero" means that the set of invertible self-adjoint elements is dense in the set of self-adjoint elements (see [54]).

Now, we investigate continuous $C^*$-algebra isomorphisms in unital $C^*$-algebras.

**Theorem 3.4.** *Let $h : A \rightarrow B$ be a continuous bijective mapping satisfying $h(0) = 0$ and $h(3^n uy) = h(3^n u)h(y)$ for all $u \in I_1(A_{sa})$, $y \in A$ and $n \in \mathbb{Z}$, for which there exists a function $\varphi : A \times A \rightarrow [0, \infty)$ satisfying (3.1), (3.2), (3.3) and (3.4). Then the bijective mapping $h : A \rightarrow B$ is a $C^*$-algebra isomorphism.*

*Proof.* It is straight forward to show that, there exists a unique $\mathbb{C}$-linear involution $H : A \rightarrow B$ satisfying (3.5). Since $h(3^n uy) = h(3^n u)h(y)$ for all $u \in I_1(A_{sa})$, $y \in A$ and $n \in \mathbb{Z}$, we have

$$H(uy) = \lim_{n\to\infty} \frac{1}{3^n} h(3^n uy) = \lim_{n\to\infty} \frac{1}{3^n} h(3^n u)h(y) = H(u)h(y) \qquad (3.11)$$

for all $u \in I_1(A_{sa})$ and $y \in A$. By the additivity of $H$ and (3.11), we have

$$3^n H(uy) = H(3^n uy) = H(u(3^n y)) = H(u)h(3^n y)$$

for all $u \in I_1(A_{sa})$ and $y \in A$. Hence we have

$$H(uy) = \frac{1}{3^n} H(u)h(3^n y) = H(u)\frac{1}{3^n} h(3^n y) \tag{3.12}$$

for all $u \in I_1(A_{sa})$ and $y \in A$. Taking $n \to \infty$ in (3.12), we obtain

$$H(uy) = H(u)H(y) \tag{3.13}$$

for all $u \in I_1(A_{sa})$ and $y \in A$. By (3.11) and (3.13), we have

$$H(e)H(y) = H(ey) = H(e)h(y)$$

for all $y \in A$. Since $\lim_{n\to\infty} \frac{h(3^n e)}{3^n} = H(e)$ is invertible,

$$H(y) = h(y)$$

for all $y \in A$. So $H : A \to B$ is continuous. But, by the assumption that $A$ has real rank zero, it is easy to show that $I_1(A_{sa})$ is dense in $\{x \in A_{sa} : \|x\| = 1\}$. Thus, for each $w \in \{z \in A_{sa} : \|z\| = 1\}$, there exists a sequence $\{\kappa_j\}$ such that $\kappa_j \to w$ as $j \to \infty$ and $\kappa_j \in I_1(A_{sa})$. Since $H : A \to B$ is continuous, it follows from (3.13) that

$$H(wy) = H(\lim_{j\to\infty} \kappa_j y) = \lim_{j\to\infty} H(\kappa_j y) = \lim_{j\to\infty} H(\kappa_j)H(y)$$
$$= H(\lim_{j\to\infty} \kappa_j)H(y) = H(w)H(y) \tag{3.14}$$

for all $w \in \{z \in A_{sa} : \|z\| = 1\}$ and $y \in A$. For each $x \in A$, $x = \frac{x+x^*}{2} + i\frac{x-x^*}{2i}$, where $x_1 := \frac{x+x^*}{2}$ and $x_2 := \frac{x-x^*}{2i}$ are self-adjoint.

First, consider the case that $x_1 \neq 0$ and $x_2 \neq 0$. Since $H : A \to B$ is $\mathbb{C}$-linear, it follows from (3.14) that

$$H(xy) = H(x_1 y + ix_2 y)$$
$$= H\left(\|x_1\|\frac{x_1}{\|x_1\|}y + i\|x_2\|\frac{x_2}{\|x_2\|}y\right)$$
$$= \|x_1\|H\left(\frac{x_1}{\|x_1\|}y\right) + i\|x_2\|H\left(\frac{x_2}{\|x_2\|}y\right)$$
$$= \|x_1\|H\left(\frac{x_1}{\|x_1\|}\right)H(y) + i\|x_2\|H\left(\frac{x_2}{\|x_2\|}\right)H(y)$$
$$= \left\{H\left(\|x_1\|\frac{x_1}{\|x_1\|}\right) + iH\left(\|x_2\|\frac{x_2}{\|x_2\|}\right)\right\}H(y)$$
$$= H(x_1 + ix_2)H(y) = H(x)H(y)$$

for all $y \in A$.

Next, consider the case that $x_1 \neq 0, x_2 = 0$. Since $H : A \to B$ is $\mathbb{C}$-linear, it follows from (3.14) that

$$
\begin{aligned}
H(xy) = H(x_1 y) &= H\left( \|x_1\| \frac{x_1}{\|x_1\|} y \right) = \|x_1\| H\left( \frac{x_1}{\|x_1\|} y \right) \\
&= \|x_1\| H\left( \frac{x_1}{\|x_1\|} \right) H(y) = H\left( \|x_1\| \frac{x_1}{\|x_1\|} \right) H(y) \\
&= H(x_1) H(y) = H(x) H(y)
\end{aligned}
$$

for all $y \in A$.

Finally, consider the case that $x_1 = 0, x_2 \neq 0$. Since $H : A \to B$ is $\mathbb{C}$-linear, it follows from (3.14) that

$$
\begin{aligned}
H(xy) = H(ix_2 y) &= H\left( i\|x_2\| \frac{x_2}{\|x_2\|} y \right) = i\|x_2\| H\left( \frac{x_2}{\|x_2\|} y \right) \\
&= i\|x_2\| H\left( \frac{x_2}{\|x_2\|} \right) H(y) = H\left( i\|x_2\| \frac{x_2}{\|x_2\|} \right) H(y) \\
&= H(ix_2) H(y) = H(x) H(y)
\end{aligned}
$$

for all $y \in A$. Hence

$$
H(xy) = H(x) H(y)
$$

for all $x, y \in A$. Therefore, the bijective mapping $h : A \to B$ is a $C^*$-algebra isomorphism. This completes the proof. $\qquad\square$

**Corollary 3.5.** *Let $h : A \to B$ be a continuous bijective mapping satisfying $h(0) = 0$ and $h(3^n u y) = h(3^n u) h(y)$ for all $u \in I_1(A_{sa})$, $y \in A$ and $n \in \mathbb{Z}$, for which there exist constants $\theta \geq 0$ and $p \in [0, 1)$ such that*

$$
\left\| 2h\left( \frac{\mu x + \mu y}{2} \right) - \mu h(x) - \mu h(y) \right\| \leq \theta(\|x\|^p + \|y\|^p)
$$

*and*

$$
\| h(3^n u^*) - h(3^n u)^* \| \leq 2 \cdot 3^{np} \theta
$$

*for all $\mu \in S^1$, $u \in I_1(A_{sa})$, $n \in \mathbb{Z}$ and $x, y \in A$. Assume that $\lim_{n \to \infty} \frac{h(3^n e)}{3^n}$ is invertible. Then the bijective mapping $h : A \to B$ is a $C^*$-algebra isomorphism.*

*Proof.* Defining

$$
\varphi(x, y) = \theta(\|x\|^p + \|y\|^p)
$$

(Rassias upper bound in the Cauchy-Rassias inequality) and applying Theorem 3.4, we get the desired result. $\qquad\square$

*Remark 3.6.* If $h : A \to B$ is a continuous bijective mapping satisfying $h(0) = 0$ and $h(3^n uy) = h(3^n u)h(y)$ for all $u \in I_1(A_{sa})$, $y \in A$ and $n \in \mathbb{Z}$, for which there exists a function $\varphi : A \times A \to [0, \infty)$ satisfying (3.1), (3.3), (3.4) and (3.10). Then the bijective mapping $h : A \to B$ is a $C^*$-algebra isomorphism.

## 3.1.2  On the Mazur-Ulam Theorem in Modules over $C^*$-Algebras

Now, we prove the Mazur-Ulam theorem in modules over $C^*$-algebras.

**Lemma 3.7 ([114]).** *If $T$ is an isometry from a normed vector space $X$ onto a normed vector space $Y$, then*

$$T(x + y) = T(x) + T(y) - T(0)$$

*and*

$$T(rx) = rT(x) + (1 - r)T(0)$$

*for all $r \in \mathbb{R}$.*

**Corollary 3.8 ([114]).** *If $T$ is an isometry from a normed vector space $X$ onto a normed vector space $Y$ and $T(0) = 0$, then $T$ is $\mathbb{R}$-linear.*

**Theorem 3.9.** *Let $X$ and $Y$ be left normed modules over a unital $C^*$-algebra $A$. If $T : X \to Y$ is a surjective isometry with $T(0) = 0$ and $T(ux) = uT(x)$ for all $u \in U(A)$ and $x \in X$, then $T : X \to Y$ is an $A$-linear isomorphism.*

*Proof.* By Corollary 3.8, $T : X \to Y$ is $\mathbb{R}$-linear. Since $i \in U(A)$, $T(ix) = iT(x)$ for all $x \in X$. For each $\lambda \in \mathbb{C}$, $\lambda = \lambda_1 + i\lambda_2$ for all $\lambda_1, \lambda_2 \in \mathbb{R}$. Thus we have

$$
\begin{aligned}
T(\lambda x) &= T(\lambda_1 x + i\lambda_2 x) = T(\lambda_1 x) + T(i \lambda_2 x) \\
&= \lambda_1 T(x) + iT(\lambda_2 x) = (\lambda_1 + i\lambda_2)T(x) \\
&= \lambda T(x)
\end{aligned}
$$

for all $x \in X$. Since each $a \in A$ is a finite linear combination of unitary elements, i.e., $a = \sum_{j=1}^{n} \lambda_j u_j$ for all $\lambda_j \in \mathbb{C}$ and $u_j \in U(A)$,

$$T(ax) = T\left( \sum_{j=1}^{n} \lambda_j u_j x \right) = \sum_{j=1}^{n} \lambda_j T(u_j x) = \sum_{j=1}^{n} \lambda_j u_j T(x) = aT(x)$$

for all $x \in X$ and so

$$T(ax + by) = T(ax) + T(by) = aT(x) + bT(y)$$

for all $a, b \in A$ and $x, y \in X$. This completes the proof.        $\square$

Now, we investigate $C^*$-algebra isomorphisms in unital $C^*$-algebras.

**Theorem 3.10.** *If $T : A \to B$ is a surjective isometry with $T(0) = 0$, $T(iu) = iT(u)$, $T(u^*) = T(u)^*$ and $T(uv) = T(u)T(v)$ for all $u, v \in U(A)$, then $T : A \to B$ is a $C^*$-algebra isomorphism.*

*Proof.* It is straight forward to show that, $T : A \to B$ is $\mathbb{R}$-linear and

$$T(\lambda u) = \lambda T(u)$$

for all $\lambda \in \mathbb{C}$ and $u \in U(A)$. Since each $a \in A$ is a finite linear combination of unitary elements, i.e., $a = \sum_{j=1}^{n} \lambda_j u_j$ for all $\lambda_j \in \mathbb{C}$ and $u_j \in U(A)$, we have

$$T(\lambda a) = T\left( \sum_{j=1}^{n} \lambda \lambda_j u_j \right) = \sum_{j=1}^{n} \lambda \lambda_j T(u_j) = \lambda \left( \sum_{j=1}^{n} \lambda_j T(u_j) \right)$$

$$= \lambda T\left( \sum_{j=1}^{n} \lambda_j u_j \right) = \lambda T(a)$$

for all $\lambda \in \mathbb{C}$ and $a \in A$. Thus $T : A \to B$ is $\mathbb{C}$-linear. Furthermore, we have

$$T(a^*) = T\left( \sum_{j=1}^{n} \overline{\lambda}_j u_j^* \right) = \sum_{j=1}^{n} \overline{\lambda}_j T(u_j^*) = \sum_{j=1}^{n} \overline{\lambda}_j T(u_j)^*$$

$$= T\left( \sum_{j=1}^{n} \lambda_j u_j \right)^* = T(a)^*$$

for all $a \in A$ and

$$T(av) = T\left( \sum_{j=1}^{n} \lambda_j u_j v \right) = \sum_{j=1}^{n} \lambda_j T(u_j v) = \sum_{j=1}^{n} \lambda_j T(u_j)T(v)$$

$$= T\left( \sum_{j=1}^{n} \lambda_j u_j \right)T(v) = T(a)T(v)$$

for all $a \in A$ and $v \in U(A)$. Since each $b \in A$ is a finite linear combination of unitary elements, i.e., $b = \sum_{j=1}^{m} v_j v_j$ for all $v_j \in \mathbb{C}$ and $v_j \in U(A)$, we have

$$T(ab) = T\left(\sum_{j=1}^{m} v_j a v_j\right) = \sum_{j=1}^{m} v_j T(a v_j) = \sum_{j=1}^{m} v_j T(a) T(v_j)$$

$$= T(a) T\left(\sum_{j=1}^{m} v_j v_j\right) = T(a) T(b)$$

for all $a, b \in A$ and so $T : A \to B$ is multiplicative. Therefore, $T : A \to B$ is a $C^*$-algebra isomorphism. This completes the proof.                     $\square$

## 3.2   Apollonius Type Additive Functional Equations

In an inner product space, the following equality:

$$\|z - x\|^2 + \|z - y\|^2 = \frac{1}{2}\|x - y\|^2 + 2\left\|z - \frac{x + y}{2}\right\|^2$$

holds, which is called the *Apollonius' identity*. The following functional equation, which was motivated by this equation,

$$Q(z - x) + Q(z - y) = \frac{1}{2}Q(x - y) + 2Q\left(z - \frac{x + y}{2}\right) \tag{3.15}$$

is quadratic. For this reason, the function equation (3.15) is called a *quadratic functional equation of Apollonius type* and each solution of the functional equation (3.15) is called a *quadratic mapping of Apollonius type*. In [144], Jun and Kim investigated the quadratic functional equation of Apollonius type.

In this section, modifying the above equality (3.15), we consider a new functional equation, which is called the *Apollonius type additive functional equation* and whose solution of the functional equation is said to be the *Apollonius type additive mapping* [209]:

$$L(z - x) + L(z - y) = -\frac{1}{2}L(x + y) + 2L\left(z - \frac{x + y}{4}\right).$$

In [126], Gilányi showed that, if $f$ has it's values in an inner product space and satisfies the functional inequality:

$$\|2f(x) + 2f(y) - f(xy^{-1})\| \le \|f(xy)\|, \tag{3.16}$$

then $f$ satisfies the *Jordan–von Neumann functional inequality*

$$2f(x) + 2f(y) = f(xy) + f(xy^{-1}).$$

In [113] and [127], Fechner and Gilányi proved the stability of the functional inequality (3.16), respectively. In [253], Park et al. proved the stability of functional inequalities associated with Jordan–von Neumann type additive functional equations.

In 1932, Jordan observed that $\mathcal{L}(\mathcal{H})$ is a (non-associative) algebra via the *anticommutator product* $x \circ y := \frac{xy+yx}{2}$. A commutative algebra $X$ with the product $x \circ y$ is called a *Jordan algebra*. A Jordan $C^*$-subalgebra of a $C^*$-algebra endowed with the anti-commutator product is called a *$JC^*$-algebra*. A $C^*$-algebra $\mathcal{C}$ endowed with the Lie product $[x, y] = \frac{xy-yx}{2}$ on $\mathcal{C}$ is called a *Lie $C^*$-algebra*.

In this section, we investigate homomorphisms and derivations in $C^*$-algebras associated with the Apollonius type additive functional equation. Also, we investigate homomorphisms and derivations on Lie $C^*$-algebras associated with the Apollonius type additive functional equation.

Finally, we investigate homomorphisms and derivations on $JC^*$-algebras associated with the Apollonius type additive functional equation.

### 3.2.1  Homomorphisms and Derivations on $C^*$-Algebras

Now, we study homomorphisms and derivations on $C^*$-algebras.

**Theorem 3.11.** *Let A be a uniquely 2-divisible Abelian group and B be a normed linear space. A mapping $f : A \to B$ satisfies the following:*

$$\left\| f(z-x) + f(z-y) + \frac{1}{2}f(x+y) \right\|_B \leq \left\| 2f\left(z - \frac{x+y}{4}\right) \right\|_B \qquad (3.17)$$

*for all $x, y, z \in A$ if and only if $f : A \to B$ is additive.*

*Proof.* Letting $x = y = z = 0$ in (3.17), we get

$$\frac{5}{2}\|f(0)\|_B \leq 2\|f(0)\|_B$$

and so $f(0) = 0$. Letting $z = 0$ and $y = -x$ in (3.17), we get

$$\|f(-x) + f(x)\|_B \leq 2\|f(0)\|_B = 0$$

for all $x \in A$ and hence $f(-x) = -f(x)$ for all $x \in A$. Letting $x = y = 2z$ in (3.17), we get

$$\left\| 2f(-z) + \frac{1}{2}f(4z) \right\|_B \leq \|2f(0)\|_B = 0$$

for all $z \in A$ and hence

$$f(4z) = -4f(-z) = 4f(z)$$

for all $z \in A$. Letting $z = \frac{x+y}{4}$ in (3.17), we get

$$\left\| f\left(\frac{-3x+y}{4}\right) + f\left(\frac{x-3y}{4}\right) + \frac{1}{2}f(x+y) \right\|_B \le \|2f(0)\|_B = 0$$

for all $x, y \in A$ and so

$$f\left(\frac{-3x+y}{4}\right) + f\left(\frac{x-3y}{4}\right) + \frac{1}{2}f(x+y) = 0 \qquad (3.18)$$

for all $x, y \in A$. Let $w_1 = \frac{-3x+y}{4}$ and $w_2 = \frac{x-3y}{4}$ in (3.18). Then we have

$$f(w_1) + f(w_2) = -\frac{1}{2}f(-2w_1 - 2w_2) = \frac{1}{2}f(2w_1 + 2w_2) = 2f\left(\frac{w_1 + w_2}{2}\right)$$

for all $w_1, w_2 \in A$ and so $f$ is additive.

It is clear that each additive mapping satisfies the inequality (3.17). This completes the proof.  $\square$

Now, we investigate $C^*$-algebra homomorphisms between $C^*$-algebras and linear derivations on $C^*$-algebras associated with the Apollonius type additive functional equation. From now on, assume that $A$ is a $C^*$-algebra with the norm $\| \cdot \|_A$ and $B$ is a $C^*$-algebra with the norm $\| \cdot \|_B$.

**Lemma 3.12.** *Let $f : A \to B$ be an additive mapping such that $f(\mu x) = \mu f(x)$ for all $x \in A$ and $\mu \in \mathbb{T}^1 = \{\lambda \in \mathbb{C} : |\lambda| = 1\}$. Then the mapping $f$ is $\mathbb{C}$-linear.*

*Proof.* Let $\lambda \in \mathbb{C}$ ($\lambda \ne 0$) and $M$ an integer greater than $4|\lambda|$.
Then we have $|\frac{\lambda}{M}| < \frac{1}{4} < 1 - \frac{2}{3} = \frac{1}{3}$. By Theorem 1 of [54], there exist three elements $\mu_1, \mu_2, \mu_3 \in \mathbb{T}^1$ such that $3\frac{\lambda}{M} = \mu_1 + \mu_2 + \mu_3$. And $f(x) = f(3 \cdot \frac{1}{3}x) = 3f(\frac{1}{3}x)$ for all $x \in A$ and so $f(\frac{1}{3}x) = \frac{1}{3}f(x)$ for all $x \in A$. Thus

$$f(\lambda x) = f\left(\frac{M}{3} \cdot 3\frac{\lambda}{M}x\right) = M \cdot f\left(\frac{1}{3} \cdot 3\frac{\lambda}{M}x\right) = \frac{M}{3}f\left(3\frac{\lambda}{M}x\right)$$

$$= \frac{M}{3}f(\mu_1 x + \mu_2 x + \mu_3 x) = \frac{M}{3}(f(\mu_1 x) + f(\mu_2 x) + f(\mu_3 x))$$

$$= \frac{M}{3}(\mu_1 + \mu_2 + \mu_3)f(x) = \frac{M}{3} \cdot 3\frac{\lambda}{M}f(x)$$

$$= \lambda f(x)$$

for all $x \in A$. Hence we have

$$f(\zeta x + \eta y) = f(\zeta x) + f(\eta y) = \zeta f(x) + \eta f(y)$$

for all $\zeta, \eta \in \mathbb{C}$ ($\zeta, \eta \ne 0$) and all $x, y \in A$. And $f(0x) = 0 = 0f(x)$ for all $x \in A$. So the additive mapping $f : A \to B$ is a $\mathbb{C}$-linear mapping. This completes the proof.  $\square$

**Theorem 3.13.** *Let $r > 1$, $\theta$ be nonnegative real numbers and $f : A \to B$ be a mapping such that*

$$\left\| f(z - \mu x) + \mu f(z - y) + \frac{1}{2} f(x + y) \right\|_B \leq \left\| 2f(z - \frac{x + y}{4}) \right\|_B, \tag{3.19}$$

$$\| f(xy) - f(x)f(y) \|_B \leq \theta \cdot \|x\|_A^r \cdot \|y\|_A^r, \tag{3.20}$$

$$\| f(x^*) - f(x)^* \|_B \leq 2\theta \|x\|_A^r \tag{3.21}$$

*for all $\mu \in \mathbb{T}^1 := \{\lambda \in \mathbb{C} : |\lambda| = 1\}$ and $x, y, z \in A$. Then the mapping $f : A \to B$ is a $C^*$-algebra homomorphism.*

*Proof.* Let $\mu = 1$ in (3.19). By Theorem 3.11, the mapping $f : A \to B$ is additive. Letting $y = -x$ and $z = 0$ in (3.19), we get

$$\| f(-\mu x) + \mu f(x) \|_B \leq \| 2f(0) \|_B = 0$$

for all $x \in A$ and $\mu \in \mathbb{T}^1$ and so

$$-f(\mu x) + \mu f(x) = f(-\mu x) + \mu f(x) = 0$$

for all $x \in A$ and $\mu \in \mathbb{T}^1$. Hence $f(\mu x) = \mu f(x)$ for all $x \in A$ and $\mu \in \mathbb{T}^1$. So, the mapping $f : A \to B$ is $\mathbb{C}$-linear. It follows from (3.20) that

$$\begin{aligned}
\| f(xy) - f(x)f(y) \|_B &= \lim_{n \to \infty} 4^n \left\| f\left(\frac{xy}{2^n \cdot 2^n}\right) - f\left(\frac{x}{2^n}\right) f\left(\frac{y}{2^n}\right) \right\|_B \\
&\leq \lim_{n \to \infty} \frac{4^n \theta}{4^{nr}} \cdot \|x\|_A^r \cdot \|y\|_A^r \\
&= 0
\end{aligned}$$

for all $x, y \in A$ and so

$$f(xy) = f(x)f(y)$$

for all $x, y \in A$. It follows from (3.21) that

$$\begin{aligned}
\| f(x^*) - f(x)^* \|_B &= \lim_{n \to \infty} 2^n \left\| f\left(\frac{x^*}{2^n}\right) - f\left(\frac{x}{2^n}\right)^* \right\|_B \\
&\leq \lim_{n \to \infty} \frac{2^{n+1} \theta}{2^{nr}} \|x\|_A^r \\
&= 0
\end{aligned}$$

for all $x \in A$ and so

$$f(x^*) = f(x)^*$$

for all $x \in A$. Therefore, the mapping $f : A \to B$ is a $C^*$-algebra homomorphism. This completes the proof. $\qquad\square$

*Remark 3.14.* Let $r < 1$, $\theta$ be positive real numbers and $f : A \to B$ be a mapping satisfying (3.19), (3.20) and (3.21). Then the mapping $f : A \to B$ is a $C^*$-algebra homomorphism.

**Theorem 3.15.** *Let $r > 1$, $\theta$ be nonnegative real numbers and $f : A \to A$ be a mapping satisfying* (3.21) *such that*

$$\|f(xy) - f(x)y - xf(y)\|_A \le \theta \cdot \|x\|_A^r \cdot \|y\|_A^r \qquad (3.22)$$

*for all $x, y \in A$. Then the mapping $f : A \to A$ is a linear derivation.*

*Proof.* By applying Lemma 3.12, the mapping $f : A \to A$ is $\mathbb{C}$-linear. It follows from (3.22) that

$$
\begin{aligned}
\|f(xy) - f(x)y - xf(y)\|_A &= \lim_{n\to\infty} 4^n \left\| f\left(\frac{xy}{4^n}\right) - f\left(\frac{x}{2^n}\right)\frac{y}{2^n} - \frac{x}{2^n}f\left(\frac{y}{2^n}\right) \right\|_A \\
&\le \lim_{n\to\infty} \frac{4^n\theta}{4^{nr}} \cdot \|x\|_A^r \cdot \|y\|_A^r \\
&= 0
\end{aligned}
$$

for all $x, y \in A$ and so

$$f(xy) = f(x)y + xf(y)$$

for all $x, y \in A$. Thus the mapping $f : A \to A$ is a linear derivation. This completes the proof. $\qquad\square$

*Remark 3.16.* Let $r < 1$, $\theta$ be positive real numbers and $f : A \to A$ be a mapping satisfying (3.19) and (3.22). Then the mapping $f : A \to A$ is a linear derivation.

### 3.2.2 Homomorphisms and Derivations in Lie $C^*$-Algebras

Assume that $A$ is a *Lie $C^*$-algebra* with the norm $\|\cdot\|_A$ and $B$ is a Lie $C^*$-algebra with the norm $\|\cdot\|_B$.

We recall that a $\mathbb{C}$-linear mapping $H : A \to B$ is called a *Lie $C^*$-algebra homomorphism* if $H : A \to B$ satisfies the following:

$$H([x, y]) = [H(x), H(y)]$$

for all $x, y \in A$. A $\mathbb{C}$-linear mapping $D : A \to A$ is called a *Lie derivation* if $D : A \to A$ satisfies the following:

$$D([x, y]) = [D(x), y] + [x, D(y)]$$

for all $x, y \in A$.

Now, we investigate Lie $C^*$-algebra homomorphisms in Lie $C^*$-algebras and Lie derivations on Lie $C^*$-algebras associated with the Apollonius type additive functional equation.

**Theorem 3.17.** *Let $r > 1$, $\theta$ be nonnegative real numbers and $f : A \to B$ be a mapping satisfying* (3.19) *such that*

$$\|f([x, y]) - [f(x), f(y)]\|_B \le \theta \cdot \|x\|_A^r \cdot \|y\|_A^r \tag{3.23}$$

*for all $x, y \in A$. Then the mapping $f : A \to B$ is a Lie $C^*$-algebra homomorphism.*

*Proof.* It is straight forward to show that, the mapping $f : A \to B$ is $\mathbb{C}$-linear. It follows from (3.23) that

$$\left\| f([x, y]) - [f(x), f(y)] \right\|_B = \lim_{n \to \infty} 4^n \left\| f\left(\frac{[x, y]}{2^n \cdot 2^n}\right) - \left[ f\left(\frac{x}{2^n}\right), f\left(\frac{y}{2^n}\right) \right] \right\|_B$$

$$\le \lim_{n \to \infty} \frac{4^n \theta}{4^{nr}} \cdot \|x\|_A^r \cdot \|y\|_A^r$$

$$= 0$$

for all $x, y \in A$ and so

$$f([x, y]) = [f(x), f(y)]$$

for all $x, y \in A$. Hence the mapping $f : A \to B$ is a Lie $C^*$-algebra homomorphism. This completes the proof.                                                                    □

*Remark 3.18.* If $r < 1$, $\theta$ is positive real numbers and $f : A \to B$ be a mapping satisfying (3.19) and (3.23). Then the mapping $f : A \to B$ is a Lie $C^*$-algebra homomorphism.

**Theorem 3.19.** *Let $r > 1$, $\theta$ be nonnegative real numbers and $f : A \to A$ be a mapping satisfying* (3.19) *such that*

$$\|f([x, y]) - [f(x), y] - [x, f(y)]\|_A \le \theta \cdot \|x\|_A^r \cdot \|y\|_A^r \tag{3.24}$$

*for all $x, y \in A$. Then the mapping $f : A \to A$ is a Lie derivation.*

*Proof.* It is straight forward to show that, the mapping $f : A \to A$ is $\mathbb{C}$-linear. It follows from (3.24) that

$$\|f([x, y]) - [f(x), y] - [x, f(y)]\|_A$$

$$= \lim_{n \to \infty} 4^n \left\| f\left(\frac{[x, y]}{4^n}\right) - \left[f\left(\frac{x}{2^n}\right), \frac{y}{2^n}\right] - \left[\frac{x}{2^n}, f\left(\frac{y}{2^n}\right)\right] \right\|_A$$

$$\leq \lim_{n \to \infty} \frac{4^n \theta}{4^{nr}} \cdot \|x\|_A^r \cdot \|y\|_A^r$$

$$= 0$$

for all $x, y \in A$ and so

$$f([x, y]) = [f(x), y] + [x, f(y)]$$

for all $x, y \in A$. Thus the mapping $f : A \to A$ is a Lie derivation. This completes the proof. □

*Remark 3.20.* If $r < 1$, $\theta$ is positive real numbers and $f : A \to A$ be a mapping satisfying (3.19) and (3.24). Then the mapping $f : A \to A$ is a Lie derivation.

### 3.2.3 Homomorphisms and Derivations in JC*-Algebras

Assume that $A$ is a $JC^*$-algebra with the norm $\| \cdot \|_A$ and $B$ is a $JC^*$-algebra with the norm $\| \cdot \|_B$.

A $\mathbb{C}$-linear mapping $H : A \to B$ is called a $JC^*$-algebra homomorphism if $H : A \to B$ satisfies the following:

$$H(x \circ y) = H(x) \circ H(y)$$

for all $x, y \in A$. A $\mathbb{C}$-linear mapping $D : A \to A$ is called a *Jordan derivation* if $D : A \to A$ satisfies the following:

$$D(x \circ y) = D(x) \circ y + x \circ D(y)$$

for all $x, y \in A$.

Now, we investigate $JC^*$-algebra homomorphisms between $JC^*$-algebras and Jordan derivations on $JC^*$-algebras associated with the Apollonius type additive functional equation.

*Remark 3.21.* Let $r > 1$, $\theta$ be nonnegative real numbers and $f : A \to B$ be a mapping satisfying (3.19) such that

$$\|f(x \circ y) - f(x) \circ f(y)\|_B \leq \theta \cdot \|x\|_A^r \cdot \|y\|_A^r \tag{3.25}$$

for all $x, y \in A$. Then the mapping $f : A \to B$ is a $JC^*$-algebra homomorphism.

*Remark 3.22.* Let $r < 1$, $\theta$ be positive real numbers and $f : A \to B$ be a mapping satisfying (3.19) and (3.25). Then the mapping $f : A \to B$ is a $JC^*$-algebra homomorphism.

*Remark 3.23.* Let $r > 1$, $\theta$ be nonnegative real numbers and $f : A \to A$ be a mapping satisfying (3.19) such that

$$\|f(x \circ y) - f(x) \circ y - x \circ f(y)\|_A \leq \theta \cdot \|x\|_A^r \cdot \|y\|_A^r \qquad (3.26)$$

for all $x, y \in A$. Then the mapping $f : A \to A$ is a Jordan derivation.

*Remark 3.24.* Let $r < 1$, $\theta$ be positive real numbers and $f : A \to A$ be a mapping satisfying (3.19) and (3.26). Then the mapping $f : A \to A$ is a Jordan derivation.

## 3.3   Stability of Jensen Type Functional Equations in $C^*$-Algebras

Using the fixed point method, we consider [250] the *Hyers-Ulam stability* of homomorphisms in $C^*$-algebras and Lie $C^*$-algebras and derivations on $C^*$-algebras and Lie $C^*$-algebras for the following *Jensen type functional equation*:

$$f\left(\frac{x+y}{2}\right) + f\left(\frac{x-y}{2}\right) = f(x).$$

### 3.3.1   Stability of Homomorphisms in $C^*$-Algebras

Assume that $A$ is a $C^*$-algebra with the norm $\| \cdot \|_A$ and $B$ is a $C^*$-algebra with the norm $\| \cdot \|_B$. For any mapping $f : A \to B$, we define

$$D_\mu f(x, y) := \mu f\left(\frac{x+y}{2}\right) + \mu f\left(\frac{x-y}{2}\right) - f(\mu x)$$

for all $\mu \in \mathbb{T}^1 := \{v \in \mathbb{C} : |v| = 1\}$ and $x, y \in A$.

Now, we prove the Hyers-Ulam stability of homomorphisms in $C^*$-algebras for the functional equation $D_\mu f(x, y) = 0$.

**Theorem 3.25.** *Let* $f : A \to B$ *be a mapping for which there exists a function* $\varphi : A^2 \to [0, \infty)$ *such that*

$$\sum_{j=0}^{\infty} 2^{-j} \varphi(2^j x, 2^j y) < \infty, \qquad (3.27)$$

$$\|D_\mu f(x, y)\|_B \le \varphi(x, y), \tag{3.28}$$

$$\|f(xy) - f(x)f(y)\|_B \le \varphi(x, y), \tag{3.29}$$

$$\|f(x^*) - f(x)^*\|_B \le \varphi(x, x) \tag{3.30}$$

for all $\mu \in \mathbb{T}^1$ and $x, y \in A$. If there exists $L < 1$ such that

$$\varphi(x, 0) \le 2L\varphi\left(\frac{x}{2}, 0\right)$$

for all $x \in A$, then there exists a unique $C^*$-algebra homomorphism $H : A \to B$ such that

$$\|f(x) - H(x)\|_B \le \frac{L}{1 - L}\varphi(x, 0) \tag{3.31}$$

for all $x \in A$.

*Proof.* Consider the set

$$X := \{g : A \to B\}$$

and introduce the *generalized metric* on $X$ defined by

$$d(g, h) = \inf\{C \in \mathbb{R}_+ : \|g(x) - h(x)\|_B \le C\varphi(x, 0), \ \forall x \in A\},$$

which $(X, d)$ is complete.

Now, we consider the linear mapping $J : X \to X$ such that

$$Jg(x) := \frac{1}{2}g(2x)$$

for all $x \in A$. Thus we have

$$d(Jg, Jh) \le Ld(g, h)$$

for all $g, h \in X$. Letting $\mu = 1$ and $y = 0$ in (3.28), we get

$$\left\|2f\left(\frac{x}{2}\right) - f(x)\right\|_B \le \varphi(x, 0) \tag{3.32}$$

for all $x \in A$ and so

$$\left\|f(x) - \frac{1}{2}f(2x)\right\|_B \le \frac{1}{2}\varphi(2x, 0) \le L\varphi(x, 0)$$

for all $x \in A$. Hence $d(f, Jf) \le L$.

By Theorem 1.3, there exists a mapping $H : A \to B$ such that

(1) $H$ is a fixed point of $J$, i.e.,

$$H(2x) = 2H(x) \tag{3.33}$$

for all $x \in A$. The mapping $H$ is a unique fixed point of $J$ in the set

$$Y = \{g \in X : d(f, g) < \infty\}.$$

This implies that $H$ is a unique mapping satisfying (3.33) such that there exists $C \in (0, \infty)$ satisfying

$$\|H(x) - f(x)\|_B \le C\varphi(x, 0)$$

for all $x \in A$;

(2) $d(J^n f, H) \to 0$ as $n \to \infty$. This implies the equality

$$\lim_{n \to \infty} \frac{f(2^n x)}{2^n} = H(x) \tag{3.34}$$

for all $x \in A$;

(3) $d(f, H) \le \frac{1}{1-L} d(f, Jf)$, which implies the inequality

$$d(f, H) \le \frac{L}{1 - L}.$$

This implies that the inequality (3.31) holds.

It follows from (3.27), (3.28) and (3.34) that

$$\left\| H\left(\frac{x+y}{2}\right) + H\left(\frac{x-y}{2}\right) - H(x) \right\|_B$$

$$= \lim_{n \to \infty} \frac{1}{2^n} \|f(2^{n-1}(x+y)) + f(2^{n-1}(x-y)) - f(2^n x)\|_B$$

$$\le \lim_{n \to \infty} \frac{1}{2^n} \varphi(2^n x, 2^n y)$$

$$= 0$$

for all $x, y \in A$ and so

$$H\left(\frac{x+y}{2}\right) + H\left(\frac{x-y}{2}\right) = H(x) \tag{3.35}$$

for all $x, y \in A$. Letting $z = \frac{x+y}{2}$ and $w = \frac{x-y}{2}$ in (3.35), we get

$$H(z) + H(w) = H(z + w)$$

for all $z, w \in A$. So the mapping $H : A \to B$ is Cauchy additive, i.e.,

$$H(z + w) = H(z) + H(w)$$

for all $z, w \in A$. Letting $y = x$ in (3.28), we get

$$\mu f(x) = f(\mu x)$$

for all $\mu \in \mathbb{T}^1$ and $x \in A$. By a similar method to above, we get

$$\mu H(x) = H(\mu x)$$

for all $\mu \in \mathbb{T}^1$ and $x \in A$. Thus one can show that the mapping $H : A \to B$ is $\mathbb{C}$-linear. It follows from (3.29) that

$$
\begin{aligned}
\|H(xy) - H(x)H(y)\|_B &= \lim_{n \to \infty} \frac{1}{4^n} \|f(4^n xy) - f(2^n x)f(2^n y)\|_B \\
&\leq \lim_{n \to \infty} \frac{1}{4^n} \varphi(2^n x, 2^n y) \\
&\leq \lim_{n \to \infty} \frac{1}{2^n} \varphi(2^n x, 2^n y) \\
&= 0
\end{aligned}
$$

for all $x, y \in A$. Then we have

$$H(xy) = H(x)H(y)$$

for all $x, y \in A$. It follows from (3.30) that

$$
\begin{aligned}
\|H(x^*) - H(x)^*\|_B &= \lim_{n \to \infty} \frac{1}{2^n} \|f(2^n x^*) - f(2^n x)^*\|_B \\
&\leq \lim_{n \to \infty} \frac{1}{2^n} \varphi(2^n x, 2^n x) \\
&= 0
\end{aligned}
$$

for all $x \in A$. Then we have

$$H(x^*) = H(x)^*$$

for all $x \in A$. Thus $H : A \to B$ is a $C^*$-algebra homomorphism satisfying (3.31). This completes the proof. $\qquad\square$

**Corollary 3.26.** *Let $r < 1$, $\theta$ be nonnegative real numbers and $f : A \rightarrow B$ be a mapping such that*

$$\|D_\mu f(x, y)\|_B \leq \theta(\|x\|_A^r + \|y\|_A^r), \tag{3.36}$$

$$\|f(xy) - f(x)f(y)\|_B \leq \theta(\|x\|_A^r + \|y\|_A^r), \tag{3.37}$$

$$\|f(x^*) - f(x)^*\|_B \leq 2\theta\|x\|_A^r \tag{3.38}$$

*for all $\mu \in \mathbb{T}^1$ and $x, y \in A$. Then there exists a unique $C^*$-algebra homomorphism $H : A \rightarrow B$ such that*

$$\|f(x) - H(x)\|_B \leq \frac{2^r\theta}{2 - 2^r}\|x\|_A^r \tag{3.39}$$

*for all $x \in A$.*

*Proof.* The proof follows from Theorem 3.25 by taking

$$\varphi(x, y) := \theta(\|x\|_A^r + \|y\|_A^r)$$

for all $x, y \in A$ and $L = 2^{r-1}$. $\qquad\qquad\qquad\qquad\qquad\qquad\qquad\qquad\square$

**Theorem 3.27.** *Let $f : A \rightarrow B$ be a mapping for which there exists a function $\varphi : A^2 \rightarrow [0, \infty)$ satisfying (3.28), (3.29) and (3.30) such that*

$$\sum_{j=0}^{\infty} 4^j \varphi\left(\frac{x}{2^j}, \frac{y}{2^j}\right) < \infty \tag{3.40}$$

*for all $x, y \in A$. If there exists $L < 1$ such that*

$$\varphi(x, 0) \leq \frac{1}{2}L\varphi(2x, 0)$$

*for all $x \in A$, then there exists a unique $C^*$-algebra homomorphism $H : A \rightarrow B$ such that*

$$\|f(x) - H(x)\|_B \leq \frac{L}{2 - 2L}\varphi(x, 0) \tag{3.41}$$

*for all $x \in A$.*

*Proof.* We consider the linear mapping $J : X \rightarrow X$ such that

$$Jg(x) := 2g\left(\frac{x}{2}\right)$$

for all $x \in A$. It follows from (3.32) that

$$\left\| f(x) - 2f\left(\frac{x}{2}\right) \right\|_B \leq \varphi\left(\frac{x}{2}, 0\right) \leq \frac{L}{2}\varphi(x, 0)$$

for all $x \in A$ and hence $d(f, Jf) \leq \frac{L}{2}$. By Theorem 1.3, there exists a mapping $H : A \to B$ such that

(1) $H$ is a fixed point of $J$, i.e.,

$$H(2x) = 2H(x) \tag{3.42}$$

for all $x \in A$. The mapping $H$ is a unique fixed point of $J$ in the set

$$Y = \{g \in X : d(f, g) < \infty\}.$$

This implies that $H$ is a unique mapping satisfying (3.42) such that there exists $C \in (0, \infty)$ satisfying

$$\|H(x) - f(x)\|_B \leq C\varphi(x, 0)$$

for all $x \in A$;

(2) $d(J^n f, H) \to 0$ as $n \to \infty$. This implies the equality

$$\lim_{n \to \infty} 2^n f\left(\frac{x}{2^n}\right) = H(x)$$

for all $x \in A$;

(3) $d(f, H) \leq \frac{1}{1-L} d(f, Jf)$, which implies the inequality

$$d(f, H) \leq \frac{L}{2 - 2L},$$

which implies that the inequality (3.41) holds.

The rest of the proof is similar to the proof of Theorem 3.25. This completes the proof. □

**Corollary 3.28.** *Let $r > 2$, $\theta$ be nonnegative real numbers and $f : A \to B$ be a mapping satisfying (3.36), (3.37) and (3.38). Then there exists a unique $C^*$-algebra homomorphism $H : A \to B$ such that*

$$\|f(x) - H(x)\|_B \leq \frac{\theta}{2^r - 2}\|x\|_A^r \tag{3.43}$$

*for all $x \in A$.*

*Proof.* The proof follows from Theorem 3.27 by taking

$$\varphi(x, y) := \theta(\|x\|_A^r + \|y\|_A^r)$$

for all $x, y \in A$ and $L = 2^{1-r}$.                                                          □

**Theorem 3.29.** *Let $f : A \to B$ be an odd mapping for which there exists a function $\varphi : A^2 \to [0, \infty)$ satisfying (3.27), (3.28), (3.29) and (3.30). If there exists $L < 1$ such that*

$$\varphi(x, 3x) \leq 2L\varphi\left(\frac{x}{2}, \frac{3x}{2}\right)$$

*for all $x \in A$, then there exists a unique $C^*$-algebra homomorphism $H : A \to B$ such that*

$$\|f(x) - H(x)\|_B \leq \frac{1}{2 - 2L}\varphi(x, 3x) \tag{3.44}$$

*for all $x \in A$.*

*Proof.* Consider the set

$$X := \{g : A \to B\}$$

and introduce the generalized metric on $X$ defined by

$$d(g, h) = \inf\{C \in \mathbb{R}_+ : \|g(x) - h(x)\|_B \leq C\varphi(x, 3x), \ \forall x \in A\},$$

which $(X, d)$ is complete.

Now, we consider the linear mapping $J : X \to X$ such that

$$Jg(x) := \frac{1}{2}g(2x)$$

for all $x \in A$. Now, we have

$$d(Jg, Jh) \leq Ld(g, h)$$

for all $g, h \in X$. Letting $\mu = 1$ and replacing $y$ by $3x$ in (3.28), we get

$$\|f(2x) - 2f(x)\|_B \leq \varphi(x, 3x) \tag{3.45}$$

for all $x \in A$ and so

$$\left\|f(x) - \frac{1}{2}f(2x)\right\|_B \leq \frac{1}{2}\varphi(x, 3x)$$

for all $x \in A$. Hence $d(f, Jf) \leq \frac{1}{2}$. By Theorem 1.3, there exists a mapping $H : A \rightarrow B$ such that

(1)  $H$ is a fixed point of $J$, i.e.,

$$H(2x) = 2H(x) \tag{3.46}$$

for all $x \in A$. The mapping $H$ is a unique fixed point of $J$ in the set

$$Y = \{g \in X : d(f, g) < \infty\}.$$

This implies that $H$ is a unique mapping satisfying (3.46) such that there exists $C \in (0, \infty)$ satisfying

$$\|H(x) - f(x)\|_B \leq C\varphi(x, 3x)$$

for all $x \in A$;

(2)  $d(J^n f, H) \rightarrow 0$ as $n \rightarrow \infty$. This implies the equality

$$\lim_{n \to \infty} \frac{f(2^n x)}{2^n} = H(x)$$

for all $x \in A$;

(3)  $d(f, H) \leq \frac{1}{1-L} d(f, Jf)$, which implies the inequality

$$d(f, H) \leq \frac{1}{2 - 2L}.$$

This implies that the inequality (3.44) holds.

The rest of the proof is similar to the proof of Theorem 3.25. This completes the proof.  □

**Corollary 3.30.** *Let $r < \frac{1}{2}$, $\theta$ be nonnegative real numbers and $f : A \rightarrow B$ be an odd mapping such that*

$$\|D_\mu f(x, y)\|_B \leq \theta \cdot \|x\|_A^r \cdot \|y\|_A^r, \tag{3.47}$$

$$\|f(xy) - f(x)f(y)\|_B \leq \theta \cdot \|x\|_A^r \cdot \|y\|_A^r, \tag{3.48}$$

$$\|f(x^*) - f(x)^*\|_B \leq \theta \|x\|_A^{2r} \tag{3.49}$$

*for all $\mu \in \mathbb{T}^1$ and $x, y \in A$. Then there exists a unique $C^*$-algebra homomorphism $H : A \rightarrow B$ such that*

$$\|f(x) - H(x)\|_B \leq \frac{3^r \theta}{2 - 2^{2r}} \|x\|_A^{2r} \tag{3.50}$$

*for all $x \in A$.*

*Proof.* The proof follows from Theorem 3.29 by taking

$$\varphi(x, y) := \theta \cdot \|x\|_A^r \cdot \|y\|_A^r$$

for all $x, y \in A$ and $L = 2^{2r-1}$.                    $\square$

**Theorem 3.31.** *Let $f : A \to B$ be an odd mapping for which there exists a function $\varphi : A^2 \to [0, \infty)$ satisfying (3.28), (3.29), (3.30) and (3.40). If there exists $L < 1$ such that*

$$\varphi(x, 3x) \leq \frac{1}{2}L\varphi(2x, 6x)$$

*for all $x \in A$, then there exists a unique $C^*$-algebra homomorphism $H : A \to B$ such that*

$$\|f(x) - H(x)\|_B \leq \frac{L}{2 - 2L}\varphi(x, 3x) \tag{3.51}$$

*for all $x \in A$.*

*Proof.* We consider the linear mapping $J : X \to X$ such that

$$Jg(x) := 2g\left(\frac{x}{2}\right)$$

for all $x \in A$. It follows from (3.45) that

$$\left\|f(x) - 2f\left(\frac{x}{2}\right)\right\|_B \leq \varphi\left(\frac{x}{2}, \frac{3x}{2}\right) \leq \frac{L}{2}\varphi(x, 3x)$$

for all $x \in A$ and hence $d(f, Jf) \leq \frac{L}{2}$. By Theorem 1.3, there exists a mapping $H : A \to B$ such that

(1) $H$ is a fixed point of $J$, i.e.,

$$H(2x) = 2H(x) \tag{3.52}$$

for all $x \in A$. The mapping $H$ is a unique fixed point of $J$ in the set

$$Y = \{g \in X : d(f, g) < \infty\}.$$

This implies that $H$ is a unique mapping satisfying (3.52) such that there exists $C \in (0, \infty)$ satisfying

$$\|H(x) - f(x)\|_B \leq C\varphi(x, 3x)$$

for all $x \in A$;

(2) $d(J^n f, H) \to 0$ as $n \to \infty$. This implies the equality

$$\lim_{n\to\infty} 2^n f\left(\frac{x}{2^n}\right) = H(x)$$

for all $x \in A$;

(3) $d(f, H) \leq \frac{1}{1-L} d(f, Jf)$, which implies the inequality

$$d(f, H) \leq \frac{L}{2 - 2L},$$

which implies that the inequality (3.51) holds.

The rest of the proof is similar to the proof of Theorem 3.25. This completes the proof.   $\square$

**Corollary 3.32.** *Let $r > 1$, $\theta$ be nonnegative real numbers and $f : A \to B$ be an odd mapping satisfying (3.47), (3.48) and (3.49). Then there exists a unique $C^*$-algebra homomorphism $H : A \to B$ such that*

$$\|f(x) - H(x)\|_B \leq \frac{\theta}{2^{2r} - 2} \|x\|_A^{2r} \tag{3.53}$$

*for all $x \in A$.*

*Proof.* The proof follows from Theorem 3.31 by taking

$$\varphi(x, y) := \theta \cdot \|x\|_A^r \cdot \|y\|_A^r$$

for all $x, y \in A$ and $L = 2^{1-2r}$.   $\square$

## 3.3.2   Stability of Derivations in $C^*$-Algebras

Now, we prove the Hyers-Ulam stability of derivations on $C^*$-algebras for the functional equation $D_\mu f(x, y) = 0$.

**Theorem 3.33.** *Let $f : A \to A$ be a mapping for which there exists a function $\varphi : A^2 \to [0, \infty)$ satisfying (3.27) such that*

$$\|D_\mu f(x, y)\|_A \leq \varphi(x, y) \tag{3.54}$$

*and*

$$\|f(xy) - f(x)y - xf(y)\|_A \leq \varphi(x, y) \tag{3.55}$$

*for all $\mu \in \mathbb{T}^1$ and $x, y \in A$. If there exists $L < 1$ such that*

$$\varphi(x,0) \le 2L\varphi\left(\frac{x}{2},0\right)$$

for all $x \in A$. Then there exists a unique derivation $\delta : A \to A$ such that

$$\|f(x) - \delta(x)\|_A \le \frac{L}{1-L}\varphi(x,0) \tag{3.56}$$

for all $x \in A$.

*Proof.* It is straight forward to show that, there exists a unique involutive $\mathbb{C}$-linear mapping $\delta : A \to A$ satisfying (3.56). The mapping $\delta : A \to A$ is given by

$$\delta(x) = \lim_{n\to\infty} \frac{f(2^n x)}{2^n}$$

for all $x \in A$. It follows from (3.55) that

$$\|\delta(xy) - \delta(x)y - x\delta(y)\|_A$$

$$= \lim_{n\to\infty} \frac{1}{4^n}\|f(4^n xy) - f(2^n x)\cdot 2^n y - 2^n xf(2^n y)\|_A$$

$$\le \lim_{n\to\infty} \frac{1}{4^n}\varphi(2^n x, 2^n y)$$

$$\le \lim_{n\to\infty} \frac{1}{2^n}\varphi(2^n x, 2^n y)$$

$$= 0$$

for all $x, y \in A$ and so

$$\delta(xy) = \delta(x)y + x\delta(y)$$

for all $x, y \in A$. Thus $\delta : A \to A$ is a derivation satisfying (3.56). This completes the proof. $\qquad\square$

**Corollary 3.34.** *Let $r < 1$, $\theta$ be nonnegative real numbers and $f : A \to A$ be a mapping such that*

$$\|D_\mu f(x,y)\|_A \le \theta(\|x\|_A^r + \|y\|_A^r) \tag{3.57}$$

*and*

$$\|f(xy) - f(x)y - xf(y)\|_A \le \theta(\|x\|_A^r + \|y\|_A^r) \tag{3.58}$$

*for all $\mu \in \mathbb{T}^1$ and $x, y \in A$. Then there exists a unique derivation $\delta : A \to A$ such that*

$$\|f(x) - \delta(x)\|_A \le \frac{2^r\theta}{2 - 2^r}\|x\|_A^r \tag{3.59}$$

*for all $x \in A$.*

*Proof.* The proof follows from Theorem 3.33 by taking

$$\varphi(x, y) := \theta(\|x\|_A^r + \|y\|_A^r)$$

for all $x, y \in A$ and $L = 2^{r-1}$.  □

*Remark 3.35.* If $f : A \to A$ is a mapping for which there exists a function $\varphi :$ $A^2 \to [0, \infty)$ satisfying (3.40), (3.54) and (3.55). If there exists $L < 1$ such that $\varphi(x, 0) \leq \frac{1}{2}L\varphi(2x, 0)$ for all $x \in A$, then there exists a unique derivation $\delta : A \to A$ such that

$$\|f(x) - \delta(x)\|_A \leq \frac{L}{2 - 2L}\varphi(x, 0) \tag{3.60}$$

for all $x \in A$.

**Corollary 3.36.** *Let $r > 2$, $\theta$ be nonnegative real numbers and $f : A \to A$ be a mapping satisfying (3.57) and (3.58).*
    *Then there exists a unique derivation $\delta : A \to A$ such that*

$$\|f(x) - \delta(x)\|_A \leq \frac{\theta}{2^r - 2}\|x\|_A^r \tag{3.61}$$

*for all $x \in A$.*

*Proof.* The proof follows from Theorem 3.35 by taking

$$\varphi(x, y) := \theta(\|x\|_A^r + \|y\|_A^r)$$

for all $x, y \in A$ and $L = 2^{1-r}$.  □

*Remark 3.37.* For the inequalities controlled by the product of powers of norms, one can obtain similar results to Theorems 3.29 and 3.31 and Corollaries 3.30 and 3.32.

### 3.3.3  Stability of Homomorphisms in Lie $C^*$-Algebras

Assume that $A$ is a Lie $C^*$-algebra with the norm $\| \cdot \|_A$ and $B$ is a Lie $C^*$-algebra with the norm $\| \cdot \|_B$.
    Now, we prove the Hyers-Ulam stability of homomorphisms in Lie $C^*$-algebras for the functional equation $D_\mu f(x, y) = 0$.

**Theorem 3.38.** *Let $f : A \to B$ be a mapping for which there exists a function $\varphi : A^2 \to [0, \infty)$ satisfying (3.27) and (3.28) such that*

$$\|f([x, y]) - [f(x), f(y)]\|_B \leq \varphi(x, y) \tag{3.62}$$

*for all $x, y \in A$. If there exists $L < 1$ such that*

$$\varphi(x, 0) \leq 2L\varphi\left(\frac{x}{2}, 0\right)$$

*for all $x \in A$, then there exists a unique Lie $C^*$-algebra homomorphism $H : A \to B$ satisfying (3.31).*

*Proof.* It is straight forward to show that, there exists a unique $\mathbb{C}$-linear mapping $H : A \to B$ satisfying (3.31). The mapping $H : A \to B$ is given by

$$H(x) = \lim_{n \to \infty} \frac{f(2^n x)}{2^n}$$

for all $x \in A$. It follows from (3.62) that

$$\|H([x, y]) - [H(x), H(y)]\|_B$$

$$= \lim_{n \to \infty} \frac{1}{4^n} \|f(4^n[x, y]) - [f(2^n x), f(2^n y)]\|_B$$

$$\leq \lim_{n \to \infty} \frac{1}{4^n} \varphi(2^n x, 2^n y)$$

$$\leq \lim_{n \to \infty} \frac{1}{2^n} \varphi(2^n x, 2^n y)$$

$$= 0$$

for all $x, y \in A$ and so

$$H([x, y]) = [H(x), H(y)]$$

for all $x, y \in A$. Thus $H : A \to B$ is a Lie $C^*$-algebra homomorphism satisfying (3.31). This completes the proof.                                        $\square$

**Corollary 3.39.** *Let $r < 1$, $\theta$ be nonnegative real numbers and $f : A \to B$ be a mapping satisfying (3.36) such that*

$$\|f([x, y]) - [f(x), f(y)]\|_B \leq \theta(\|x\|_A^r + \|y\|_A^r) \tag{3.63}$$

*for all $x, y \in A$.*

*Then there exists a unique Lie $C^*$-algebra homomorphism $H : A \to B$ satisfying (3.39).*

*Proof.* The proof follows from Theorem 3.38 by taking

$$\varphi(x, y) := \theta(\|x\|_A^r + \|y\|_A^r)$$

for all $x, y \in A$ and $L = 2^{r-1}$.                                        $\square$

*Remark 3.40.* If $f : A \rightarrow B$ is a mapping for which there exists a function $\varphi : A^2 \rightarrow [0, \infty)$ satisfying (3.28), (3.40) and (3.62). If there exists $L < 1$ such that $\varphi(x, 0) \leq \frac{1}{2} L \varphi(2x, 0)$ for all $x \in A$, then there exists a unique Lie $C^*$-algebra homomorphism $H : A \rightarrow B$ satisfying (3.41).

**Corollary 3.41.** *Let $r > 2$, $\theta$ be nonnegative real numbers and $f : A \rightarrow B$ be a mapping satisfying (3.36) and (3.63). Then there exists a unique Lie $C^*$-algebra homomorphism $H : A \rightarrow B$ satisfying (3.43).*

*Proof.* The proof follows from Theorem 3.40 by taking

$$\varphi(x, y) := \theta(\|x\|_A^r + \|y\|_A^r)$$

for all $x, y \in A$ and $L = 2^{1-r}$. $\qquad\qquad\qquad\qquad\qquad\qquad\qquad\qquad\square$

*Remark 3.42.* For the inequalities controlled by the product of powers of norms, one can obtain similar results to Theorems 3.29 and 3.31 and their corollaries.

### 3.3.4 Stability of Lie Derivations in $C^*$-Algebras

Assume that $A$ is a Lie $C^*$-algebra with the norm $\| \cdot \|_A$.

Now, we prove the Hyers-Ulam stability of derivations on Lie $C^*$-algebras for the functional equation $D_\mu f(x, y) = 0$.

**Theorem 3.43.** *Let $f : A \rightarrow A$ be a mapping for which there exists a function $\varphi : A^2 \rightarrow [0, \infty)$ satisfying (3.27) and (3.54) such that*

$$\|f([x, y]) - [f(x), y] - [x, f(y)]\|_A \leq \varphi(x, y) \qquad (3.64)$$

*for all $x, y \in A$. If there exists $L < 1$ such that*

$$\varphi(x, 0) \leq 2L\varphi\left(\frac{x}{2}, 0\right)$$

*for all $x \in A$. Then there exists a unique Lie derivation $\delta : A \rightarrow A$ satisfying (3.56).*

*Proof.* It is easy to show that, there exists a unique involution $\mathbb{C}$-linear mapping $\delta : A \rightarrow A$ satisfying (3.56). The mapping $\delta : A \rightarrow A$ is given by

$$\delta(x) = \lim_{n \to \infty} \frac{f(2^n x)}{2^n}$$

for all $x \in A$. It follows from (3.62) that

$$\|\delta([x, y]) - [\delta(x), y] - [x, \delta(y)]\|_A$$

$$= \lim_{n \to \infty} \frac{1}{4^n} \|f(4^n[x, y]) - [f(2^n x), 2^n y] - [2^n x, f(2^n y)]\|_A$$

$$\leq \lim_{n\to\infty} \frac{1}{4^n} \varphi(2^n x, 2^n y)$$

$$\leq \lim_{n\to\infty} \frac{1}{2^n} \varphi(2^n x, 2^n y)$$

$$= 0$$

for all $x, y \in A$ and so

$$\delta([x, y]) = [\delta(x), y] + [x, \delta(y)]$$

for all $x, y \in A$. Thus $\delta : A \to A$ is a derivation satisfying (3.56). This completes the proof.                                                                                                            □

**Corollary 3.44.** *Let $r < 1$, $\theta$ be nonnegative real numbers and $f : A \to A$ be a mapping satisfying (3.57) such that*

$$\|f([x, y]) - [f(x), y] - [x, f(y)]\|_A \leq \theta(\|x\|_A^r + \|y\|_A^r) \tag{3.65}$$

*for all $x, y \in A$. Then there exists a unique Lie derivation $\delta : A \to A$ satisfying (3.59).*

*Proof.* The proof follows from Theorem 3.38 by taking

$$\varphi(x, y) := \theta(\|x\|_A^r + \|y\|_A^r)$$

for all $x, y \in A$ and $L = 2^{r-1}$.                                                                         □

*Remark 3.45.* If $f : A \to A$ is a mapping for which there exists a function $\varphi : A^2 \to [0, \infty)$ satisfying (3.40), (3.54) and (3.64). Whenever there exists $L < 1$ such that

$$\varphi(x, 0) \leq \frac{1}{2} L \varphi(2x, 0)$$

for all $x \in A$, then there exists a unique Lie derivation $\delta : A \to A$ satisfying (3.60).

**Corollary 3.46.** *Let $r > 2$, $\theta$ be nonnegative real numbers and $f : A \to A$ be a mapping satisfying (3.57) and (3.65). Then there exists a unique Lie derivation $\delta : A \to A$ satisfying (3.61).*

*Proof.* The proof follows from Theorem 3.45 by taking

$$\varphi(x, y) := \theta(\|x\|_A^r + \|y\|_A^r)$$

for all $x, y \in A$ and $L = 2^{1-r}$.                                                                         □

*Remark 3.47.* For the inequalities controlled by the product of powers of norms, one can obtain similar results to Theorems 3.29 and 3.31 and their corollaries.

## 3.4  Generalized Additive Mapping

Recently, Park and Park [243] introduced and investigated the following generalized additive functional equation

$$\sum_{i=1}^{n} r_i L\left(\sum_{j=1}^{n} r_j(x_i - x_j)\right) + \left(\sum_{i=1}^{n} r_i\right) L\left(\sum_{i=1}^{n} r_i x_i\right)$$

$$= \left(\sum_{i=1}^{n} r_i\right) \sum_{i=1}^{n} r_i L(x_i) \tag{3.66}$$

for all $r_1, \cdots, r_n \in (0, \infty)$ whose solution is called a *generalized additive mapping*.

In this section, we consider [210] the following additive functional equation which is somewhat different from (3.66):

$$\sum_{j=1}^{n} f\left(\frac{1}{2} \sum_{1 \le i \le n, i \neq j} r_i x_i - \frac{1}{2} r_j x_j\right) + \sum_{i=1}^{n} r_i f(x_i) = nf\left(\frac{1}{2} \sum_{i=1}^{n} r_i x_i\right) \tag{3.67}$$

for all $r_1, \cdots, r_n \in \mathbb{R}$. Every solution of the functional equation (3.67) is said to be a *generalized additive mapping*.

Using the fixed point method, we investigate the Hyers–Ulam stability of the functional equation (3.67) in Banach modules over a $C^*$-algebra. These results are applied to investigate $C^*$-algebra homomorphisms in unital $C^*$-algebras.

Throughout this section, assume that $A$ is a unital $C^*$-algebra with the norm $\| \cdot \|_A$ and the unit $e$ that $B$ is a unital $C^*$-algebra with the norm $\| \cdot \|_B$ and $X$ and $Y$ are left Banach modules over a unital $C^*$-algebra $A$ with the norms $\| \cdot \|_X$ and $\| \cdot \|_Y$, respectively. Let $U(A)$ be the group of unitary elements in $A$ and let $r_1, \cdots, r_n \in \mathbb{R}$.

### 3.4.1  Hyers–Ulam Stability of Functional Equations in Banach Modules over a $C^*$-Algebra

For any mapping $f : X \rightarrow Y$, $u \in U(A)$ and $\mu \in \mathbb{C}$, we define $D_{u, r_1, \cdots, r_n} f$ and $D_{\mu, r_1, \cdots, r_n} f : X^n \rightarrow Y$ by

$$D_{u, r_1, \cdots, r_n} f(x_1, \cdots, x_n)$$

$$:= \sum_{j=1}^{n} f\left(\frac{1}{2} \sum_{1 \le i \le n, i \neq j} r_i u x_i - \frac{1}{2} r_j u x_j\right) + \sum_{i=1}^{n} r_i u f(x_i) - nf\left(\frac{1}{2} \sum_{i=1}^{n} r_i u x_i\right)$$

and

$$D_{\mu,r_1,\cdots,r_n}f(x_1,\cdots,x_n)$$

$$:= \sum_{j=1}^{n} f\Big(\frac{1}{2}\sum_{1\leq i\leq n, i\neq j}\mu r_i x_i - \frac{1}{2}\mu r_j x_j\Big) + \sum_{i=1}^{n}\mu r_i f(x_i) - nf\Big(\frac{1}{2}\sum_{i=1}^{n}\mu r_i x_i\Big)$$

for all $x_1,\cdots,x_n \in X$, respectively.

**Lemma 3.48.** *Let $X$ and $Y$ be linear spaces and $r_1,\cdots,r_n$ be real numbers with $\sum_{k=1}^{n} r_k \neq 0$ and $r_i \neq 0$, $r_j \neq 0$ for some $1 \leq i < j \leq n$. Assume that a mapping $L : X \to Y$ satisfies the functional equation (3.67) for all $x_1,\cdots,x_n \in X$. Then the mapping $L$ is additive. Moreover, $L(r_k x) = r_k L(x)$ for all $x \in X$ and $1 \leq k \leq n$.*

*Proof.* Since $\sum_{k=1}^{n} r_k \neq 0$, putting $x_1 = \cdots = x_n = 0$ in (3.67), we get $L(0) = 0$. Without loss of generality, we assume that $r_1, r_2 \neq 0$. Letting $x_3 = \cdots = x_n = 0$ in (3.67), we get

$$L\Big(\frac{-r_1 x_1 + r_2 x_2}{2}\Big) + L\Big(\frac{r_1 x_1 - r_2 x_2}{2}\Big) + r_1 L(x_1) + r_2 L(x_2)$$

$$= 2L\Big(\frac{r_1 x_1 + r_2 x_2}{2}\Big) \tag{3.68}$$

for all $x_1, x_2 \in X$. Letting $x_2 = 0$ in (3.68), we get

$$r_1 L(x_1) = L\Big(\frac{r_1 x_1}{2}\Big) - L\Big(\frac{-r_1 x_1}{2}\Big) \tag{3.69}$$

for all $x_1 \in X$. Similarly, by putting $x_1 = 0$ in (3.68), we get

$$r_2 L(x_2) = L\Big(\frac{r_2 x_2}{2}\Big) - L\Big(\frac{-r_2 x_2}{2}\Big) \tag{3.70}$$

for all $x_2 \in X$. It follows from (3.68), (3.69) and (3.70) that

$$L\Big(\frac{-r_1 x_1 + r_2 x_2}{2}\Big) + L\Big(\frac{r_1 x_1 - r_2 x_2}{2}\Big)$$

$$+ L\Big(\frac{r_1 x_1}{2}\Big) + L\Big(\frac{r_2 x_2}{2}\Big) - L\Big(\frac{-r_1 x_1}{2}\Big) - L\Big(\frac{-r_2 x_2}{2}\Big) \tag{3.71}$$

$$= 2L\Big(\frac{r_1 x_1 + r_2 x_2}{2}\Big)$$

for all $x_1, x_2 \in X$. Replacing $x_1$ and $x_2$ by $\frac{2x}{r_1}$ and $\frac{2x}{r_2}$ in (3.71), we get

$$L(-x + y) + L(x - y) + L(x) + L(y) - L(-x) - L(-y)$$

$$= 2L(x + y) \tag{3.72}$$

for all $x, y \in X$. Letting $y = -x$ in (3.72), we get that

$$L(-2x) + L(2x) = 0$$

for all $x \in X$. Then the mapping $L$ is odd. Therefore, it follows from (3.72) that the mapping $L$ is additive. Moreover, let $x \in X$ and $1 \leq k \leq n$. Setting $x_k = x$ and $x_l = 0$ for all $1 \leq l \leq n$ with $l \neq k$ in (3.67) and, using the oddness of $L$, it follows that $L(r_k x) = r_k L(x)$. This completes the proof. $\qquad\square$

Using the same method as in the proof of Lemma 3.48, we have an alternative result of Lemma 3.48 when $\sum_{k=1}^{n} r_k = 0$.

**Lemma 3.49.** *Let $X$ and $Y$ be linear spaces and $r_1, \cdots, r_n$ be real numbers with $r_i \neq 0, r_j \neq 0$ for some $1 \leq i < j \leq n$. Assume that a mapping $L : X \to Y$ with $L(0) = 0$ satisfies the functional equation (3.67) for all $x_1, \cdots, x_n \in X$. Then the mapping $L$ is additive. Moreover, $L(r_k x) = r_k L(x)$ for all $x \in X$ and all $1 \leq k \leq n$.*

Now, we investigate the Hyers-Ulam stability of a generalized additive mapping in Banach modules over a unital $C^*$-algebra. Here $r_1, \cdots, r_n$ are real numbers such that $r_i \neq 0$ and $r_j \neq 0$ for fixed $1 \leq i < j \leq n$.

**Theorem 3.50.** *Let $f : X \to Y$ be a mapping satisfying $f(0) = 0$ for which there exists a function $\varphi : X^n \to [0, \infty)$ such that*

$$\|D_{e,r_1,\cdots,r_n} f(x_1, \cdots, x_n)\|_Y \leq \varphi(x_1, \cdots, x_n) \tag{3.73}$$

*for each $x_1, \cdots, x_n \in X$. Let*

$$\varphi_{ij}(x, y) := \varphi\Big(0, \cdots, 0, \underbrace{x}_{i\,th}, 0, \cdots, 0, \underbrace{y}_{j\,th}, 0, \cdots, 0\Big)$$

*for all $x, y \in X$ and $1 \leq i < j \leq n$. If there exists $0 < C < 1$ such that*

$$\varphi(2x_1, \cdots, 2x_n) \leq 2C\varphi(x_1, \cdots, x_n)$$

*for all $x_1, \cdots, x_n \in X$, then there exists a unique generalized additive mapping $L : X \to Y$ such that*

$$\begin{aligned}
\|f(x) - L(x)\|_Y \leq \frac{1}{4 - 4C} \Big\{ &\varphi_{ij}\Big(\frac{2x}{r_i}, \frac{2x}{r_j}\Big) + 2\varphi_{ij}\Big(\frac{x}{r_i}, -\frac{x}{r_j}\Big) \\
&+ \varphi_{ij}\Big(\frac{2x}{r_i}, 0\Big) + 2\varphi_{ij}\Big(\frac{x}{r_i}, 0\Big) \\
&+ \varphi_{ij}\Big(0, \frac{2x}{r_j}\Big) + 2\varphi_{ij}\Big(0, -\frac{x}{r_j}\Big) \Big\}
\end{aligned} \tag{3.74}$$

*for all $x \in X$. Moreover, $L(r_k x) = r_k L(x)$ for all $x \in X$ and $1 \leq k \leq n$.*

*Proof.* For each $1 \leq k \leq n$ with $k \neq i,j$, let $x_k = 0$ in (3.73). Then we get the following inequality:

$$\left\| f\left(\frac{-r_ix_i + r_jx_j}{2}\right) + f\left(\frac{r_ix_i - r_jx_j}{2}\right) - 2f\left(\frac{r_ix_i + r_jx_j}{2}\right) \right.$$
$$\left. + r_if(x_i) + r_jf(x_j) \right\|_Y$$
$$\leq \varphi\Big(0,\cdots,0,\underbrace{x_i}_{i\,th},0,\cdots,0,\underbrace{x_j}_{j\,th},0,\cdots,0\Big) \tag{3.75}$$

for all $x_i, x_j \in X$. Letting $x_i = 0$ in (3.75), we get

$$\left\| f\left(-\frac{r_jx_j}{2}\right) - f\left(\frac{r_jx_j}{2}\right) + r_jf(x_j) \right\|_Y \leq \varphi_{ij}(0, x_j) \tag{3.76}$$

for all $x_j \in X$. Similarly, letting $x_j = 0$ in (3.75), we get

$$\left\| f\left(-\frac{r_ix_i}{2}\right) - f\left(\frac{r_ix_i}{2}\right) + r_if(x_i) \right\|_Y \leq \varphi_{ij}(x_i, 0) \tag{3.77}$$

for all $x_i \in X$. It follows from (3.75), (3.76) and (3.77) that

$$\left\| f\left(\frac{-r_ix_i + r_jx_j}{2}\right) + f\left(\frac{r_ix_i - r_jx_j}{2}\right) - 2f\left(\frac{r_ix_i + r_jx_j}{2}\right) \right.$$
$$\left. + f\left(\frac{r_ix_i}{2}\right) + f\left(\frac{r_jx_j}{2}\right) - f\left(-\frac{r_ix_i}{2}\right) - f\left(-\frac{r_jx_j}{2}\right) \right\|_Y$$
$$\leq \varphi_{ij}(x_i, x_j) + \varphi_{ij}(x_i, 0) + \varphi_{ij}(0, x_j) \tag{3.78}$$

for all $x_i, x_j \in X$. Replacing $x_i$ and $x_j$ by $\frac{2x}{r_i}$ and $\frac{2y}{r_j}$ in (3.78), it follows that

$$\| f(-x + y) + f(x - y) - 2f(x + y)$$
$$+ f(x) + f(y) - f(-x) - f(-y) \|_Y$$
$$\leq \varphi_{ij}\left(\frac{2x}{r_i}, \frac{2y}{r_j}\right) + \varphi_{ij}\left(\frac{2x}{r_i}, 0\right) + \varphi_{ij}\left(0, \frac{2y}{r_j}\right) \tag{3.79}$$

for all $x, y \in X$. Putting $y = x$ in (3.79), we get

$$\| 2f(x) - 2f(-x) - 2f(2x) \|_Y$$
$$\leq \varphi_{ij}\left(\frac{2x}{r_i}, \frac{2x}{r_j}\right) + \varphi_{ij}\left(\frac{2x}{r_i}, 0\right) + \varphi_{ij}\left(0, \frac{2x}{r_j}\right) \tag{3.80}$$

for all $x \in X$. Replacing $x$ and $y$ by $\frac{x}{2}$ and $-\frac{x}{2}$ in (3.79), respectively, we get

$$\|f(x) + f(-x)\|_Y$$
$$\leq \varphi_{ij}\left(\frac{x}{r_i}, -\frac{x}{r_j}\right) + \varphi_{ij}\left(\frac{x}{r_i}, 0\right) + \varphi_{ij}\left(0, -\frac{x}{r_j}\right) \tag{3.81}$$

for all $x \in X$. It follows from (3.80) and (3.81) that

$$\left\|\frac{1}{2}f(2x) - f(x)\right\|_Y \leq \frac{1}{4}\psi(x) \tag{3.82}$$

for all $x \in X$, where

$$\psi(x) := \varphi_{ij}\left(\frac{2x}{r_i}, \frac{2x}{r_j}\right) + 2\varphi_{ij}\left(\frac{x}{r_i}, -\frac{x}{r_j}\right)$$
$$+ \varphi_{ij}\left(\frac{2x}{r_i}, 0\right) + 2\varphi_{ij}\left(\frac{x}{r_i}, 0\right) + \varphi_{ij}\left(0, \frac{2x}{r_j}\right) + 2\varphi_{ij}\left(0, -\frac{x}{r_j}\right).$$

Consider the set $\mathcal{W} := \{g : X \to Y\}$ and introduce the generalized metric on $\mathcal{W}$ defined by

$$d(g, h) = \inf\{C \in \mathbb{R}_+ : \|g(x) - h(x)\|_Y \leq C\psi(x), \ \forall x \in X\}.$$

It is easy to show that $(\mathcal{W}, d)$ is complete. Now, we consider the linear mapping $J : \mathcal{W} \to \mathcal{W}$ such that

$$Jg(x) := \frac{1}{2}g(2x) \tag{3.83}$$

for all $x \in X$. It follows that $d(Jg, Jh) \leq Cd(g, h)$ for all $g, h \in \mathcal{W}$ and so $d(f, Jf) \leq \frac{1}{4}$. By Theorem 1.3, there exists a mapping $L : X \to Y$ such that

(1) $L$ is a fixed point of $J$, i.e.,

$$L(2x) = 2L(x) \tag{3.84}$$

for all $x \in X$. The mapping $L$ is a unique fixed point of $J$ in the set

$$Z = \{g \in \mathcal{W} : d(f, g) < \infty\}.$$

This implies that $L$ is a unique mapping satisfying (3.84) such that there exists $C \in (0, \infty)$ satisfying

$$\|L(x) - f(x)\|_Y \leq C\psi(x)$$

for all $x \in X$;

(2) $d(J^n f, L) \to 0$ as $n \to \infty$. This implies the equality

$$\lim_{n \to \infty} \frac{f(2^n x)}{2^n} = L(x)$$

for all $x \in X$;

(3) $d(f, L) \leq \frac{1}{1-C} d(f, Jf)$, which implies the inequality $d(f, L) \leq \frac{1}{4-4C}$. This implies that the inequality (3.74) holds.

Since $\varphi(2x_1, \cdots, 2x_n) \leq 2C\varphi(x_1, \cdots, x_n)$, we have

$$
\begin{aligned}
\|D_{e,r_1,\cdots,r_n} L(x_1, \cdots, x_n)\|_Y &= \lim_{k \to \infty} \frac{1}{2^k} \|D_{e,r_1,\cdots,r_n} f(2^k x_1, \cdots, 2^k x_n)\|_Y \\
&\leq \lim_{k \to \infty} \frac{1}{2^k} \varphi(2^k x_1, \cdots, 2^k x_n) \\
&\leq \lim_{k \to \infty} C^k \varphi(x_1, \cdots, x_n) = 0
\end{aligned}
$$

for all $x_1, \cdots, x_n \in X$. Therefore, the mapping $L : X \to Y$ satisfies the Eq. (3.67) and $L(0) = 0$. Hence, by Lemma 3.49, $L$ is a generalized additive mapping and $L(r_k x) = r_k L(x)$ for all $x \in X$ and all $1 \leq k \leq n$. This completes the proof. $\square$

**Theorem 3.51.** *Let $f : X \to Y$ be a mapping satisfying $f(0) = 0$ for which there exists a function $\varphi : X^n \to [0, \infty)$ satisfying*

$$\|D_{u,r_1,\cdots,r_n} f(x_1, \cdots, x_n)\| \leq \varphi(x_1, \cdots, x_n) \tag{3.85}$$

*for all $x_1, \cdots, x_n \in X$ and $u \in U(A)$. If there exists $0 < C < 1$ such that*

$$\varphi(2x_1, \cdots, 2x_n) \leq 2C\varphi(x_1, \cdots, x_n)$$

*for all $x_1, \cdots, x_n \in X$, then there exists a unique A-linear generalized additive mapping $L : X \to Y$ satisfying (3.74) for all $x \in X$. Moreover, $L(r_k x) = r_k L(x)$ for all $x \in X$ and $1 \leq k \leq n$.*

*Proof.* By Theorem 3.50, there exists a unique generalized additive mapping $L : X \to Y$ satisfying (3.74) and, moreover, $L(r_k x) = r_k L(x)$ for all $x \in X$ and $1 \leq k \leq n$. By the assumption, for all $u \in U(A)$, we get

$$
\left\| D_{u,r_1,\cdots,r_n} L(0, \cdots, 0, \underbrace{x}_{i\,\text{th}}, 0 \cdots, 0) \right\|_Y
$$

$$
= \lim_{k \to \infty} \frac{1}{2^k} \left\| D_{u,r_1,\cdots,r_n} f(0, \cdots, 0, \underbrace{2^k x}_{i\,\text{th}}, 0 \cdots, 0) \right\|_Y
$$

$$\leq \lim_{k\to\infty} \frac{1}{2^k}\varphi\Big(0,\cdots,0,\underbrace{2^k x}_{i\,\text{th}},0\cdots,0\Big)$$

$$\leq \lim_{k\to\infty} C^k\varphi\Big(0,\cdots,0,\underbrace{x}_{i\,\text{th}},0\cdots,0\Big)$$

$$= 0$$

for all $x \in X$ and so

$$r_i u L(x) = L(r_i u x)$$

for all $u \in U(A)$ and $x \in X$. Since $L(r_i x) = r_i L(x)$ for all $x \in X$ and $r_i \neq 0$, we have

$$L(ux) = uL(x)$$

for all $u \in U(A)$ and $x \in X$. Now, we have

$$L(ax + by) = L(ax) + L(by) = aL(x) + bL(y)$$

for all $a, b \in A$ $(a, b \neq 0)$ and $x, y \in X$. Since $L(0x) = 0 = 0L(x)$ for all $x \in X$, the unique generalized additive mapping $L : X \to Y$ is an $A$-linear mapping. This completes the proof. $\qquad\square$

**Theorem 3.52.** *Let $f : X \to Y$ be a mapping satisfying $f(0) = 0$ for which there exists a function $\varphi : X^n \to [0, \infty)$ such that*

$$\|D_{e,r_1,\cdots,r_n} f(x_1, \cdots, x_n)\|_Y \leq \varphi(x_1, \cdots, x_n) \tag{3.86}$$

*for all $x_1, \cdots, x_n \in X$. If there exists $0 < C < 1$ such that*

$$\varphi(x_1, \cdots, 2_n) \leq \frac{C}{2}\varphi(2x_1, \cdots, 2x_n)$$

*for all $x_1, \cdots, x_n \in X$, then there exists a unique generalized additive mapping $L : X \to Y$ such that*

$$\|f(x) - L(x)\|_Y$$
$$\leq \frac{C}{4 - 4C}\Big\{\varphi_{ij}\Big(\frac{2x}{r_i}, \frac{2x}{r_j}\Big) + 2\varphi_{ij}\Big(\frac{x}{r_i}, -\frac{x}{r_j}\Big)$$
$$+ \varphi_{ij}\Big(\frac{2x}{r_i}, 0\Big) + 2\varphi_{ij}\Big(\frac{x}{r_i}, 0\Big) \quad + \varphi_{ij}\Big(0, \frac{2x}{r_j}\Big) + 2\varphi_{ij}\Big(0, -\frac{x}{r_j}\Big)\Big\} \tag{3.87}$$

*for all $x \in X$, where $\varphi_{ij}$ is defined in the statement of Theorem 3.50. Moreover, $L(r_k x) = r_k L(x)$ for all $x \in X$ and $1 \leq k \leq n$.*

*Proof.* It follows from (3.82) that

$$\left\| f(x) - f\left(\frac{x}{2}\right) \right\|_Y \leq \frac{1}{2}\psi\left(\frac{x}{2}\right) \leq \frac{C}{4}\psi(x)$$

for all $x \in X$, where $\psi$ is defined in the proof of Theorem 3.50. The rest of the proof is similar to the proof of Theorem 3.50. This completes the proof. $\qquad\square$

*Remark 3.53.* Let $f : X \to Y$ be a mapping with $f(0) = 0$ for which there exists a function $\varphi : X^n \to [0, \infty)$ satisfying

$$\|D_{u,r_1,\cdots,r_n}f(x_1,\cdots,x_n)\| \leq \varphi(x_1,\cdots,x_n) \tag{3.88}$$

for all $x_1,\cdots,x_n \in X$ and $u \in U(A)$. If there exists $0 < C < 1$ such that

$$\varphi(x_1,\cdots,2_n) \leq \frac{C}{2}\varphi(2x_1,\cdots,2x_n)$$

for all $x_1,\cdots,x_n \in X$, then there exists a unique $A$-linear generalized additive mapping $L : X \to Y$ satisfying (3.87) for all $x \in X$. Moreover, $L(r_k x) = r_k L(x)$ for all $x \in X$ and $1 \leq k \leq n$.

*Remark 3.54.* In Theorems 3.52 and 3.53, one can assume that $\sum_{k=1}^{n} r_k \neq 0$ instead of $f(0) = 0$.

### 3.4.2   Homomorphisms in Unital $C^*$-Algebras

Now, we investigate $C^*$-algebra homomorphisms in unital $C^*$-algebras.

We use the following lemma in the proof of the following theorem.

**Lemma 3.55 ([229]).** *Let $f : A \to B$ be an additive mapping such that $f(\mu x) = \mu f(x)$ for all $x \in A$ and $\mu \in \mathbb{S}^1 := \{\lambda \in \mathbb{C} : |\lambda| = 1\}$. Then the mapping $f : A \to B$ is $\mathbb{C}$-linear.*

**Theorem 3.56.** *Let $f : A \to B$ be a mapping with $f(0) = 0$ for which there exists a function $\varphi : A^n \to [0, \infty)$ satisfying*

$$\left\| D_{\mu,r_1,\cdots,r_n}f(x_1,\cdots,x_n) \right\|_B \leq \varphi(x_1,\cdots,x_n), \tag{3.89}$$

$$\left\| f(2^k u^*) - f(2^k u)^* \right\|_B \leq \varphi\big( \underbrace{2^k u,\cdots,2^k u}_{n \text{ times}} \big), \tag{3.90}$$

$$\left\| f(2^k ux) - f(2^k u)f(x) \right\|_B \leq \varphi\big( \underbrace{2^k ux,\cdots,2^k ux}_{n \text{ times}} \big) \tag{3.91}$$

for all $x, x_1, \cdots, x_n \in A$, $u \in U(A)$, $k \in \mathbb{N}$ and $\mu \in \mathbb{S}^1$. If there exists $0 < C < 1$ such that

$$\varphi(2x_1, \cdots, 2x_n) \leq 2C\varphi(x_1, \cdots, x_n)$$

for all $x_1, \cdots, x_n \in A$, then the mapping $f : A \to B$ is a $C^*$-algebra homomorphism.

*Proof.* Since $|J| \geq 3$, letting $\mu = 1$ and $x_k = 0$ for all $1 \leq k \leq n$ with $k \neq i, j$ in (3.89), we get

$$f\left(\frac{-r_i x_i + r_j x_j}{2}\right) + f\left(\frac{r_i x_i - r_j x_j}{2}\right) + r_i f(x_i) + r_j f(x_j)$$
$$= 2f\left(\frac{r_i x_i + r_j x_j}{2}\right)$$

for all $x_i, x_j \in A$. By Lemma 3.48, the mapping $f$ is additive and $f(r_k x) = r_k f(x)$ for all $x \in A$ and $k = i, j$. So, by letting $x_i = x$ and $x_k = 0$ for all $1 \leq k \leq n$ with $k \neq i$ in (3.89), it follows that $f(\mu x) = \mu f(x)$ for all $x \in A$ and $\mu \in \mathbb{S}^1$. Therefore, by Lemma 3.55, the mapping $f$ is $\mathbb{C}$-linear. Hence it follows from (3.90) and (3.91) that

$$\|f(u^*) - f(u)^*\|_B = \lim_{k \to \infty} \frac{1}{2^k} \|f(2^k u^*) - f(2^k u)^*\|_B$$
$$\leq \lim_{k \to \infty} \frac{1}{2^k} \varphi\Big(\underbrace{2^k u, \cdots, 2^k u}_{n \text{ times}}\Big)$$
$$\leq \lim_{k \to \infty} C^k \varphi\Big(\underbrace{u, \cdots, u}_{n \text{ times}}\Big)$$
$$= 0$$

and

$$\|f(ux) - f(u)f(x)\|_B = \lim_{k \to \infty} \frac{1}{2^k} \|f(2^k ux) - f(2^k u)f(x)\|_B$$
$$\leq \lim_{k \to \infty} \frac{1}{2^k} \varphi\Big(\underbrace{2^k ux, \cdots, 2^k ux}_{n \text{ times}}\Big)$$
$$\leq \lim_{k \to \infty} C^k \varphi\Big(\underbrace{ux, \cdots, ux}_{n \text{ times}}\Big)$$
$$= 0$$

for all $x \in A$ and $u \in U(A)$ and so

$$f(u^*) = f(u)^*, \quad f(ux) = f(u)f(x)$$

for all $x \in A$ and $u \in U(A)$. Since $f$ is $\mathbb{C}$-linear and each $x \in A$ is a finite linear combination of unitary elements (see [168]), i.e., $x = \sum_{k=1}^{m} \lambda_k u_k$, where $\lambda_k \in \mathbb{C}$ and $u_k \in U(A)$ for all $1 \leq k \leq n$, we have

$$f(x^*) = f\left(\sum_{k=1}^{m} \overline{\lambda_k} u_k^*\right) = \sum_{k=1}^{m} \overline{\lambda_k} f\left(u_k^*\right) = \sum_{k=1}^{m} \overline{\lambda_k} f(u_k)^*$$

$$= \left(\sum_{k=1}^{m} \lambda_k f(u_k)\right)^* = f\left(\sum_{k=1}^{m} \lambda_k u_k\right)^* = f(x)^*$$

and

$$f(xy) = f\left(\sum_{k=1}^{m} \lambda_k u_k y\right) = \sum_{k=1}^{m} \lambda_k f(u_k y)$$

$$= \sum_{k=1}^{m} \lambda_k f(u_k) f(y) = f\left(\sum_{k=1}^{m} \lambda_k u_k\right) f(y) = f(x)f(y)$$

for all $x, y \in A$. Therefore, the mapping $f : A \to B$ is a $C^*$-algebra homomorphism. This completes the proof.  □

*Remark 3.57.* Let $f : A \to B$ be a mapping with $f(0) = 0$ for which there exists a function $\varphi : A^n \to [0, \infty)$ satisfying

$$\|D_{\mu, r_1, \cdots, r_n} f(x_1, \cdots, x_n)\|_B \leq \varphi\left(x_1, \cdots, x_n\right),$$

$$\left\|f\left(\frac{u^*}{2^k}\right) - f\left(\frac{u}{2^k}\right)^*\right\|_B \leq \phi\left(\underbrace{\frac{u}{2^k}, \cdots, \frac{u}{2^k}}_{n \text{ times}}\right), \tag{3.92}$$

$$\left\|f\left(\frac{ux}{2^k}\right) - f\left(\frac{u}{2^k}\right)f(x)\right\|_B \leq \phi\left(\underbrace{\frac{ux}{2^k}, \cdots, \frac{ux}{2^k}}_{n \text{ times}}\right) \tag{3.93}$$

for all $x, x_1, \cdots, x_n \in A$, $u \in U(A)$, $k \in \mathbb{N}$ and $\mu \in \mathbb{S}^1$. If there exists $0 < C < 1$ such that

$$\varphi(x_1, \cdots, 2_n) \leq \frac{C}{2} \varphi(2x_1, \cdots, 2x_n)$$

for all $x_1, \cdots, x_n \in A$, then the mapping $f : A \to B$ is a $C^*$-algebra homomorphism.

*Remark 3.58.* In Theorem 3.56 and last remark, one can assume that $\sum_{k=1}^{n} r_k \neq 0$ instead of $f(0) = 0$.

**Theorem 3.59.** *Let* $f : A \to B$ *be a mapping with* $f(0) = 0$ *for which there exists a function* $\varphi : A^n \to [0, \infty)$ *satisfying* (3.90), (3.91) *and*

$$\|D_{\mu, r_1, \cdots, r_n} f(x_1, \cdots, x_n)\|_B \leq \varphi(x_1, \cdots, x_n), \tag{3.94}$$

*for all* $x_1, \cdots, x_n \in A$ *and* $\mu \in \mathbb{S}^1$. *Assume that* $\lim_{k \to \infty} \frac{1}{2^k} f(2^k e)$ *is invertible. If there exists* $0 < C < 1$ *such that*

$$\varphi(2x_1, \cdots, 2x_n) \leq 2C\varphi(x_1, \cdots, x_n)$$

*for all* $x_1, \cdots, x_n \in A$, *then the mapping* $f : A \to B$ *is a* $C^*$-*algebra homomorphism.*

*Proof.* Consider the $C^*$-algebras $A$ and $B$ as left Banach modules over the unital $C^*$-algebra $\mathbb{C}$. By Theorem 3.51, there exists a unique $\mathbb{C}$-linear generalized additive mapping $H : A \to B$ defined by

$$H(x) = \lim_{k \to \infty} \frac{1}{2^k} f(2^k x)$$

for all $x \in A$. By (3.90) and (3.91), we get

$$
\begin{aligned}
\left\| H\left(u^*\right) - H(u)^* \right\|_B &= \lim_{k \to \infty} \frac{1}{2^k} \left\| f\left(2^k u^*\right) - f\left(2^k u\right)^* \right\|_B \\
&\leq \lim_{k \to \infty} \frac{1}{2^k} \varphi\left( \underbrace{2^k u, \cdots, 2^k u}_{n \text{ times}} \right) \\
&= 0
\end{aligned}
$$

and

$$
\begin{aligned}
\|H(ux) - H(u)f(x)\|_B &= \lim_{k \to \infty} \frac{1}{2^k} \left\| f\left(2^k ux\right) - f(2^k u)f(x) \right\|_B \\
&\leq \lim_{k \to \infty} \frac{1}{2^k} \varphi\left( \underbrace{2^k ux, \cdots, 2^k ux}_{n \text{ times}} \right) \\
&= 0
\end{aligned}
$$

for all $u \in U(A)$ and $x \in A$ and so

$$H\left(u^*\right) = H(u)^*, \quad H(ux) = H(u)f(x)$$

for all $u \in U(A)$ and $x \in A$. Therefore, by the additivity of $H$, we have

$$H(ux) = \lim_{k\to\infty} \frac{1}{2^k} H\left(2^k ux\right) = H(u) \lim_{k\to\infty} \frac{1}{2^k} f\left(2^k x\right) = H(u)H(x) \qquad (3.95)$$

for all $u \in U(A)$ and $x \in A$. Since $H$ is $\mathbb{C}$-linear and each $x \in A$ is a finite linear combination of unitary elements, i.e., $x = \sum_{k=1}^{m} \lambda_k u_k$, where $\lambda_k \in \mathbb{C}$ and $u_k \in U(A)$ for all $1 \leq k \leq n$, it follows from (3.95) that

$$H(xy) = H\left(\sum_{k=1}^{m} \lambda_k u_k y\right) = \sum_{k=1}^{m} \lambda_k H(u_k y)$$

$$= \sum_{k=1}^{m} \lambda_k H(u_k) H(y) = H\left(\sum_{k=1}^{m} \lambda_k u_k\right) H(y)$$

$$= H(x)H(y)$$

and

$$H\left(x^*\right) = H\left(\sum_{k=1}^{m} \overline{\lambda_k} u_k^*\right) = \sum_{k=1}^{m} \overline{\lambda_k} H(u_k^*)$$

$$= \sum_{k=1}^{m} \overline{\lambda_k} H(u_k)^* = \left(\sum_{k=1}^{m} \lambda_k H(u_k)\right)^*$$

$$= H\left(\sum_{k=1}^{m} \lambda_k u_k\right)^* = H(x)^*$$

for all $x, y \in A$. Since $H(e) = \lim_{k\to\infty} \frac{1}{2^k} f(2^k e)$ is invertible and

$$H(e)H(y) = H(ey) = H(e)f(y), \quad H(y) = f(y)$$

for all $y \in A$. Therefore, the mapping $f : A \to B$ is a $C^*$-algebra homomorphism. This completes the proof.                                                                                            $\square$

*Remark 3.60.* Let $f : A \to B$ be a mapping with $f(0) = 0$ for which there exists a function $\varphi : A^n \to [0, \infty)$ satisfying (3.92), (3.93) and

$$\|D_{\mu, r_1, \cdots, r_n} f(x_1, \cdots, x_n)\|_B \leq \varphi(x_1, \cdots, x_n),$$

for all $x_1, \cdots, x_n \in A$ and $\mu \in \mathbb{S}^1$. Assume that $\lim_{k\to\infty} 2^k f(\frac{e}{2^k})$ is invertible. If there exists $0 < C < 1$ such that

$$\varphi(x_1, \cdots, 2_n) \leq \frac{C}{2} \varphi(2x_1, \cdots, 2x_n)$$

for all $x_1, \cdots, x_n \in A$, then the mapping $f : A \to B$ is a $C^*$-algebra homomorphism.

In the last Remark, one can assume that $\sum_{k=1}^{n} r_k \neq 0$ instead of $f(0) = 0$.

**Theorem 3.61.** *Let $f : A \to B$ be a mapping with $f(0) = 0$ for which there exists a function $\varphi : A^n \to [0, \infty)$ satisfying (3.90), (3.91) and*

$$\|D_{\mu,r_1,\cdots,r_n}f(x_1,\cdots,x_n)\|_B \leq \varphi(x_1,\cdots,x_n) \tag{3.96}$$

*for $\mu = i, 1$ and $x_1, \cdots, x_n \in A$. Assume that $\lim_{k\to\infty} \frac{1}{2^k}f(2^k e)$ is invertible and, for each fixed $x \in A$, the mapping $t \mapsto f(tx)$ is continuous in $t \in \mathbb{R}$. If there exists $0 < C < 1$ such that*

$$\varphi(2x_1, \cdots, 2x_n) \leq 2C\varphi(x_1, \cdots, x_n)$$

*for all $x_1, \cdots, x_n \in A$, then the mapping $f : A \to B$ is a $C^*$-algebra homomorphism.*

*Proof.* Put $\mu = 1$ in (3.96). By the same reasoning as in the proof of Theorem 3.50, there exists a unique generalized additive mapping $H : A \to B$ defined by

$$H(x) = \lim_{k\to\infty} \frac{f(2^k x)}{2^k}$$

for all $x \in A$. It is straight forward to show that, the generalized additive mapping $H : A \to B$ is $\mathbb{R}$-linear. By the same method as in the proof of Theorem 3.51, we have

$$\left\| D_{\mu,r_1,\cdots,r_n}H(0,\cdots,0,\underbrace{x}_{j\text{ th}},0\cdots,0) \right\|_Y$$

$$= \lim_{k\to\infty} \frac{1}{2^k}\left\| D_{\mu,r_1,\cdots,r_n}f(0,\cdots,0,\underbrace{2^k x}_{j\text{ th}},0\cdots,0) \right\|_Y$$

$$\leq \lim_{k\to\infty} \frac{1}{2^k}\varphi(0,\cdots,0,\underbrace{2^k x}_{j\text{ th}},0\cdots,0)$$

$$= 0$$

for all $x \in A$ and so

$$r_j\mu H(x) = H(r_j\mu x)$$

for all $x \in A$. Since $H(r_j x) = r_j H(x)$ for all $x \in X$ and $r_j \neq 0$, we have

$$H(\mu x) = \mu H(x)$$

for all $x \in A$ and $\mu = i, 1$. For each element $\lambda \in \mathbb{C}$, we have $\lambda = s + it$, where $s, t \in \mathbb{R}$. Thus it follows that

$$H(\lambda x) = H(sx + itx) = sH(x) + tH(ix) = sH(x) + itH(x)$$
$$= (s + it)H(x) = \lambda H(x)$$

for all $\lambda \in \mathbb{C}$ and $x \in A$. So, we have

$$H(\zeta x + \eta y) = H(\zeta x) + H(\eta y) = \zeta H(x) + \eta H(y)$$

for all $\zeta, \eta \in \mathbb{C}$ and $x, y \in A$. Hence the generalized additive mapping $H : A \rightarrow B$ is $\mathbb{C}$-linear.

The rest of the proof is the same as in the proof of Theorem 3.59. This completes the proof.                                                                            $\square$

The following is an alternative result of Theorem 3.61.

*Remark 3.62.* Let $f : A \rightarrow B$ be a mapping with $f(0) = 0$ for which there exists a function $\varphi : A^n \rightarrow [0, \infty)$ satisfying (3.92), (3.93) and

$$\|D_{\mu, r_1, \cdots, r_n} f(x_1, \cdots, x_n)\|_B \leq \varphi(x_1, \cdots, x_n)$$

for all $x, x_1, \cdots, x_n \in A$ and $\mu = i, 1$. Assume that $\lim_{k \rightarrow \infty} 2^k f(\frac{e}{2^k})$ is invertible and, for each fixed $x \in A$, the mapping $t \mapsto f(tx)$ is continuous in $t \in \mathbb{R}$. If there exists $0 < C < 1$ such that

$$\varphi(x_1, \cdots, 2_n) \leq \frac{C}{2} \varphi(2x_1, \cdots, 2x_n)$$

for all $x_1, \cdots, x_n \in A$, then the mapping $f : A \rightarrow B$ is a $C^*$-algebra homomorphism.

*Remark 3.63.* Also, one can assume that $\sum_{k=1}^{n} r_k \neq 0$ instead of $f(0) = 0$.

## 3.5   Generalized Additive Mappings in Banach Modules

Let $X, Y$ be vector spaces. It is shown that, if an odd mapping $f : X \rightarrow Y$ satisfies the functional equation

$$rf\left(\frac{\sum_{j=1}^{d} x_j}{r}\right) + \sum_{\substack{\iota(j) \ = 0, 1 \\ \sum_{j=1}^{d} \iota(j) \ = l}} rf\left(\frac{\sum_{j=1}^{d} (-1)^{\iota(j)} x_j}{r}\right)$$

$$= (_{d-1}C_l - _{d-1}C_{l-1} + 1) \sum_{j=1}^{d} f(x_j), \tag{3.97}$$

then the odd mapping $f : X \to Y$ is additive. Also, we consider the Hyers-Ulam stability of the functional equation (3.97) in Banach modules over a unital $C^*$-algebra. As an application, we show that every almost linear bijection $h : A \to B$ of a unital $C^*$-algebra $A$ onto a unital $C^*$-algebra $B$ is a $C^*$-algebra isomorphism when $h(\frac{2^n}{r^n}uy) = h(\frac{2^n}{r^n}u)h(y)$ for all unitaries $u \in A$, $y \in A$ and $n \geq 0$ [20].

Throughout this section, assume that $r$ is a positive rational number and $d, l$ are integers with $1 < l < \frac{d}{2}$.

### 3.5.1 Odd Functional Equations in $d$–Variables

Now, assume that $X$ and $Y$ are real linear spaces.

**Lemma 3.64.** *An odd mapping $f : X \to Y$ satisfies (3.97) for all $x_1, x_2, \cdots, x_d \in X$ if and only if $f$ is Cauchy additive.*

*Proof.* Assume that $f : X \to Y$ satisfies (3.97) for all $x_1, x_2, \cdots, x_d \in X$. Note that $f(0) = 0$ and $f(-x) = -f(x)$ for all $x \in X$ since $f$ is an odd mapping. Putting $x_1 = x, x_2 = y$ and $x_3 = \cdots = x_d = 0$ in (3.97), we get

$$(_{d-2}C_l - _{d-2}C_{l-2} + 1) r f\left(\frac{x+y}{r}\right)$$
$$= (_{d-1}C_l - _{d-1}C_{l-1} + 1)(f(x) + f(y)) \tag{3.98}$$

for all $x, y \in X$. Since $_{d-2}C_l - _{d-2}C_{l-2} + 1 = _{d-1}C_l - _{d-1}C_{l-1} + 1$, we have

$$rf\left(\frac{x+y}{r}\right) = f(x) + f(y)$$

for all $x, y \in X$. Letting $y = 0$ in (3.98), we get $rf(\frac{x}{r}) = f(x)$ for all $x \in X$. Hence we have

$$f(x+y) = rf\left(\frac{x+y}{r}\right) = f(x) + f(y)$$

for all $x, y \in X$. Thus $f$ is Cauchy additive.

The converse is obviously true. This completes the proof. $\qquad\square$

In the proof of Lemma 3.64, we prove the following.

**Corollary 3.65.** *An odd mapping $f : X \to Y$ satisfies*

$$rf\left(\frac{x+y}{r}\right) = f(x) + f(y)$$

*for all $x, y \in X$ if and only if $f$ is Cauchy additive.*

### 3.5.2  Stability of Odd Functional Equations in Banach Modules over a $C^*$-Algebra

Assume that $A$ is a unital $C^*$-algebra with the norm $|\cdot|$ and a unitary group $U(A)$ and $X$, $Y$ are left Banach modules over a unital $C^*$-algebra $A$ with norms $\|\cdot\|$ and $\|\cdot\|$, respectively.

For any mapping $f : X \to Y$, we set

$$D_u f(x_1, \cdots, x_d)$$

$$:= rf\left(\frac{\sum_{j=1}^d ux_j}{r}\right) + \sum_{\substack{\iota(j) \ = 0,1 \\ \sum_{j=1}^d \iota(j) \ = l}} rf\left(\frac{\sum_{j=1}^d (-1)^{\iota(j)} ux_j}{r}\right)$$

$$- (_{d-1}C_l - _{d-1}C_{l-1} + 1) \sum_{j=1}^d uf(x_j)$$

for all $u \in U(A)$ and $x_1, \cdots, x_d \in X$.

**Theorem 3.66.** *Let $r \neq 2$. Let $f : X \to Y$ be an odd mapping for which there exists a function $\varphi : X^d \to [0, \infty)$ such that*

$$\tilde{\varphi}(x_1, \cdots, x_d) := \sum_{j=0}^\infty \frac{r^j}{2^j} \varphi\left(\frac{2^j}{r^j} x_1, \cdots, \frac{2^j}{r^j} x_d\right) < \infty \qquad (3.99)$$

*and*

$$\|D_u f(x_1, \cdots, x_d)\| \leq \varphi(x_1, \cdots, x_d) \qquad (3.100)$$

*for all $u \in U(A)$ and $x_1, \cdots, x_d \in X$. Then there exists a unique $A$-linear generalized additive mapping $L : X \to Y$ such that*

$$\|f(x) - L(x)\| \leq \frac{1}{2(_{d-2}C_l - _{d-2}C_{l-2} + 1)} \tilde{\varphi}\big(x, x, \underbrace{0, \cdots, 0}_{d-2\ times}\big) \qquad (3.101)$$

*for all $x \in X$.*

*Proof.* Note that $f(0) = 0$ and $f(-x) = -f(x)$ for all $x \in X$ since $f$ is an odd mapping. Let $u = 1 \in U(A)$. Putting $x_1 = x_2 = x$ and $x_3 = \cdots = x_d = 0$ in (3.100), we have

$$\left\| rf\left(\frac{2}{r}x\right) - 2f(x) \right\| \leq \frac{1}{_{d-2}C_l - _{d-2}C_{l-2} + 1} \varphi\big(x, x, \underbrace{0, \cdots, 0}_{d-2\ times}\big)$$

for all $x \in X$. Letting $t :=_{d-2} C_l -_{d-2} C_{l-2} + 1$, we get

$$\left\| f(x) - \frac{r}{2}f\left(\frac{2}{r}x\right) \right\| \le \frac{1}{2t}\varphi\big(x, x, \underbrace{0, \cdots, 0}_{d-2 \text{ times}}\big)$$

for all $x \in X$. Hence we have

$$\left\| \frac{r^n}{2^n}f\left(\frac{2^n}{r^n}x\right) - \frac{r^{n+1}}{2^{n+1}}f\left(\frac{2^{n+1}}{r^{n+1}}x\right) \right\|$$

$$= \frac{r^n}{2^n}\left\| f\left(\frac{2^n}{r^n}x\right) - \frac{r}{2}f\left(\frac{2}{r} \cdot \frac{2^n}{r^n}x\right) \right\|$$

$$\le \frac{r^n}{2^{n+1}t}\varphi\big(\frac{2^n}{r^n}x, \frac{2^n}{r^n}x, \underbrace{0, \cdots, 0}_{d-2 \text{ times}}\big) \tag{3.102}$$

for all $x \in X$ and $n \ge 1$. By (3.102), we have

$$\left\| \frac{r^m}{2^m}f\left(\frac{2^m}{r^m}x\right) - \frac{r^n}{2^n}f\left(\frac{2^n}{r^n}x\right) \right\|$$

$$\le \sum_{k=m}^{n-1} \frac{r^k}{2^{k+1} t}\varphi\big(\frac{2^k}{r^k}x, \frac{2^k}{r^k}x, \underbrace{0, \cdots, 0}_{d-2 \text{ times}}\big) \tag{3.103}$$

for all $x \in X$ and $m, n \ge 1$ with $m < n$. This shows that the sequence $\{\frac{r^n}{2^n}f(\frac{2^n}{r^n}x)\}$ is a Cauchy sequence for all $x \in X$. Since $Y$ is complete, the sequence $\{\frac{r^n}{2^n}f(\frac{2^n}{r^n}x)\}$ converges for all $x \in X$. So we can define a mapping $L : X \to Y$ by

$$L(x) := \lim_{n \to \infty} \frac{r^n}{2^n}f\left(\frac{2^n}{r^n}x\right)$$

for all $x \in X$. Since $f(-x) = -f(x)$ for all $x \in X$, we have $L(-x) = -L(x)$ for all $x \in X$. Also, we get

$$\|D_1 L(x_1, \cdots, x_d)\| = \lim_{n \to \infty} \frac{r^n}{2^n}\left\| D_1 f\left(\frac{2^n}{r^n}x_1, \cdots, \frac{2^n}{r^n}x_d\right) \right\|$$

$$\le \lim_{n \to \infty} \frac{r^n}{2^n}\varphi\left(\frac{2^n}{r^n}x_1, \cdots, \frac{2^n}{r^n}x_d\right)$$

$$= 0$$

for all $x_1, \cdots, x_d \in X$. So $L$ is a generalized additive mapping. Putting $m = 0$ and letting $n \to \infty$ in (3.103), we get (3.101).

Now, let $L' : X \to Y$ be another additive mapping satisfying (3.101). By Lemma 3.64, $L$ and $L'$ are additive. So we have

$$\|L(x) - L'(x)\| = \frac{r^n}{2^n}\left\|L\left(\frac{2^n}{r^n}x\right) - L'\left(\frac{2^n}{r^n}x\right)\right\|$$

$$\leq \frac{r^n}{2^n}\left(\left\|L\left(\frac{2^n}{r^n}x\right) - f\left(\frac{2^n}{r^n}x\right)\right\| + \left\|L'\left(\frac{2^n}{r^n}x\right) - f\left(\frac{2^n}{r^n}x\right)\right\|\right)$$

$$\leq \frac{2r^n}{2^{n+1}t}\tilde{\varphi}\Big(\frac{2^n}{r^n}x, \frac{2^n}{r^n}x, \underbrace{0, \cdots, 0}_{d-2 \text{ times}}\Big),$$

which tends to zero as $n \to \infty$ for all $x \in X$. So we can conclude that $L(x) = L'(x)$ for all $x \in X$. This proves the uniqueness of $L$.

By the assumption, for each $u \in U(A)$, we get

$$\|D_u L(x, \underbrace{0, \cdots, 0}_{d-1 \text{ times}})\| = \lim_{n\to\infty}\frac{r^n}{2^n}\Big\|D_u f\Big(\frac{2^n}{r^n}x, \underbrace{0, \cdots, 0}_{d-1 \text{ times}}\Big)\Big\|$$

$$\leq \lim_{n\to\infty}\frac{r^n}{2^n}\varphi\Big(\frac{2^n}{r^n}x, \underbrace{0, \cdots, 0}_{d-1 \text{ times}}\Big)$$

$$= 0$$

for all $x \in X$ and so

$$(_{d-1}C_l - _{d-1}C_{l-1} + 1)rL\Big(\frac{ux}{r}\Big) = (_{d-1}C_l - _{d-1}C_{l-1} + 1)uL(x)$$

for all $u \in U(A)$ and $x \in X$. Since $L$ is additive,

$$L(ux) = rL\Big(\frac{ux}{r}\Big) = uL(x) \tag{3.104}$$

for all $u \in U(A)$ and $x \in X$.

Now, let $a \in A$ $(a \neq 0)$ and $M$ be an integer greater than $4|a|$. Then we have

$$\Big|\frac{a}{M}\Big| < \frac{1}{4} < 1 - \frac{2}{3} = \frac{1}{3}.$$

By Kadison and Pederson [167], there exist three elements $u_1, u_2, u_3 \in U(A)$ such that $3\frac{a}{M} = u_1 + u_2 + u_3$. So, by (3.104), we have

$$L(ax) = L\Big(\frac{M}{3}\cdot 3\frac{a}{M}x\Big) = M\cdot L\Big(\frac{1}{3}\cdot 3\frac{a}{M}x\Big) = \frac{M}{3}L\Big(3\frac{a}{M}x\Big)$$

$$= \frac{M}{3}L(u_1 x + u_2 x + u_3 x) = \frac{M}{3}(L(u_1 x) + L(u_2 x) + L(u_3 x))$$

$$= \frac{M}{3}(u_1 + u_2 + u_3)L(x) = \frac{M}{3}\cdot 3\frac{a}{M}L(x)$$

$$= aL(x)$$

for all $a \in A$ and $x \in X$. Hence we have

$$L(ax + by) = L(ax) + L(by) = aL(x) + bL(y)$$

for all $a, b \in A$ with $a, b \neq 0$ and $x, y \in X$ and $L(0x) = 0 = 0L(x)$ for all $x \in X$. So the unique generalized additive mapping $L : X \to Y$ is an $A$-linear mapping. This completes the proof. □

**Corollary 3.67.** *Let $r > 2$ and $\theta, p > 1$ be positive real numbers or let $r < 2$ and $\theta, p < 1$ be positive real numbers. Let $f : X \to Y$ be an odd mapping such that*

$$\|D_u f(x_1, \cdots, x_d)\| \leq \theta \sum_{j=1}^{d} ||x_j||^p$$

*for all $u \in U(A)$ and $x_1, \cdots, x_d \in X$. Then there exists a unique $A$-linear generalized additive mapping $L : X \to Y$ such that*

$$\|f(x) - L(x)\| \leq \frac{r^{p-1}\theta}{(r^{p-1} - 2^{p-1})(_{d-2}C_l -_{d-2} C_{l-2} + 1)} \|x\|^p$$

*for all $x \in X$.*

*Proof.* Defining $\varphi(x_1, \cdots, x_d) = \theta \sum_{j=1}^{d} ||x_j||^p$ and applying Theorem 3.66, we get the desired result. □

**Theorem 3.68.** *Let $r \neq 2$. Let $f : X \to Y$ be an odd mapping for which there exists a function $\varphi : X^d \to [0, \infty)$ satisfying (3.100) such that*

$$\tilde{\varphi}(x_1, \cdots, x_d) := \sum_{j=1}^{\infty} \frac{2^j}{r^j} \varphi\left(\frac{r^j}{2^j}x_1, \cdots, \frac{r^j}{2^j}x_d\right) < \infty$$

*for all $u \in U(A)$ and $x_1, \cdots, x_d \in X$. Then there exists a unique $A$-linear generalized additive mapping $L : X \to Y$ such that*

$$\|f(x) - L(x)\| \leq \frac{1}{2(_{d-2}C_l -_{d-2} C_{l-2} + 1)} \tilde{\varphi}\left(x, x, \underbrace{0, \cdots, 0}_{d-2 \ times}\right)$$

*for all $x \in X$.*

*Proof.* Note that $f(0) = 0$ and $f(-x) = -f(x)$ for all $x \in X$ since $f$ is an odd mapping. Let $u = 1 \in U(A)$. Putting $x_1 = x_2 = x$ and $x_3 = \cdots = x_d = 0$ in (3.100), we have

$$\left\|rf\left(\frac{2}{r}x\right) - 2f(x)\right\| \leq \frac{1}{_{d-2}C_l -_{d-2} C_{l-2} + 1} \varphi\left(x, x, \underbrace{0, \cdots, 0}_{d-2 \ times}\right)$$

for all $x \in X$. Letting $t :=_{d-2} C_l -_{d-2} C_{l-2} + 1$, we get

$$\left\| f(x) - \frac{2}{r} f\left(\frac{r}{2}x\right) \right\| \le \frac{1}{rt} \varphi\left(\frac{r}{2}x, \frac{r}{2}x, \underbrace{0, \cdots, 0}_{d-2 \text{ times}}\right)$$

for all $x \in X$.

The rest of the proof is similar to the proof of Theorem 3.66. This completes the proof.                                                                    □

**Corollary 3.69.** *Let $r < 2$ and $\theta$, $p > 1$ be positive real numbers or let $r > 2$ and $\theta$, $p < 1$ be positive real numbers. Let $f : X \to Y$ be an odd mapping such that*

$$\|D_u f(x_1, \cdots, x_d)\| \le \theta \sum_{j=1}^{d} \|x_j\|^p$$

*for all $u \in U(A)$ and $x_1, \cdots, x_d \in X$. Then there exists a unique $A$-linear generalized additive mapping $L : X \to Y$ such that*

$$\|f(x) - L(x)\| \le \frac{r^{p-1}\theta}{(2^{p-1} - r^{p-1})(_{d-2}C_l -_{d-2} C_{l-2} + 1)} \|x\|^p$$

*for all $x \in X$.*

*Proof.* Defining

$$\varphi(x_1, \cdots, x_d) = \theta \sum_{j=1}^{d} ||x_j||^p$$

and applying Theorem 3.68, we get the desired result.                           □

Now, we investigate the Hyers–Ulam stability of linear mappings for the case $d = 2$.

**Theorem 3.70.** *Let $r \ne 2$. Let $f : X \to Y$ be an odd mapping for which there exists a function $\varphi : X^2 \to [0, \infty)$ such that*

$$\tilde{\varphi}(x, y) := \sum_{j=0}^{\infty} \frac{r^j}{2^j} \varphi\left(\frac{2^j}{r^j}x, \frac{2^j}{r^j}y\right) < \infty$$

*and*

$$\left\| rf\left(\frac{ux + uy}{r}\right) - uf(x) - uf(y) \right\| \le \varphi(x, y) \tag{3.105}$$

*for all $u \in U(A)$ and $x, y \in X$. Then there exists a unique A-linear mapping*
$L : X \to Y$ *such that*

$$\|f(x) - L(x)\| \leq \frac{1}{2}\tilde{\varphi}(x, x)$$

*for all $x \in X$.*

*Proof.* Let $u = 1 \in U(A)$. Putting $x = y$ in (3.105), we have

$$\left\| rf\left(\frac{2}{r}x\right) - 2f(x) \right\| \leq \varphi(x, x)$$

for all $x \in X$ and so

$$\left\| f(x) - \frac{r}{2}f\left(\frac{2}{r}x\right) \right\| \leq \frac{1}{2}\varphi(x, x)$$

for all $x \in X$.

The rest of the proof is the same as in the proof of Theorem 3.66. This completes
the proof. □

**Corollary 3.71.** *Let $r > 2$ and $\theta$, $p > 1$ be positive real numbers or let $r < 2$ and*
$\theta$, $p < 1$ *be positive real numbers. Let $f : X \to Y$ be an odd mapping such that*

$$\left\| rf\left(\frac{ux + uy}{r}\right) - uf(x) - uf(y) \right\| \leq \theta(\|x\|^p + \|y\|^p)$$

*for all $u \in U(A)$ and $x, y \in X$. Then there exists a unique A-linear mapping*
$L : X \to Y$ *such that*

$$\|f(x) - L(x)\| \leq \frac{r^{p-1}\theta}{r^{p-1} - 2^{p-1}}\|x\|^p$$

*for all $x \in X$.*

*Proof.* Defining $\varphi(x, y) = \theta(\|x\|^p + \|y\|^p)$ and applying Theorem 3.70, we get the
desired results. □

**Theorem 3.72.** *Let $r \neq 2$. Let $f : X \to Y$ be an odd mapping for which there exists*
*a function $\varphi : X^2 \to [0, \infty)$ satisfying (3.105) such that*

$$\tilde{\varphi}(x, y) := \sum_{j=1}^{\infty} \frac{2^j}{r^j}\varphi\left(\frac{r^j}{2^j}x, \frac{r^j}{2^j}y\right) < \infty$$

*for all $u \in U(A)$ and all $x, y \in X$. Then there exists a unique A-linear mapping*
$L : X \to Y$ *such that*

$$\|f(x) - L(x)\| \leq \frac{1}{2}\tilde{\varphi}(x,x)$$

*for all $x \in X$.*

*Proof.* Let $u = 1 \in U(A)$. Putting $x = y$ in (3.105), we have

$$\left\| rf\left(\frac{2}{r}x\right) - 2f(x) \right\| \leq \varphi(x,x)$$

for all $x \in X$ and so

$$\left\| f(x) - \frac{2}{r}f\left(\frac{r}{2}x\right) \right\| \leq \frac{1}{r}\varphi\left(\frac{r}{2}x, \frac{r}{2}x\right)$$

for all $x \in X$.

The rest of the proof is similar to the proof of Theorem 3.66. This completes the proof.                                                                                            □

**Corollary 3.73.** *Let $r < 2$ and $\theta$, $p > 1$ be positive real numbers or let $r > 2$ and $\theta$, $p < 1$ be positive real numbers. Let $f : X \to Y$ be an odd mapping such that*

$$\left\| rf\left(\frac{ux + uy}{r}\right) - uf(x) - uf(y) \right\| \leq \theta(\|x\|^p + \|y\|^p)$$

*for all $u \in U(A)$ and $x, y \in X$. Then there exists a unique A-linear mapping $L : X \to Y$ such that*

$$\|f(x) - L(x)\| \leq \frac{r^{p-1}\theta}{2^{p-1} - r^{p-1}}\|x\|^p$$

*for all $x \in X$.*

*Proof.* Defining $\varphi(x,y) = \theta(\|x\|^p + \|y\|^p)$ and applying Theorem 3.72, we get the desired results.                                                                         □

### 3.5.3  Isomorphisms in Unital $C^*$-Algebras

Assume that $A$ is a unital $C^*$-algebra with the norm $\|\cdot\|$ and the unit $e$ and $B$ is a unital $C^*$-algebra with the norm $\|\cdot\|$. Let $U(A)$ be the set of unitary elements in $A$.

Now, we investigate $C^*$-algebra isomorphisms in unital $C^*$-algebras.

**Theorem 3.74.** *Let $r \neq 2$. Let $h : A \to B$ be an odd bijective mapping satisfying $h(\frac{2^n}{r^n}uy) = h(\frac{2^n}{r^n}u)h(y)$ for all $u \in U(A)$, $y \in A$ and $n \geq 0$ for which there exists a function $\varphi : A^d \to [0, \infty)$ such that*

$$\sum_{j=0}^{\infty} \frac{r^j}{2^j} \varphi\Big(\frac{2^j}{r^j}x_1, \cdots, \frac{2^j}{r^j}x_d\Big) < \infty, \tag{3.106}$$

$$\|D_\mu h(x_1, \cdots, x_d)\| \le \varphi(x_1, \cdots, x_d),$$

$$\Big\|h\Big(\frac{2^n}{r^n}u^*\Big) - h\Big(\frac{2^n}{r^n}u\Big)^*\Big\| \le \varphi\Big(\underbrace{\frac{2^n}{r^n}u, \cdots, \frac{2^n}{r^n}u}_{d \text{ times}}\Big) \tag{3.107}$$

*for all $\mu \in S^1 := \{\lambda \in \mathbb{C} : |\lambda| = 1\}$, $u \in U(A)$, $n \ge 0$ and $x_1, \cdots, x_d \in A$. Assume that $\lim_{n\to\infty} \frac{r^n}{2^n} h(\frac{2^n}{r^n}e)$ is invertible. Then the odd bijective mapping $h : A \to B$ is a $C^*$-algebra isomorphism.*

*Proof.* Consider the $C^*$-algebras $A$ and $B$ as left Banach modules over the unital $C^*$-algebra $\mathbb{C}$. By Theorem 3.66, there exists a unique $\mathbb{C}$-linear generalized additive mapping $H : A \to B$ such that

$$\|h(x) - H(x)\| \le \frac{1}{2(_{d-2}C_l -_{d-2} C_{l-2} + 1)} \tilde{\varphi}\Big(x, x, \underbrace{0, \cdots, 0}_{d-2 \text{ times}}\Big) \tag{3.108}$$

for all $x \in A$. The generalized additive mapping $H : A \to B$ is given by

$$H(x) = \lim_{n\to\infty} \frac{r^n}{2^n} h\Big(\frac{2^n}{r^n}x\Big)$$

for all $x \in A$. By (3.106) and (3.107), we get

$$H(u^*) = \lim_{n\to\infty} \frac{r^n}{2^n} h\Big(\frac{2^n}{r^n}u^*\Big) = \lim_{n\to\infty} \frac{r^n}{2^n} h\Big(\frac{2^n}{r^n}u\Big)^*$$

$$= \Big(\lim_{n\to\infty} \frac{r^n}{2^n} h\Big(\frac{2^n}{r^n}u\Big)\Big)^* = H(u)^*$$

for all $u \in U(A)$. Since $H$ is $\mathbb{C}$-linear and each $x \in A$ is a finite linear combination of unitary elements (see [168]), i.e., $x = \sum_{j=1}^{m} \lambda_j u_j$ for all $\lambda_j \in \mathbb{C}$ and $u_j \in U(A)$, we have

$$H(x^*) = H\Big(\sum_{j=1}^{m} \overline{\lambda_j}u_j^*\Big) = \sum_{j=1}^{m} \overline{\lambda_j}H(u_j^*) = \sum_{j=1}^{m} \overline{\lambda_j}H(u_j)^*$$

$$= \Big(\sum_{j=1}^{m} \lambda_j H(u_j)\Big)^* = H\Big(\sum_{j=1}^{m} \lambda_j u_j\Big)^* = H(x)^*$$

for all $x \in A$. Since $h(\frac{2^n}{r^n}uy) = h(\frac{2^n}{r^n}u)h(y)$ for all $u \in U(A)$, $y \in A$ and $n \ge 0$, we have

$$H(uy) = \lim_{n \to \infty} \frac{r^n}{2^n} h\left(\frac{2^n}{r^n} uy\right) = \lim_{n \to \infty} \frac{r^n}{2^n} h\left(\frac{2^n}{r^n} u\right) h(y)$$
$$= H(u)h(y) \qquad (3.109)$$

for all $u \in U(A)$ and $y \in A$. By the additivity of $H$ and (3.109), we have

$$\frac{2^n}{r^n} H(uy) = H\left(\frac{2^n}{r^n} uy\right) = H\left(u\left(\frac{2^n}{r^n} y\right)\right) = H(u)h\left(\frac{2^n}{r^n} y\right)$$

for all $u \in U(A)$ and $y \in A$. Hence we have

$$H(uy) = \frac{r^n}{2^n} H(u) h\left(\frac{2^n}{r^n} y\right) = H(u) \frac{r^n}{2^n} h\left(\frac{2^n}{r^n} y\right) \qquad (3.110)$$

for all $u \in U(A)$ and $y \in A$. Taking $n \to \infty$ in (3.110), we obtain

$$H(uy) = H(u)H(y) \qquad (3.111)$$

for all $u \in U(A)$ and $y \in A$. Since $H$ is $\mathbb{C}$-linear and each $x \in A$ is a finite linear combination of unitary elements, i.e., $x = \sum_{j=1}^{m} \lambda_j u_j$ for all $\lambda_j \in \mathbb{C}$ and $u_j \in U(A)$, it follows from (3.111) that

$$H(xy) = H\left(\sum_{j=1}^{m} \lambda_j u_j y\right) = \sum_{j=1}^{m} \lambda_j H(u_j y) = \sum_{j=1}^{m} \lambda_j H(u_j) H(y)$$

$$= H(\sum_{j=1}^{m} \lambda_j u_j) H(y) = H(x) H(y)$$

for all $x, y \in A$. By (3.109) and (3.111), we have

$$H(e)H(y) = H(ey) = H(e)h(y)$$

for all $y \in A$. Since $\lim_{n \to \infty} \frac{r^n}{2^n} h(\frac{2^n}{r^n} e) = H(e)$ is invertible,

$$H(y) = h(y)$$

for all $y \in A$. Therefore, the odd bijective mapping $h : A \to B$ is a $C^*$-algebra isomorphism. This completes the proof. $\qquad \square$

**Corollary 3.75.** *Let $r > 2$ and $\theta, p > 1$ be positive real numbers or let $r < 2$ and $\theta, p < 1$ be positive real numbers. Let $h : A \to B$ be an odd bijective mapping satisfying $h(\frac{2^n}{r^n} uy) = h(\frac{2^n}{r^n} u)h(y)$ for all $u \in U(A)$, $y \in A$ and $n \geq 0$ such that*

$$\|D_\mu h(x_1, \cdots, x_d)\| \leq \theta \sum_{j=1}^{d} \|x_j\|^p$$

*and*

$$\left\| h\left(\frac{2^n}{r^n} u^*\right) - h\left(\frac{2^n}{r^n} u\right)^* \right\| \leq d\frac{2^{pn}}{r^{pn}} \theta$$

*for all* $\mu \in S^1$, $u \in U(A)$, $n \geq 0$ *and* $x_1, \cdots, x_d \in A$. *Assume that* $\lim_{n\to\infty} \frac{r^n}{2^n} h(\frac{2^n}{r^n} e)$ *is invertible. Then the odd bijective mapping* $h : A \to B$ *is a* $C^*$-*algebra isomorphism.*

*Proof.* Defining

$$\varphi(x_1, \cdots, x_d) = \theta \sum_{j=1}^{d} \|x_j\|^p$$

and applying Theorem 3.74, we get the desired results.                                    $\square$

**Theorem 3.76.** *Let* $r \neq 2$. *Let* $h : A \to B$ *be an odd bijective mapping satisfying* $h(\frac{2^n}{r^n} uy) = h(\frac{2^n}{r^n} u)h(y)$ *for all* $u \in U(A)$, $y \in A$ *and* $n \geq 0$ *for which there exists a function* $\varphi : A^d \to [0, \infty)$ *satisfying* (3.106), (3.107). *Assume that* $\lim_{n\to\infty} \frac{r^n}{2^n} h(\frac{2^n}{r^n} e)$ *is invertible such that*

$$\|D_\mu h(x_1, \cdots, x_d)\| \leq \varphi(x_1, \cdots, x_d) \tag{3.112}$$

*for all* $x_1, \cdots, x_d \in A$ *and* $\mu = 1, i$. *If* $h(tx)$ *is continuous in* $t \in \mathbb{R}$ *for each fixed* $x \in A$, *then the odd bijective mapping* $h : A \to B$ *is a* $C^*$-*algebra isomorphism.*

*Proof.* Put $\mu = 1$ in (3.112). By the same reasoning as in the proof of Theorem 3.74, there exists a unique generalized additive mapping $H : A \to B$ satisfying (3.108). By (3.112), the generalized additive mapping $H : A \to B$ is $\mathbb{R}$-linear. Put $\mu = i$ in (3.112). By the same method as in the proof of Theorem 3.66, one can obtain that

$$H(ix) = \lim_{n\to\infty} \frac{r^n}{2^n} h\left(\frac{2^n}{r^n} ix\right) = \lim_{n\to\infty} \frac{i\, r^n}{2^n} h\left(\frac{2^n}{r^n} x\right) = iH(x)$$

for all $x \in A$. For each element $\lambda \in \mathbb{C}$, let $\lambda = s + it$, where $s, t \in \mathbb{R}$. Then we have

$$H(\lambda x) = H(sx + itx) = sH(x) + tH(ix) = sH(x) + itH(x)$$
$$= (s + it)H(x) = \lambda H(x)$$

for all $\lambda \in \mathbb{C}$ and $x \in A$. Thus we have

$$H(\zeta x + \eta y) = H(\zeta x) + H(\eta y) = \zeta H(x) + \eta H(y)$$

for all $\zeta, \eta \in \mathbb{C}$ and $x, y \in A$. Hence the generalized additive mapping $H : A \to B$ is $\mathbb{C}$-linear.

The rest of the proof is the same as in the proof of Theorem 3.66. This completes the proof.                                                                                        $\square$

## 3.6   Jordan *-Derivations and Quadratic Jordan *-Derivations

Jordan *-derivations were introduced in [307, 308] for the first time and the structure of such derivations has been investigated in [52]. The reason for introducing these mappings was the fact that the problem of representing quadratic forms by sesquilinear ones is closely connected with the structure of Jordan *-derivations. In [13], An et al. investigated Jordan *-derivations on $C^*$-algebras and Jordan *-derivations on $JC^*$-algebras associated with a special functional inequality.

In this section, we consider the Hyers-Ulam stability of Jordan *-derivations and quadratic Jordan *-derivations on real $C^*$-algebras and real $JC^*$-algebras. We also prove the superstability of Jordan *-derivations and quadratic Jordan *-derivations on real $C^*$-algebras and real $JC^*$-algebras under some conditions [50].

### 3.6.1   Stability of Jordan *-Derivations

Here we prove the Hyers–Ulam stability of Jordan *-derivations on real $C^*$-algebras and real $JC^*$-algebras.

**Definition 3.77.** Let $\mathcal{A}$ be a real $C^*$-algebra. An $\mathbb{R}$-linear mapping $D : \mathcal{A} \to \mathcal{A}$ is called a *Jordan *-derivation* if

$$D(a^2) = a^* D(a) + D(a)a^*$$

for all $a \in \mathcal{A}$.

The mapping $D_x : \mathcal{A} \to \mathcal{A}, a \mapsto a^* x - x a^*$, where $x$ is a fixed element in $\mathcal{A}$, is a Jordan *-derivation. A real $C^*$-algebra $\mathcal{A}$ endowed with the Jordan product $a \circ b := \frac{ab+ba}{2}$ on $\mathcal{A}$ is called a real $JC^*$-algebra (see [13, 225]).

**Definition 3.78.** Let $\mathcal{A}$ be a real $JC^*$-algebra. An $\mathbb{R}$-linear mapping $\delta : \mathcal{A} \to \mathcal{A}$ is called a *Jordan *-derivation* if

$$\delta(a^2) = a^* \circ D(a) + D(a) \circ a^*$$

for all $a \in \mathcal{A}$.

**Theorem 3.79.** *Let $A$ be a real $C^*$-algebra. Suppose that $f : A \to A$ is a mapping with $f(0) = 0$ for which there exists a function $\varphi : A^3 \to [0, \infty)$ such that*

$$\tilde{\varphi}(a, b, c) := \sum_{n=0}^{\infty} \frac{1}{2^{n+1}} \varphi(2^n a 2^n b, 2^n c) < \infty \tag{3.113}$$

*and*

$$\|f(\lambda a + b + c^2) - \lambda f(a) - f(b) - f(c)c^* - c^* f(c)\|$$
$$\leq \varphi(a, b, c) \tag{3.114}$$

*for all $\lambda \in \mathbb{R}$ and $a, b, c \in A$. Then there exists a unique Jordan $*$-derivation $\delta$ on $A$ satisfying*

$$\|f(a) - \delta(a)\| \leq \tilde{\varphi}(a, a, 0) \tag{3.115}$$

*for all $a \in A$.*

*Proof.* Setting $a = b, c = 0$ and $\lambda = 1$ in (3.114), we have

$$\|f(2a) - 2f(a)\| \leq \varphi(a, a, 0)$$

for all $a \in A$. One can use induction to show that

$$\left\| \frac{f(2^n a)}{2^n} - \frac{f(2^m a)}{2^m} \right\| = \sum_{k=m}^{n-1} \frac{1}{2^{k+1}} \varphi(2^k a, 2^k a, 0) \tag{3.116}$$

for all $n > m \geq 0$ and $a \in A$. It follows from (3.113) and (3.116) that the sequence $\{\frac{f(2^n a)}{2^n}\}$ is a Cauchy sequence. Due to the completeness of $A$, this sequence is convergent. Define

$$d(a) := \lim_{n \to \infty} \frac{f(2^n a)}{2^n} \tag{3.117}$$

for all $a \in A$. Then we have

$$\delta\left(\frac{1}{2^k} a\right) = \lim_{n \to \infty} \frac{1}{2^k} \frac{f(2^{n-k} a)}{2^{n-k}} = \frac{1}{2^k} d(a)$$

for each $k \in \mathbb{N}$. Putting $c = 0$ and replacing $a$ and $b$ by $2^n a$ and $2^n b$, respectively, in (3.114), we get

$$\left\| \frac{1}{2^n} f(2^n(\lambda a + b)) - \lambda \frac{1}{2^n} f(2^n a) - \frac{1}{2^n} f(2^n b) \right\| \leq \frac{1}{2^n} \varphi(2^n a, 2^n b, 0).$$

Taking $n \to \infty$, we obtain

$$\delta(\lambda a + b) = \lambda \delta(a) + \delta(b)$$

for all $a, b \in \mathcal{A}$ and $\lambda \in \mathbb{R}$. So $\delta$ is $\mathbb{R}$-linear. Putting $a = b = 0$ and substituting $c$ by $2^n c$ in (3.114), we get

$$\left\| \frac{1}{2^{2n}} f(2^{2n} c^2) - \frac{1}{2^{2n}} f(2^n c)(2^n c^*) - \frac{1}{2^{2n}} (2^n c^*) f(2^n c) \right\|$$

$$\leq \frac{1}{2^{2n}} \varphi(0, 0, 2^n c)$$

$$\leq \frac{1}{2^n} \varphi(0, 0, 2^n c).$$

Taking $n \to \infty$, we obtain

$$\delta(c^2) = \delta(c)c^* + c^*\delta(c)$$

for all $c \in \mathcal{A}$. Moreover, it follows from (3.116) with $m = 0$ and (3.117) that

$$\|\delta(a) - f(a)\| \leq \tilde{\varphi}(a, a, 0)$$

for all $a \in \mathcal{A}$.

For the uniqueness of $\delta$, let $\tilde{\delta} : A \longrightarrow B$ be another Jordan $*$-derivation satisfying (3.115). Then we have

$$\|\delta(a) - \tilde{\delta}(a)\| = \frac{1}{2^n} \|\delta(2^n a) - \tilde{\delta}(2^n a)\|$$

$$\leq \frac{1}{2^n} (\|\delta(2^n a) - f(2^n a)\| + \|f(2^n a) - \tilde{\delta}(2^n a)\|)$$

$$\leq 2 \sum_{j=1}^{\infty} \frac{1}{2^{n+j}} \varphi(2^{n+j} a, 2^{n+j} a, 0)$$

$$= 2 \sum_{j=n}^{\infty} \frac{1}{4^j} \varphi(2^j a, 2^j a, 0),$$

which tends to zero as $n \to \infty$ for all $a \in \mathcal{A}$. So $\delta$ is unique. Therefore, $\delta$ is a Jordan $*$-derivation on $\mathcal{A}$. This completes the proof.  $\square$

*Remark 3.80.* Let $\mathcal{A}$ be a real $C^*$-algebra. Suppose that $f : \mathcal{A} \to \mathcal{A}$ is a mapping with $f(0) = 0$ for which there exists a function $\varphi : \mathcal{A}^3 \to [0, \infty)$ satisfying (3.114) and

$$\tilde{\varphi}(a,b,c) := \sum_{n=1}^{\infty} 2^{n-1} \varphi\left(\frac{a}{2^n}, \frac{b}{2^n}, \frac{c}{2^n}\right) < \infty$$

for all $a, b, c \in \mathcal{A}$. Then there exists a unique Jordan *-derivation $\delta$ on $\mathcal{A}$ satisfying

$$\|f(a) - \delta(a)\| \leq \tilde{\varphi}(a, a, 0)$$

for all $a \in \mathcal{A}$.

**Corollary 3.81.** *Let $\mathcal{A}$ be a real $C^*$-algebra and $\varepsilon, p$ be positive real numbers with $p \neq 1$. Suppose that $f : \mathcal{A} \to \mathcal{A}$ is a mapping satisfying*

$$\|f(\lambda a + b + c^2) - \lambda f(a) - f(b) - cf(c) - f(c)c^*\|$$
$$\leq \varepsilon(\|a\|^p + \|b\|^p + \|c\|^p)$$

*for all $\lambda \in \mathbb{R}$ and $a, b, c \in \mathcal{A}$. Then there exists a unique Jordan *-derivation $\delta$ on $\mathcal{A}$ satisfying*

$$\|f(a) - \delta(a)\| \leq \frac{2\varepsilon}{|2 - 2^p|}\|a\|^p \tag{3.118}$$

*for all $a \in \mathcal{A}$.*

*Proof.* Putting

$$\varphi(a, b, c) = \varepsilon(\|a\|^p + \|b\|^p + \|c\|^p)$$

in Theorem 3.79, we get the desired result.                    □

Now, we consider the Hyers-Ulam stability of Jordan *-derivations on a real $JC^*$-algebra $\mathcal{A}$. Since the proofs are similar to the above results, here we omit them.

*Remark 3.82.* Let $\mathcal{A}$ be a real $JC^*$-algebra. Suppose that $f : \mathcal{A} \to \mathcal{A}$ is a mapping with $f(0) = 0$ for which there exists a function $\varphi : \mathcal{A}^3 \to [0, \infty)$ such that

$$\tilde{\varphi}(a, b, c) := \sum_{n=0}^{\infty} \frac{1}{2^{n+1}} \varphi(2^n a, 2^n b, 2^n c) < \infty$$

and

$$\|f(\lambda a + b + c^2) - \lambda f(a) - f(b) - f(c) \circ c^* - c^* \circ f(c)\|$$
$$\leq \varphi(a, b, c) \tag{3.119}$$

for all $\lambda \in \mathbb{R}$ and $a, b, c \in \mathcal{A}$. Then there exists a unique Jordan $*$-derivation $\delta$ on $\mathcal{A}$ satisfying

$$\|f(a) - \delta(a)\| \leq \tilde{\varphi}(a, a, 0)$$

for all $a \in \mathcal{A}$.

*Remark 3.83.* Let $\mathcal{A}$ be a real $JC^*$-algebra. Suppose that $f : \mathcal{A} \to \mathcal{A}$ is a mapping with $f(0) = 0$ for which there exists a function $\varphi : \mathcal{A}^3 \to [0, \infty)$ satisfying (3.119) and

$$\tilde{\varphi}(a, b, c) := \sum_{n=1}^{\infty} 2^{n-1} \varphi\left(\frac{a}{2^n}, \frac{b}{2^n}, \frac{c}{2^n}\right) < \infty$$

for all $a, b, c \in \mathcal{A}$. Then there exists a unique Jordan $*$-derivation $\delta$ on $\mathcal{A}$ satisfying

$$\|f(a) - \delta(a)\| \leq \tilde{\varphi}(a, a, 0)$$

for all $a \in \mathcal{A}$.

**Corollary 3.84.** *Let $\mathcal{A}$ be a real $JC^*$-algebra and $\varepsilon$, $p$ be positive real numbers with $p \neq 1$. Suppose that $f : \mathcal{A} \to \mathcal{A}$ is a mapping satisfying*

$$\|f(\lambda a + b + c^2) - \lambda f(a) - f(b) - c^* \circ f(c) - f(c) \circ c^*\|$$
$$\leq \varepsilon(\|a\|^p + \|b\|^p + \|c\|^p)$$

*for all $\lambda \in \mathbb{R}$ and $a, b, c \in \mathcal{A}$. Then there exists a unique Jordan $*$-derivation $\delta$ on $\mathcal{A}$ satisfying*

$$\|f(a) - \delta(a)\| \leq \frac{2\varepsilon}{|2 - 2^p|} \|a\|^p$$

*for all $a \in \mathcal{A}$.*

*Proof.* The result follows from Remarks 3.82 and 3.83 by putting

$$\varphi(a, b, c) = \varepsilon(\|a\|^p + \|b\|^p + \|c\|^p). \qquad \square$$

### 3.6.2   Stability of Quadratic Jordan $*$-Derivations

Now, we prove the Hyers-Ulam stability of quadratic Jordan $*$-derivations on real $C^*$-algebras and real $JC^*$-algebras.

**Definition 3.85.** Let $\mathcal{A}$ be a real $C^*$-algebra. A mapping $D : \mathcal{A} \to \mathcal{A}$ is called a *quadratic Jordan ∗-derivation* if $D$ is a quadratic $\mathbb{R}$-homogeneous mapping, that is, $D$ is quadratic, $D(\lambda a) = \lambda^2 D(a)$ for all $a \in \mathcal{A}$ and $\lambda \in \mathbb{R}$ and

$$D(a^2) = (a^*)^2 D(a) + D(a)(a^*)^2$$

for all $a \in \mathcal{A}$.

The mapping $D_x : \mathcal{A} \to \mathcal{A}$, $a \mapsto (a^*)^2 x - x(a^*)^2$, where $x$ is a fixed element in $\mathcal{A}$, is a quadratic Jordan ∗-derivation.

**Definition 3.86.** Let $\mathcal{A}$ be a real $JC^*$-algebra. A mapping $\delta : \mathcal{A} \to \mathcal{A}$ is called a *quadratic Jordan ∗-derivation* if $\delta$ is a quadratic $\mathbb{R}$-homogeneous mapping and

$$\delta(a^2) = (a^*)^2 \circ D(a) + D(a) \circ (a^*)^2$$

for all $a \in \mathcal{A}$.

**Theorem 3.87.** *Let $\mathcal{A}$ be a real $C^*$-algebra. Suppose that $f : \mathcal{A} \to \mathcal{A}$ is a mapping with $f(0) = 0$ for which there exists a function $\varphi : \mathcal{A}^2 \to [0, \infty)$ such that*

$$\tilde{\varphi}(a, b) := \sum_{k=0}^{\infty} \frac{1}{4^k} \varphi(2^k a, 2^k b) < \infty,$$

$$\|f(\lambda a + \lambda b) + f(\lambda a - \lambda b) - 2\lambda^2 f(a) - 2\lambda^2 f(b)\|$$
$$\leq \varphi(a, b), \tag{3.120}$$

$$\|f(a^2) - f(a)(a^*)^2 - (a^*)^2 f(a)\| \leq \varphi(a, a) \tag{3.121}$$

*for all $a, b \in \mathcal{A}$ and $\lambda \in \mathbb{R}$. Then there exists a unique quadratic Jordan ∗-derivation $\delta$ on $\mathcal{A}$ satisfying*

$$\|f(a) - \delta(a)\| \leq \frac{1}{4}\tilde{\varphi}(a, a) \tag{3.122}$$

*for all $a \in \mathcal{A}$.*

*Proof.* Putting $a = b$ and $\lambda = 1$ in (3.120), we have

$$\|f(2a) - 4f(a)\| \leq \varphi(a, a)$$

for all $a \in \mathcal{A}$. One can use induction to show that

$$\left\|\frac{f(2^n a)}{4^n} - \frac{f(2^m a)}{4^m}\right\| \leq \frac{1}{4}\sum_{k=m}^{n-1}\frac{\varphi(2^k a, 2^k a)}{4^k} \tag{3.123}$$

for all $n > m \geq 0$ and $a \in \mathcal{A}$. It follows from (3.123) that the sequence $\left\{\dfrac{f(2^n a)}{4^n}\right\}$ is a Cauchy sequence. Since $\mathcal{A}$ is complete, this sequence is convergent. Define

$$\delta(a) := \lim_{n\to\infty}\frac{f(2^n a)}{4^n}$$

for all $a \in \mathcal{A}$. Since $f(0) = 0$, we have $\delta(0) = 0$. Replacing $a$ and $b$ by $2^n a$ and $2^n b$, respectively, in (3.120), we get

$$\left\|\frac{f(2^n(\lambda a + \lambda b))}{4^n} + \frac{f(2^n(\lambda a - \lambda b))}{4^n} - 2\lambda^2\frac{f(2^n a)}{4^n} - 2\lambda^2\frac{f(2^n b)}{4^n}\right\|$$
$$\leq \frac{\varphi(2^n a, 2^n b)}{4^n}.$$

Taking $n \to \infty$, we obtain

$$\delta(\lambda a + \lambda b) + \delta(\lambda a - \lambda b) = 2\lambda^2\delta(a) + 2\lambda^2\delta(b) \tag{3.124}$$

for all $a, b \in \mathcal{A}$ and $\lambda \in \mathbb{R}$. Putting $\lambda = 1$ in (3.124), we obtain that $\delta$ is a quadratic mapping. It is easy to check that the quadratic mapping $\delta$ satisfying (3.122) is unique (see the proof of Theorem 3.79). Setting $b := a$ in (3.124), we get $\delta(2\lambda a) = 4\lambda^2\delta(a)$ for all $a \in \mathcal{A}$ and $\lambda \in \mathbb{R}$. Hence $\delta(\lambda a) = \lambda^2\delta(a)$ for all $a \in \mathcal{A}$ and $\lambda \in \mathbb{R}$. Replacing $a$ by $2^n a$ in (3.121), we get

$$\left\|\frac{f(2^n a \cdot 2^n a)}{4^{2n}} - \frac{2^{2n}(a^*)^2 f(2^n a)}{4^{2n}} - \frac{f(2^n a)2^{2n}(a^*)^2}{4^{2n}}\right\|$$
$$= \left\|\frac{f(2^{2n}a^2)}{4^{2n}} - \frac{2^{2n}(a^*)^2 f(2^n a)}{2^{2n}\ 4^n} - \frac{f(2^n a)\ 2^{2n}(a^*)^2}{4^n\ 2^{2n}}\right\|$$
$$\leq \frac{\varphi(2^n a, 2^n a)}{4^{2n}}$$
$$\leq \frac{\varphi(2^n a, 2^n a)}{4^n}$$

for all $a \in \mathcal{A}$. Thus we have

$$\|\delta(a^2) - (a^*)^2\delta(a) - \delta(a)(a^*)^2\| \leq \lim_{n\to\infty}\frac{\varphi(2^n a, 2^n a)}{4^n} = 0.$$

Therefore, $\delta$ is a quadratic Jordan $*$-derivation on $\mathcal{A}$. This completes the proof. $\quad\square$

*Remark 3.88.* Let $\mathcal{A}$ be a real $C^*$-algebra. Suppose that $f : \mathcal{A} \rightarrow \mathcal{A}$ is a mapping with $f(0) = 0$ for which there exists a function $\varphi : \mathcal{A}^2 \rightarrow [0, \infty)$ satisfying (3.120), (3.121) and

$$\tilde{\varphi}(a, b) := \sum_{k=1}^{\infty} 4^k \varphi\left(\frac{a}{2^k}, \frac{b}{2^k}\right) < \infty$$

for all $a, b \in \mathcal{A}$. Then there exists a unique quadratic Jordan *-derivation $\delta$ on $\mathcal{A}$ satisfying

$$\|f(a) - \delta(a)\| \leq \frac{1}{4}\tilde{\varphi}(a, a)$$

for all $a \in \mathcal{A}$.

**Corollary 3.89.** *Let $\mathcal{A}$ be a real $C^*$-algebra and $\varepsilon$, $p$ be positive real numbers with $p \neq 2$. Suppose that $f : \mathcal{A} \rightarrow \mathcal{A}$ is a mapping such that*

$$\|f(\lambda a + \lambda b) + f(\lambda a - \lambda b) - 2\lambda^2 f(a) - 2\lambda^2 f(b)\| \leq \varepsilon(\|a\|^p + \|b\|^p)$$

*and*

$$\|f(a^2) - a^2 f(a) - f(a)(a^*)^2\| \leq 2\varepsilon\|a\|^p$$

*for all $a, b \in \mathcal{A}$ and $\lambda \in \mathbb{R}$. Then there exists a unique quadratic Jordan *-derivation $\delta$ on $\mathcal{A}$ satisfying*

$$\|f(a) - \delta(a)\| \leq \frac{2\varepsilon}{|4 - 2^p|}\|a\|^p \tag{3.125}$$

*for all $a \in \mathcal{A}$.*

*Proof.* Putting

$$\varphi(a, b) = \varepsilon(\|a\|^p + \|b\|^p)$$

in Theorem 3.87, we get the desired result.                    □

Here we assume that $\mathcal{A}$ is a real $JC^*$-algebra. Then we can get the Hyers-Ulam stability of quadratic Jordan *-derivations on $\mathcal{A}$.

*Remark 3.90.* Suppose that $f : \mathcal{A} \rightarrow \mathcal{A}$ is a mapping with $f(0) = 0$ for which there exists a function $\varphi : \mathcal{A}^2 \rightarrow [0, \infty)$ such that

$$\tilde{\varphi}(a, b) := \sum_{k=0}^{\infty} \frac{1}{4^k}\varphi(2^k a, 2^k b) < \infty,$$

$$\|f(\lambda a + \lambda b) + f(\lambda a - \lambda b) - 2\lambda^2 f(a) - 2\lambda^2 f(b)\| \leq \varphi(a, b), \quad (3.126)$$

$$\|f(a^2) - (a^*)^2 \circ f(a) - f(a) \circ (a^*)^2\| \leq \varphi(a, a) \quad (3.127)$$

for all $a, b \in \mathcal{A}$ and $\lambda \in \mathbb{R}$. Then there exists a unique quadratic Jordan $*$-derivation $\delta$ on $\mathcal{A}$ satisfying

$$\|f(a) - \delta(a)\| \leq \frac{1}{4}\tilde{\varphi}(a, a)$$

for all $a \in \mathcal{A}$.

*Remark 3.91.* Suppose that $f : \mathcal{A} \to \mathcal{A}$ is a mapping with $f(0) = 0$ for which there exists a function $\varphi : \mathcal{A}^2 \to [0, \infty)$ satisfying (3.126), (3.127) and

$$\tilde{\varphi}(a, b) := \sum_{k=1}^{\infty} 4^k \varphi\left(\frac{a}{2^k}, \frac{b}{2^k}\right) < \infty$$

for all $a, b \in \mathcal{A}$. Then there exists a unique quadratic Jordan $*$-derivation $\delta$ on $\mathcal{A}$ satisfying

$$\|f(a) - \delta(a)\| \leq \frac{1}{4}\tilde{\varphi}(a, a)$$

for all $a \in \mathcal{A}$.

We can obtain the following Remark by letting $\varphi(a, b) = \varepsilon(\|a\|^p + \|b\|^p)$ in Remarks 3.90 and 3.91.

*Remark 3.92.* Let $\varepsilon$ and $p$ be positive real numbers with $p \neq 2$. Suppose that $f : \mathcal{A} \to \mathcal{A}$ is a mapping such that

$$\|f(\lambda a + \lambda b) + f(\lambda a - \lambda b) - 2\lambda^2 f(a) - 2\lambda^2 f(b)\| \leq \varepsilon(\|a\|^p + \|b\|^p)$$

and

$$\|f(a^2) - (a^*)^2 \circ f(a) - f(a) \circ (a^*)^2\| \leq 2\varepsilon\|a\|^p$$

for all $a, b \in \mathcal{A}$ and $\lambda \in \mathbb{R}$. Then there exists a unique quadratic Jordan $*$-derivation $\delta$ on $\mathcal{A}$ satisfying

$$\|f(a) - \delta(a)\| \leq \frac{2\varepsilon}{|4 - 2^p|}\|a\|^p$$

for all $a \in \mathcal{A}$.

### 3.6.3 Stability of Jordan ∗-Derivations: The Fixed Point Method

Now, we assume that $\mathcal{A}$ is a real $C^*$-algebra and prove the stability of Jordan $\ast$-derivations by the fixed point method.

**Theorem 3.93.** *Let $f : \mathcal{A} \to \mathcal{A}$ be a mapping with $f(0) = 0$ and $\varphi : \mathcal{A}^3 \to [0, \infty)$ be a function such that*

$$\|f(\lambda a + b + c^2) - \lambda f(a) - f(b) - f(c)c^* - c^* f(c)\|$$
$$\leq \varphi(a, b, c) \tag{3.128}$$

*for all $\lambda \in \mathbb{R}$ and $a, b, c \in \mathcal{A}$. If there exists a constant $k \in (0, 1)$ such that*

$$\varphi(2a, 2b, 2c) \leq 2k\varphi(a, b, c) \tag{3.129}$$

*for all $a, b, c \in \mathcal{A}$, then there exists a unique Jordan $\ast$-derivation $\delta : \mathcal{A} \to \mathcal{A}$ satisfying*

$$\|f(a) - \delta(a)\| \leq \frac{1}{2(1-k)}\varphi(a, a, 0) \tag{3.130}$$

*for all $a \in \mathcal{A}$.*

*Proof.* It follows from (3.129) that

$$\lim_{j \to \infty} \frac{\varphi(2^j a, 2^j b, 2^j c)}{2^j} = 0$$

for all $a, b, c \in \mathcal{A}$. Putting $\lambda = 1, a = b$ and $c = 0$ in (3.128), we have

$$\|f(2a) - 2f(a)\| \leq \varphi(a, a, 0)$$

for all $a \in \mathcal{A}$ and so

$$\left\| f(a) - \frac{1}{2}f(2a) \right\| \leq \frac{1}{2}\varphi(a, a, 0) \tag{3.131}$$

for all $a \in A$. We consider the set $\Omega := \{h : \mathcal{A} \to \mathcal{A} : h(0) = 0\}$ and introduce the generalized metric on $\Omega$ defined as follows:

$$d(h_1, h_2) := \inf\{C \in (0, \infty) : \|h_1(a) - h_2(a)\| \leq C\varphi(a, a, 0), \ \forall a \in \mathcal{A}\}$$

if there exists such a constant $C$ and, otherwise, $d(h_1, h_2) = \infty$. We know that $d$ is a generalized metric on $\Omega$ and the metric space $(\Omega, d)$ is complete. We now define the linear mapping $T : \Omega \to \Omega$ by

$$Th(a) = \frac{1}{2}h(2a) \qquad (3.132)$$

for all $a \in \mathcal{A}$. For any $h_1, h_2 \in \Omega$, let $C \in \mathbb{R}^+$ be an arbitrary constant with $d(h_1, h_2) \leq C$, that is,

$$\|h_1(a) - h_2(a)\| \leq C\varphi(a, a, 0) \qquad (3.133)$$

for all $a \in A$. Substituting $a$ by $2a$ in the inequality (3.133) and using the equalities (3.129) and (3.132), we have

$$\|Th_1(a) - Th_2(a)\| = \frac{1}{2}\|h_1(2a) - h_2(2a)\|$$

$$\leq \frac{1}{2}C\varphi(2a, 2a, 0)$$

$$\leq Ck\varphi(2a, 2a, 0)$$

for all $a \in A$ and so $d(Th_1, Th_2) \leq Ck$. Therefore, we conclude that $d(Th_1, Th_2) \leq kd(h_1, h_2)$ for all $h_1, h_2 \in \Omega$. It follows from (3.131) that

$$d(Tf, f) \leq \frac{1}{2}. \qquad (3.134)$$

By Theorem 1.3, the sequence $\{T^n f\}$ converges to a unique fixed point $\delta : \mathcal{A} \to \mathcal{A}$ in the set $\Omega_1 = \{h \in \Omega : d(f, h) < \infty\}$, that is,

$$\lim_{n \to \infty} \frac{f(2^n a)}{2^n} = \delta(a)$$

for all $a \in A$. By Theorem 1.3 and (3.134), we have

$$d(f, \delta) \leq \frac{d(Tf, f)}{1 - k} \leq \frac{1}{2(1 - k)}.$$

The above inequalities show that (3.130) holds for all $a \in \mathcal{A}$. Thus, by the same proof of Theorem 3.79, we can deduce that $\delta$ is $\mathbb{R}$-linear by letting $c = 0$ and replacing $a$ and $b$ by $2^n a$ and $2^n b$, respectively, in (3.128). Also, we have

$$\delta(c^2) = \delta(c)c^* + c^*\delta(c)$$

for all $c \in \mathcal{A}$. This completes the proof.                                        $\square$

The following shows that we can obtain a more accurate approximation of (3.118) in the case $p < 1$.

**Corollary 3.94.** *Let $p$, $\theta$ be non-negative real numbers with $p < 1$ and $f : A \to A$ be a mapping with $f(0) = 0$ such that*

$$\|f(\lambda a + b + c^2) - \lambda f(a) - f(b) - f(c)c^* - c^* f(c)\|$$
$$\leq \theta(\|a\|^p + \|b\|^p + \|c\|^p)$$

*for all $\lambda \in \mathbb{R}$ and $a, b, c \in A$. Then there exists a unique Jordan ∗-derivation $\delta : A \to A$ satisfying*

$$\|f(a) - \delta(a)\| \leq \frac{\theta}{2 - 2^p} \|a\|^p$$

*for all $a \in A$.*

*Proof.* The result follows from Theorem 3.93 by taking

$$\varphi(a, b, c) = \theta(\|a\|^p + \|b\|^p + \|c\|^p).$$

$\square$

In the following, we show that, under some conditions, the superstability for Jordan ∗-derivations on real $C^*$-algebras.

**Corollary 3.95.** *Let $p$, $q$, $r$, $\theta$ be non-negative real numbers such that $p + q + r \in (1, \infty)$. Suppose that a mapping $f : A \to A$ satisfies the following:*

$$\|f(\lambda a + b + c^2) - \lambda f(a) - f(b) - f(c)c^* - c^* f(c)\|$$
$$\leq \theta(\|a\|^p \|b\|^q \|c\|^r) \tag{3.135}$$

*for all $a, b, c \in A$. Then $f$ is a Jordan ∗-derivation on $A$.*

*Proof.* Letting $a = b = c = 0$ in (3.135), we have $f(0) = 0$. Once more, if we put $\lambda = 1, c = 0$ and $a = b$ in (3.135), then we get $f(2a) = 2f(a)$ for all $a \in A$. By induction, it is easy to see that $f(2^n a) = 2^n f(a)$ and so $f(a) = \frac{f(2^n a)}{2^n}$ for all $a \in A$ and $n \in \mathbb{N}$. Now, it follows from Theorem 3.93 that $f$ is a Jordan ∗-derivation. $\square$

Note that, in Corollary 3.95, if $p + q + r \in (0, 1)$ and $p > 0$ such that the inequality (3.135) holds, then, by applying $\varphi(x, y) = \theta(\|a\|^p \|b\|^q \|c\|^r)$ in Theorem 3.93, $f$ is again a Jordan ∗-derivation.

**Theorem 3.96.** *Let $f : A \to A$ be a mapping with $f(0) = 0$ and $\varphi : A^2 \to [0, \infty)$ be a function such that*

$$\|f(\lambda a + \lambda b) + f(\lambda a - \lambda b) - 2\lambda^2 f(a) - 2\lambda^2 f(b)\| \leq \varphi(a, b) \tag{3.136}$$

*and*

$$\|f(a^2) - f(a)(a^*)^2 - (a^*)^2 f(a)\| \leq \varphi(a, a)$$

*for all $a, b \in \mathcal{A}$ and $\lambda \in \mathbb{R}$. If there exists a constant $k \in (0, 1)$ such that*

$$\varphi(2a, 2b) \leq 4k\varphi(a, b) \tag{3.137}$$

*for all $a, b \in \mathcal{A}$, then there exists a unique quadratic Jordan $*$-derivation $\delta : \mathcal{A} \to \mathcal{A}$ satisfying*

$$\|f(a) - \delta(a)\| \leq \frac{1}{4(1-k)} \varphi(a, a) \tag{3.138}$$

*for all $a \in \mathcal{A}$.*

*Proof.* By the same proof of Theorem 3.93, we consider the set $\Omega = \{g : \mathcal{A} \to \mathcal{A} : g(0) = 0\}$ and define the mapping $d$ on $\Omega \times \Omega$ as follows:

$$d(g, h) := \inf\{c \in (0, \infty) : \|g(a) - h(a)\| \leq c\phi(a, a), \ \forall a \in \mathcal{A}\}$$

if there exists such a constant $c$ and, otherwise, $d(g, h) = \infty$. One can easily show that $(\Omega, d)$ is complete. Now, we consider the mapping $T : \Omega \to \Omega$ defined by

$$Tg(a) = \frac{1}{4} g(2a)$$

for all $a \in \mathcal{A}$. For any $g, h \in \Omega$ with $d(g, h) < c$, by the definition of $d$ and $T$, we get

$$\left\| \frac{1}{4} g(2a) - \frac{1}{4} h(2a) \right\| \leq \frac{1}{4} c\varphi(2a, 2a)$$

for all $a \in \mathcal{A}$. Using (3.137), we have

$$\left\| \frac{1}{4} g(2a) - \frac{1}{4} h(2a) \right\| \leq ck\varphi(a, a)$$

for all $a \in \mathcal{A}$. The above inequality shows that $d(Tg, Th) \leq kd(g, h)$ for all $g, h \in \Omega$. Hence $T$ is a strictly contractive mapping on $\Omega$ with the Lipschitz constant $k$.

Now, we prove that $d(Tf, f) < \infty$. Putting $a = b$ and $\lambda = 1$ in (3.136), we obtain

$$\|f(2a) - 4f(a)\| \leq \varphi(a, a)$$

for all $a \in \mathcal{A}$ and so

$$\left\| \frac{1}{4} f(2a) - f(a) \right\| \leq \frac{1}{4} \varphi(a, a) \tag{3.139}$$

for all $a \in A$. We deduce from (3.139) that $d(Tf, f) \leq \frac{1}{4}$. It follows from Theorem 1.3 that $d(T^n g, T^{n+1} g) < \infty$ for all $n \geq 0$ and so, in Theorem 1.3, we have $n_0 = 0$. Thus Theorem 1.3 hold on the whole $\Omega$. Hence there exists a unique mapping $\delta : A \to A$ such that $\delta$ is a fixed point of $T$ and $T^n f \to \delta$ as $n \to \infty$. Thus we have

$$\lim_{n \to \infty} \frac{f(2^n a)}{4^n} = \delta(a)$$

for all $a \in A$ and so

$$d(f, \delta) \leq \frac{1}{1 - k} d(Tf, f) \leq \frac{1}{4(1 - k)}.$$

The above equalities show that (3.138) is true for all $a \in A$. Now, it follows from (3.137) that

$$\lim_{n \to \infty} \frac{\varphi(2^n a, 2^n b)}{4^n} = 0.$$

The rest of the proof is easy.    □

In the following, we find a more accurate approximation relative to Corollary 3.89 with the same conditions on the mapping $f$ when $p < 2$. In fact, we obtain a refinement of the inequality (3.125).

**Corollary 3.97.** *Let $\theta$ and $p$ be positive real numbers with $p < 2$. Suppose that $f : A \to A$ is a mapping such that*

$$\|f(\lambda a + \lambda b) + f(\lambda a - \lambda b) - 2\lambda^2 f(a) - 2\lambda^2 f(b)\| \leq \theta(\|a\|^p + \|b\|^p)$$

*and*

$$\|f(a^2) - (a^*)^2 f(a) - f(a)(a^*)^2\| \leq 2\theta \|a\|^p$$

*for all $a, b \in A$ and $\lambda \in \mathbb{R}$. Then there exists a unique quadratic Jordan ∗-derivation $\delta$ on $A$ satisfying*

$$\|f(a) - \delta(a)\| \leq \frac{\theta}{4 - 2^p} \|a\|^p$$

*for all $a \in A$.*

*Proof.* If we put

$$\varphi(a, b) = \theta(\|a\|^p + \|b\|^p)$$

in Theorem 3.96, then we obtain the desired result.    □

The following shows that, under which conditions, a quadratic Jordan $*$-derivation on a real $C^*$-algebra is superstable.

**Corollary 3.98.** *Let $\theta$, $p$, $q$ be positive real numbers with $p + q \neq 2$. Suppose that $f : A \to A$ is a mapping such that*

$$\|f(\lambda a + \lambda b) + f(\lambda a - \lambda b) - 2\lambda^2 f(a) - 2\lambda^2 f(b)\|$$
$$\leq \theta(\|a\|^p \|b\|^q) \tag{3.140}$$

*and*

$$\|f(a^2) - (a^*)^2 f(a) - f(a)(a^*)^2\| \leq \theta \|a\|^{p+q} \tag{3.141}$$

*for all $a, b \in A$ and $\lambda \in \mathbb{R}$. Then $f$ is a quadratic Jordan $*$-derivation on $A$.*

*Proof.* Putting $a = b = 0$ in (3.140), we get $f(0) = 0$. Now, if we put $a = b$ and $\lambda = 1$ in (3.140), then we have $f(2a) = 4f(a)$ for all $a \in A$. It is easy to see, by induction, that $f(2^n a) = 4^n f(a)$ and so $f(a) = \frac{f(2^n a)}{4^n}$ for all $a \in A$ and $n \in \mathbb{N}$. It follows from Theorem 3.96 that $f$ is a quadratic homogeneous mapping. Letting $\varphi(a, b) = \theta(\|a\|^p \|b\|^q)$ in Theorem 3.96, the we can obtain the desired result. $\square$

## 3.7   $(\alpha, \beta, \gamma)$-Derivations on Lie $C^*$-Algebras: The Direct Method

Let $A$ be a Lie $C^*$-algebra. A $\mathbb{C}$-linear mapping $D : A \to A$ is called an $(\alpha, \beta, \gamma)$-*derivation* of $A$ if there exist $\alpha, \beta, \gamma \in \mathbb{C}$ such that

$$\alpha D([x, y]) = \beta [D(x), y] + \gamma [x, D(y)]$$

for all $x, y \in A$.

In this section, we review some works of Eshaghi Gordji et al. [98, 100] on the Hyers-Ulam stability of $(\alpha, \beta, \gamma)$-derivations on Lie $C^*$-algebras associated with the following functional equation:

$$f\left(\frac{x_2 - x_1}{3}\right) + f\left(\frac{x_1 - 3x_3}{3}\right) + f\left(\frac{3x_1 + 3x_3 - x_2}{3}\right) = f(x_1).$$

In fact, we investigate the superstability and the Hyers–Ulam stability of $(\alpha, \beta, \gamma)$-derivations on Lie $C^*$-algebras.

Assume that $A$ is a Lie $C^*$-algebra with the norm $\| \cdot \|$. Now, we remember a lemma to be used in the last sections.

**Lemma 3.99.** *Let X, Y be linear spaces and $f : X \to Y$ be an additive mapping such that $f(\mu x) = \mu f(x)$ for all $x \in X$ and $\mu \in \mathbb{T}^1 := \{\lambda \in \mathbb{C} : |\lambda| = 1\}$. Then the mapping $f : X \to Y$ is $\mathbb{C}$-linear.*

**Lemma 3.100.** *Let $f : A \to A$ be a mapping such that*

$$\left\| f\left(\frac{x_2 - x_1}{3}\right) + f\left(\frac{x_1 - 3x_3}{3}\right) + f\left(\frac{3x_1 + 3x_3 - x_2}{3}\right) \right\|_A$$
$$\leq \|f(x_1)\| \tag{3.142}$$

*for all $x_1, x_2, x_3 \in A$. Then the mapping $f : A \to A$ is additive.*

*Proof.* Letting $x_1 = x_2 = x_3 = 0$ in (3.142), we get $\|3f(0)\|_A \leq \|f(0)\|_A = 0$ and so $f(0) = 0$. Letting $x_1 = x_2 = 0$ in (3.142), we get

$$\|f(-x_3) + f(x_3)\|_A \leq \|f(0)\|_A = 0$$

for all $x_3 \in A$. Hence $f(-x_3) = -f(x_3)$ for all $x_3 \in A$. Letting $x_1 = 0$ and $x_2 = 6x_3$ in (3.142), we get

$$\|f(2x_3) - 2f(x_3)\|_A \leq \|f(0)\|_A = 0$$

for all $x_3 \in A$ and so $f(2x_3) = 2f(x_3)$ for all $x_3 \in A$. Letting $x_1 = 0$ and $x_2 = 9x_3$ in (3.142), we get

$$\|f(3x_3) - 3f(x_3)\|_A \leq \|f(0)\|_A = 0$$

for all $x_3 \in A$ and so $f(3x_3) = 3f(x_3)$ for all $x_3 \in A$. Letting $x_1 = 0$ in (3.142), we get

$$\left\| f\left(\frac{x_2}{3}\right) + f(-x_3) + f\left(x_3 - \frac{x_2}{3}\right) \right\|_A \leq \|f(0)\|_A = 0$$

for all $x_2, x_3 \in A$. Then we have

$$f\left(\frac{x_2}{3}\right) + f(-x_3) + f\left(x_3 - \frac{x_2}{3}\right) = 0 \tag{3.143}$$

for all $x_2, x_3 \in A$. Letting $t_1 = x_3 - \frac{x_2}{3}$ and $t_2 = \frac{x_2}{3}$ in (3.143), we get

$$f(t_2) - f(t_1 + t_2) + f(t_1) = 0$$

for all $t_1, t_2 \in A$ and so $f$ is additive. This completes the proof. $\qquad \square$

Now, we prove the superstability of $(\alpha, \beta, \gamma)$-derivations in Lie $C^*$-algebra $A$.

**Theorem 3.101.** *Let $p \neq 1$, $\theta$ be nonnegative real numbers and $f : A \to A$ be a mapping such that, for some $\alpha, \beta, \gamma \in \mathbb{C}$,*

$$\left\| f\left(\frac{\mu x_2 - x_1}{3}\right) + f\left(\frac{x_1 - 3\mu x_3}{3}\right) + \mu f\left(\frac{3x_1 + 3x_3 - x_2}{3}\right) \right\|$$
$$\leq \|f(x_1)\| \tag{3.144}$$

*and*

$$\|\alpha f([x_1, x_2]) - \beta[f(x_1), x_2] - \gamma[x_1, f(x_2)]\|$$
$$\leq \theta(\|x_1\|^{2p} + \|x_2\|^{2p}) \tag{3.145}$$

*for all $\mu \in \mathbb{T}^1$ and $x_1, x_2, x_3 \in A$. Then the mapping $f : A \to A$ is an $(\alpha, \beta, \gamma)$-derivation.*

*Proof.* Assume $p > 1$. Let $\mu = 1$ in (3.144). By Lemma 3.100, the mapping $f : A \to A$ is additive. Letting $x_1 = x_2 = 0$ in (3.144), we get

$$\|f(-\mu x_3) + \mu f(x_3)\|_A \leq \|f(0)\|_A = 0$$

for all $x_3 \in A$ and $\mu \in \mathbb{T}^1$. Then we have

$$-f(\mu x_3) + \mu f(x_3) = f(-\mu x_3) + \mu f(x_3) = 0$$

for all $x_3 \in A$ and $\mu \in \mathbb{T}^1$. Hence $f(\mu x_3) = \mu f(x_3)$ for all $x_3 \in A$ and $\mu \in \mathbb{T}^1$. By Lemma 3.99, the mapping $f : A \to A$ is $\mathbb{C}$-linear. Since $f$ is additive, it follows from (3.145) that

$$\|\alpha f([x_1, x_2]) - \beta[f(x_1), x_2] - \gamma[x_1, f(x_2)]\|_A$$
$$= \lim_{n \to \infty} 4^n \left\| \alpha f\left(\frac{[x_1, x_2]}{4^n}\right) - \beta\left[f(\frac{x_1}{2^n}), \frac{x_2}{2^n}\right] - \gamma\left[\frac{x_1}{2^n}, f(\frac{x_2}{2^n})\right] \right\|_A$$
$$\leq \lim_{n \to \infty} \frac{4^n \theta}{4^{np}}(\|x_1\|_A^{2p} + \|x_2\|_A^{2p})$$
$$= 0$$

for all $x_1, x_2 \in A$. Thus for some $\alpha, \beta, \gamma \in \mathbb{C}$

$$\alpha f([x_1, x_2]) = \beta[f(x_1), x_2] + \gamma[x_1, f(x_2)]$$

for all $x_1, x_2 \in A$. Hence the mapping $f : A \to A$ is an $(\alpha, \beta, \gamma)$-derivation. Similarly, one obtains the results for the case $p < 1$. This completes the proof. $\square$

Now, we prove the Hyers-Ulam stability of $(\alpha, \beta, \gamma)$-derivations on Lie $C^*$-algebras.

**Theorem 3.102.** *Let $p > 1$, $\theta$ be nonnegative real numbers and $f : A \to A$ be a mapping with $f(0) = 0$ such that, for some $\alpha, \beta, \gamma \in \mathbb{C}$,*

$$\left\| f\left(\frac{\mu x_2 - x_1}{3}\right) + f\left(\frac{x_1 - 3\mu x_3}{3}\right) + \mu f\left(\frac{3x_1 + 3x_3 - x_2}{3}\right) - \mu f(x_1) \right\|$$
$$\leq \theta(\|x_1\|^p + \|x_2\|^p + \|x_3\|^p) \tag{3.146}$$

*and*

$$\|\alpha f([x_1, x_2]) - \beta[f(x_1), x_2] - \gamma[x_1, f(x_2)]\|$$
$$\leq \theta(\|x_1\|^{2p} + \|x_2\|^{2p}) \tag{3.147}$$

*for all $\mu \in \mathbb{T}^1$ and $x_1, x_2, x_3 \in A$. Then there exists a unique $(\alpha, \beta, \gamma)$-derivation $D : A \to A$ such that*

$$\|D(x_1) - f(x_1)\| \leq \frac{3^p(1 + 2^p)\theta \|x_1\|^p}{3^p - 3} \tag{3.148}$$

*for all $x_1 \in A$.*

*Proof.* Letting $\mu = 1$, $x_2 = 2x_1$ and $x_3 = 0$ in (3.146), we get

$$\left\| 3f\left(\frac{x_1}{3}\right) - f(x_1) \right\| \leq (1 + 2^p)\theta \|x_1\|^p \tag{3.149}$$

for all $x_1 \in A$. By induction, we have

$$\left\| 3^n f\left(\frac{x_1}{3^n}\right) - f(x_1) \right\| \leq (1 + 2^p)\theta \|x_1\|^p \sum_{i=0}^{n-1} 3^{i(1-p)}$$

for all $x_1 \in A$. Hence

$$\left\| 3^{n+m} f\left(\frac{x_1}{3^{n+m}}\right) - 3^m f\left(\frac{x_1}{3^m}\right) \right\|$$
$$\leq (1 + 2^p)\theta \|x_1\|^p \sum_{i=0}^{n-1} 3^{(i+m)(1-p)}$$
$$\leq (1 + 2^p)\theta \|x_1\|^p \sum_{i=m}^{n+m-1} 3^{i(1-p)} \tag{3.150}$$

for all $m, n \geq 1$ and $x_1 \in A$. This implies that the sequence $\{3^n f(\frac{x_1}{3^n})\}$ is a Cauchy sequence for all $x_1 \in A$. Since $A$ is complete, the sequence $\{3^n f(\frac{x_1}{3^n})\}$ converges. Thus one can define the mapping $D : A \to A$ by

$$D(x_1) := \lim_{n \to \infty} 3^n f\left(\frac{x_1}{3^n}\right)$$

for all $x_1 \in A$. Moreover, letting $m = 0$ and $n \to \infty$ in (3.150), we get (3.148). It follows from (3.146) that

$$\left\| D\left(\frac{\mu x_2 - x_1}{3}\right) + D\left(\frac{x_1 - 3\mu x_3}{3}\right) + \mu D\left(\frac{3x_1 + 3x_3 - x_2}{3}\right) - \mu D(x_1) \right\|$$

$$= \lim_{n \to \infty} 3^n \left( \left\| f\left(\frac{\mu x_2 - x_1}{3^{n+1}}\right) + f\left(\frac{x_1 - 3\mu x_3}{3^{n+1}}\right) \right. \right.$$

$$\left. \left. + \mu f\left(\frac{3x_1 + 3x_3 - x_2}{3^{n+1}}\right) - \mu f\left(\frac{x_1}{3^n}\right) \right\| \right)$$

$$\leq \lim_{n \to \infty} \frac{3^n \theta}{3^{np}} (\|x_1\|^p + \|x_2\|^p + \|x_3\|^p)$$

$$= 0$$

for all $\mu \in \mathbb{T}^1$ and $x_1, x_2, x_3 \in A$ and so

$$D\left(\frac{\mu x_2 - x_1}{3}\right) + D\left(\frac{x_1 - 3\mu x_3}{3}\right) + \mu D\left(\frac{3x_1 + 3x_3 - x_2}{3}\right)$$

$$= \mu D(x_1) \tag{3.151}$$

for all $\mu \in \mathbb{T}^1$ and $x_1, x_2, x_3 \in A$. Let $\mu = 1$ in (3.151). Then the mapping $D : A \to A$ satisfies the inequality (3.142). By Lemma 3.100, the mapping $D : A \to A$ is additive. Letting $x_1 = x_2 = 0$ in (3.151), we get $D\left(\frac{-3\mu x_3}{3}\right) + \mu D\left(\frac{3x_3}{3}\right) = 0$ and so $D(\mu x_3) = \mu D(x_3)$ for all $\mu \in \mathbb{T}^1$ and all $x_3 \in A$. By Lemma 3.99, $D$ is $\mathbb{C}$-linear. It follows from (3.147) that

$$\|\alpha D([x_1, x_2]) - \beta[D(x_1), x_2] - \gamma[x_1, D(x_2)]\|$$

$$= \lim_{n \to \infty} 9^n \left\| \alpha f\left(\frac{[x_1, x_2]}{9^n}\right) - \beta\left[f\left(\frac{x_1}{3^n}\right), \frac{x_2}{3^n}\right] - \gamma\left[\frac{x_1}{3^n}, f\left(\frac{x_2}{3^n}\right)\right] \right\|$$

$$\leq \lim_{n \to \infty} \frac{9^n \theta}{9^{np}} (\|x_1\|^{2p} + \|x_2\|^{2p}) = 0$$

for all $x_1, x_2 \in A$. So, for some $\alpha, \beta, \gamma \in A$, we have

$$\alpha D([x_1, x_2]) = \beta[D(x_1), x_2] + \gamma[x_1, D(x_2)]$$

for all $x_1, x_2 \in A$. Thus $D$ is an $(\alpha, \beta, \gamma)$-derivation.

Now, let $D' : A \to A$ be another $(\alpha, \beta, \gamma)$-derivation satisfying (3.148). Then we have

$$\|D(x_1) - D'(x_1)\|$$

$$= 3^n \left\| D\left(\frac{x_1}{3^n}\right) - D'\left(\frac{x_1}{3^n}\right) \right\|$$

$$\leq 3^n \left( \left\| D\left(\frac{x_1}{3^n}\right) - f\left(\frac{x_1}{3^n}\right) \right\| + \left\| D'\left(\frac{x_1}{3^n}\right) - f\left(\frac{x_1}{3^n}\right) \right\| \right)$$

$$\leq \frac{2 \cdot 3^n (1 + 2^p) 3^p}{3^{np}(3^p - 3)} \theta \|x\|^p \, ,$$

which tends to zero as $n \to \infty$ for all $x_1 \in A$. So we can conclude that $D(x_1) = D'(x_1)$ for all $x_1 \in A$. This proves the uniqueness of $D$. Therefore, the mapping $D : A \to A$ is a unique $(\alpha, \beta, \gamma)$-derivation satisfying (3.148). This completes the proof.  □

**Theorem 3.103.** *Let $p < 1$, $\theta$ be nonnegative real numbers and $f : A \to A$ with $f(0) = 0$ be a mapping satisfying (3.146) and (3.147). Then there exists a unique $(\alpha, \beta, \gamma)$-derivation $D : A \to A$ such that*

$$\|D(x_1) - f(x_1)\| \leq \frac{3^p (1 + 2^p) \theta \|x_1\|^p}{3 - 3^p} \tag{3.152}$$

*for all $x_1 \in A$.*

*Proof.* It follows from (3.149) that

$$\left\| f(x_1) - \frac{1}{3} f(3x_1) \right\| \leq \frac{3^p}{3} (1 + 2^p) \theta \|x_1\|^p$$

for all $x_1 \in A$. By induction, we have

$$\left\| f(x_1) - \frac{1}{3^n} f(3^n x_1) \right\| \leq (1 + 2^p) \theta \|x_1\|^p \sum_{i=1}^{n} 3^{i(p-1)}$$

for all $x_1 \in A$. Hence we have

$$\left\| \frac{1}{3^m} f(3^m x_1) - \frac{1}{3^{n+m}} f(3^{n+m} x_1) \right\|$$

$$\leq (1 + 2^p) \theta \|x_1\|^p \sum_{i=1}^{n} 3^{(i+m)(p-1)}$$

$$\leq (1 + 2^p) \theta \|x_1\|^p \sum_{i=m+1}^{n+m} 3^{i(p-1)} \tag{3.153}$$

for all $m \geq 1$ and $x_1 \in A$. This implies that the sequence $\{\frac{1}{3^n} f(3^n x_1)\}$ is a Cauchy sequence for all $x_1 \in A$. Since $A$ is complete, the sequence $\{\frac{1}{3^n} f(3^n x_1)\}$ converges. Thus one can define the mapping $D : A \to A$ by

$$D(x_1) := \lim_{n \to \infty} \frac{1}{3^n} f(3^n x_1)$$

for all $x_1 \in A$. Moreover, letting $m = 0$ and $n \to \infty$ in (3.153), we get (3.152).

The rest of the proof is similar to the proof of Theorem 3.102. This completes the proof.                                                                       □

**Corollary 3.104.** *Let $\theta$ be a nonnegative real number. Let $f : A \to A$ with be a mapping $f(0) = 0$ such that, for some $\alpha, \beta, \gamma \in \mathbb{C}$,*

$$\left\| f\left(\frac{\mu x_2 - x_1}{3}\right) + f\left(\frac{x_1 - 3\mu x_3}{3}\right) + \mu f\left(\frac{3x_1 + 3x_3 - x_2}{3}\right) - \mu f(x_1) \right\|$$
$$\leq \theta$$

*and*

$$\| \alpha f([x_1, x_2]) - \beta [f(x_1), x_2] - \gamma [x_1, f(x_2)] \| \leq \theta$$

*for all $\mu \in \mathbb{T}^1$ and $x_1, x_2, x_3 \in A$. Then there exists a unique $(\alpha, \beta, \gamma)$-derivation $D : A \to A$ such that*

$$\| D(x_1) - f(x_1) \| \leq \frac{\theta}{2}$$

*for all $x_1 \in A$.*

## 3.8   Square Roots and 3rd Root Functional Equations: The Direct Method

In this section, we introduce a square root functional equation and a 3rd root functional equation. We prove the Hyers-Ulam stability of the square root functional equation and of the 3rd root functional equation in $C^*$-algebras.

**Definition 3.105 ([96]).** Let $A$ be a $C^*$-algebra and $x \in A$ be a self-adjoint element, i.e., $x^* = x$. Then $x$ is said to be *positive* if it is of the form $yy^*$ for some $y \in A$. The set of positive elements of $A$ is denoted by $A^+$.

Note that $A^+$ is a closed convex cone (see [96]). It is well-known that, for a positive element $x$ and a positive integer $n$, there exists a unique positive element $y \in A^+$ such that $x = y^n$. We denote $y$ by $x^{\frac{1}{n}}$ (see [128]).

In this section, we introduce a *square root functional equation*:

$$S\left(x + y + x^{\frac{1}{4}} y^{\frac{1}{2}} x^{\frac{1}{4}} + y^{\frac{1}{4}} x^{\frac{1}{2}} y^{\frac{1}{4}}\right) = S(x) + S(y) \tag{3.154}$$

and a *3rd root functional equation*:

$$T\left(x + y + 3x^{\frac{1}{3}}y^{\frac{1}{3}}x^{\frac{1}{3}} + 3y^{\frac{1}{3}}x^{\frac{1}{3}}y^{\frac{1}{3}}\right) = T(x) + T(y) \tag{3.155}$$

for all $x, y \in A^+$. Each solution of the square root functional equation is called a *square root mapping* and each solution of the 3rd root functional equation is called a *3rd root mapping*.

Note that the functions $S(x) = \sqrt{x} = x^{\frac{1}{2}}$ and $T(x) = \sqrt[3]{x} = x^{\frac{1}{3}}$ in the set of non-negative real numbers are solutions of the functional equations (3.154) and (3.155), respectively.

Throughout this section, let $A^+$ and $B^+$ be the sets of positive elements in $C^*$-algebras $A$ and $B$, respectively.

### 3.8.1   Stability of the Square Root Functional Equation

Here we investigate the square root functional equation in $C^*$-algebras.

**Lemma 3.106.** *Let $S : A^+ \to B^+$ be a square root mapping satisfying (3.154). Then $S$ satisfies*

$$S(4^n x) = 2^n S(x) \tag{3.156}$$

*for all $x \in A^+$ and $n \in \mathbb{Z}$.*

*Proof.* Putting $x = y = 0$ in (3.154), we obtain $S(0) = 0$. Letting $y = 0$ in (3.154), we obtain

$$S(4^0 x) = S(x) = 2^0 S(x)$$

for all $x \in A^+$.

First of all, we use the induction on $n$ to prove the equality (3.155) for all $n \geq 1$. Replacing $y$ by $x$ in (3.154), we get

$$S(4x) = 2S(x) \tag{3.157}$$

for all $x \in A^+$. So the equality (3.156) holds for $n = 1$. Assume that

$$S(4^k x) = 2^k S(x) \tag{3.158}$$

holds for a positive integer $k$. Replacing $x$ by $4x$ in (3.158) and using (3.157), we obtain

$$S(4^{k+1} x) = S(4^k \cdot 4x) = 2^k S(4x) = 2^{k+1} S(x)$$

for all $x \in A^+$. So the equality (3.156) holds for $n = k + 1$. Thus, by induction, we have

$$S(4^n x) = 2^n S(x) \tag{3.159}$$

for all $x \in A^+$ and $n \geq 1$.

Next, replacing $x$ by $4^{-n}x$ in (3.159), we obtain

$$S(x) = S(4^n \cdot 4^{-n}x) = 2^n S(4^{-n}x)$$

for all $x \in A^+$ and $n \geq 1$ and so

$$S(4^n x) = 2^n S(x)$$

for all $x \in A^+$ and $n \geq 1$. Therefore, we have

$$S(4^n x) = 2^n S(x)$$

for all $x \in A^+$ and $n \in \mathbb{Z}$. This completes the proof. $\qquad\qquad\square$

Now, we prove the Hyers-Ulam stability of the square root functional equation in $C^*$-algebras.

**Theorem 3.107.** *Let* $f : A^+ \to B^+$ *be a mapping for which there exists a function* $\varphi : A^+ \times A^+ \to [0, \infty)$ *such that*

$$\tilde{\varphi}(x, y) := \sum_{j=1}^{\infty} 2^j \varphi\left(\frac{x}{4^j}, \frac{y}{4^j}\right) < \infty \tag{3.160}$$

*and*

$$\left\| f\left(x + y + x^{\frac{1}{4}}y^{\frac{1}{2}}x^{\frac{1}{4}} + y^{\frac{1}{4}}x^{\frac{1}{2}}y^{\frac{1}{4}}\right) - f(x) - f(y) \right\| \leq \varphi(x, y) \tag{3.161}$$

*for all* $x, y \in A^+$. *Then there exists a unique square root mapping* $S : A^+ \to A^+$ *satisfying* (3.154) *and*

$$\|f(x) - S(x)\| \leq \frac{1}{2}\tilde{\varphi}(x, y) \tag{3.162}$$

*for all* $x \in A^+$.

*Proof.* Letting $y = x$ in (3.161), we get

$$\left\| f(4x) - 2f(x) \right\| \leq \varphi(x, x) \tag{3.163}$$

for all $x \in A^+$. It follows from (3.163) that

$$\left\| f(x) - 2f\left(\frac{x}{4}\right) \right\| \leq \varphi\left(\frac{x}{4}, \frac{x}{4}\right)$$

for all $x \in A^+$. Hence

$$\left\| 2^l f\left(\frac{x}{4^l}\right) - 2^m f\left(\frac{x}{4^m}\right) \right\| \leq \frac{1}{2} \sum_{j=l+1}^{m} 2^j \varphi\left(\frac{x}{4^j}, \frac{x}{4^j}\right) \tag{3.164}$$

for all $m, l \geq 1$ with $m > l$ and $x \in A^+$. It follows from (3.160) and (3.164) that the sequence $\{2^k f\left(\frac{x}{4^k}\right)\}$ is a Cauchy sequence for all $x \in A^+$. Since $B^+$ is complete, the sequence $\{2^k f\left(\frac{x}{4^k}\right)\}$ converges and so one can define the mapping $S : A^+ \to B^+$ by

$$S(x) := \lim_{k \to \infty} 2^k f\left(\frac{x}{4^k}\right)$$

for all $x \in A^+$. By (3.163) and (3.164), we have

$$\left\| S\left(x + y + x^{\frac{1}{4}} y^{\frac{1}{2}} x^{\frac{1}{4}} + y^{\frac{1}{4}} x^{\frac{1}{2}} y^{\frac{1}{4}}\right) - S(x) - S(y) \right\|$$
$$= \lim_{k \to \infty} 2^k \left\| f\left(\frac{x + y + x^{\frac{1}{4}} y^{\frac{1}{2}} x^{\frac{1}{4}} + y^{\frac{1}{4}} x^{\frac{1}{2}} y^{\frac{1}{4}}}{4^k}\right) - f\left(\frac{x}{4^k}\right) - f\left(\frac{y}{4^k}\right) \right\|$$
$$\leq \lim_{k \to \infty} 2^k \varphi\left(\frac{x}{4^k}, \frac{y}{4^k}\right)$$
$$= 0$$

for all $x, y \in A^+$ and so

$$S\left(x + y + x^{\frac{1}{4}} y^{\frac{1}{2}} x^{\frac{1}{4}} + y^{\frac{1}{4}} x^{\frac{1}{2}} y^{\frac{1}{4}}\right) - S(x) - S(y) = 0.$$

Hence the mapping $S : A^+ \to B^+$ is a square root mapping. Moreover, letting $l = 0$ and $m \to \infty$ in (3.164), we get (3.162). So, there exists a square root mapping $S : A^+ \to B^+$ satisfying (3.154) and (3.162).

Now, let $S' : A^+ \to B^+$ be another square root mapping satisfying (3.154) and (3.162). Then we have

$$\| S(x) - S'(x) \| = 2^q \left\| S\left(\frac{x}{4^q}\right) - S'\left(\frac{x}{4^q}\right) \right\|$$
$$\leq 2^q \left\| S\left(\frac{x}{4^q}\right) - f\left(\frac{x}{4^q}\right) \right\| + 2^q \left\| S'\left(\frac{x}{4^q}\right) - f\left(\frac{x}{4^q}\right) \right\|$$
$$\leq \frac{2 \cdot 2^q}{2} \tilde{\varphi}\left(\frac{x}{4^q}, \frac{x}{4^q}\right),$$

which tends to zero as $q \to \infty$ for all $x \in A^+$. So we can conclude that $S(x) = S'(x)$ for all $x \in A^+$. This proves the uniqueness of $S$. This completes the proof.  $\square$

**Corollary 3.108.** *Let $p > \frac{1}{2}$ and $\theta_1, \theta_2$ be non-negative real numbers, and let $f : A^+ \to B^+$ be a mapping such that*

$$\left\| f\left(x + y + x^{\frac{1}{4}}y^{\frac{1}{2}}x^{\frac{1}{4}} + y^{\frac{1}{4}}x^{\frac{1}{2}}y^{\frac{1}{4}}\right) - f(x) - f(y) \right\|$$

$$\leq \theta_1 (\|x\|^p + \|y\|^p) + \theta_2 \cdot \|x\|^{\frac{p}{2}} \cdot \|y\|^{\frac{p}{2}} \tag{3.165}$$

*for all $x, y \in A^+$. Then there exists a unique square root mapping $S : A^+ \to B^+$ satisfying (3.154) and*

$$\|f(x) - S(x)\| \leq \frac{2\theta_1 + \theta_2}{4^p - 2} \|x\|^p$$

*for all $x \in A^+$.*

*Proof.* Define

$$\varphi(x, y) = \theta_1 (\|x\|^p + \|y\|^p) + \theta_2 \cdot \|x\|^{\frac{p}{2}} \cdot \|y\|^{\frac{p}{2}}$$

and apply Theorem 3.107. Then we get the desired result.                                □

**Theorem 3.109.** *Let $f : A^+ \to B^+$ be a mapping for which there exists a function $\varphi : A^+ \times A^+ \to [0, \infty)$ satisfying (3.161) such that*

$$\tilde{\varphi}(x, y) := \sum_{j=0}^{\infty} 2^{-j} \varphi(4^j x, 4^j y) < \infty$$

*for all $x, y \in A^+$. Then there exists a unique square root mapping $S : A^+ \to B^+$ satisfying (3.154) and*

$$\|f(x) - S(x)\| \leq \frac{1}{2} \tilde{\varphi}(x, x)$$

*for all $x \in A^+$.*

*Proof.* It follows from (3.163) that

$$\left\| f(x) - \frac{1}{2} f(4x) \right\| \leq \frac{1}{2} \varphi(x, x)$$

for all $x \in A^+$. The rest of the proof is similar to the proof of Theorem 3.107.      □

**Corollary 3.110.** *Let $0 < p < \frac{1}{2}$, $\theta_1, \theta_2$ be non-negative real numbers and $f : A^+ \to B^+$ be a mapping satisfying (3.165). Then there exists a unique square root mapping $S : A^+ \to B^+$ satisfying (3.154) and*

$$\|f(x) - S(x)\| \leq \frac{2\theta_1 + \theta_2}{2 - 4^p} \|x\|^p$$

*for all $x \in A^+$.*

*Proof.* Define

$$\varphi(x, y) = \theta_1(\|x\|^p + \|y\|^p) + \theta_2 \cdot \|x\|^{\frac{p}{2}} \cdot \|y\|^{\frac{p}{2}}$$

and apply Theorem 3.109. Then we get the desired result.  □

### 3.8.2  Stability of the 3rd Root Functional Equation

Now, we investigate the 3rd root functional equation in $C^*$-algebras.

**Lemma 3.111.** *Let $T : A^+ \to B^+$ be a 3rd root mapping satisfying* (3.155). *Then $T$ satisfies*

$$T(8^n x) = 2^n T(x)$$

*for all $x \in A^+$ and $n \in \mathbb{Z}$.*

*Proof.* The proof is similar to the proof of Lemma 3.106.  □

Now, we prove the Hyers-Ulam stability of the 3rd root functional equation in $C^*$-algebras.

**Theorem 3.112.** *Let $f : A^+ \to B^+$ be a mapping for which there exists a function $\varphi : A^+ \times A^+ \to [0, \infty)$ such that*

$$\tilde{\varphi}(x, y) := \sum_{j=1}^{\infty} 2^j \varphi\left(\frac{x}{8^j}, \frac{y}{8^j}\right) < \infty$$

*and*

$$\left\| f\left(x + y + 3x^{\frac{1}{3}}y^{\frac{1}{3}}x^{\frac{1}{3}} + 3y^{\frac{1}{3}}x^{\frac{1}{3}}y^{\frac{1}{3}}\right) - f(x) - f(y) \right\|$$
$$\leq \varphi(x, y) \tag{3.166}$$

*for all $x, y \in A^+$. Then there exists a unique 3rd root mapping $T : A^+ \to A^+$ satisfying* (3.155) *and*

$$\|f(x) - T(x)\| \leq \frac{1}{2}\tilde{\varphi}(x, y)$$

*for all $x \in A^+$.*

*Proof.* Letting $y = x$ in (3.166), we get

$$\|f(8x) - 2f(x)\| \le \varphi(x, x) \tag{3.167}$$

for all $x \in A^+$. The rest of the proof is similar to the proof of Theorem 3.107.  □

**Corollary 3.113.** *Let $p > \frac{1}{3}$, $\theta_1$, $\theta_2$ be non-negative real numbers and $f : A^+ \to B^+$ be a mapping such that*

$$\left\| f\left(x + y + 3x^{\frac{1}{3}}y^{\frac{1}{3}}x^{\frac{1}{3}} + 3y^{\frac{1}{3}}x^{\frac{1}{3}}y^{\frac{1}{3}}\right) - f(x) - f(y) \right\|$$

$$\le \theta_1(\|x\|^p + \|y\|^p) + \theta_2 \cdot \|x\|^{\frac{p}{2}} \cdot \|y\|^{\frac{p}{2}} \tag{3.168}$$

*for all $x, y \in A^+$. Then there exists a unique 3rd root mapping $T : A^+ \to B^+$ satisfying (3.155) and*

$$\|f(x) - T(x)\| \le \frac{2\theta_1 + \theta_2}{8^p - 2} \|x\|^p$$

*for all $x \in A^+$.*

*Proof.* Define

$$\varphi(x, y) = \theta_1(\|x\|^p + \|y\|^p) + \theta_2 \cdot \|x\|^{\frac{p}{2}} \cdot \|y\|^{\frac{p}{2}}$$

and apply Theorem 3.112. Then we get the desired result.  □

**Theorem 3.114.** *Let $f : A^+ \to B^+$ be a mapping for which there exists a function $\varphi : A^+ \times A^+ \to [0, \infty)$ satisfying (3.166) such that*

$$\tilde{\varphi}(x, y) := \sum_{j=0}^{\infty} 2^{-j} \varphi(8^j x, 8^j y) < \infty$$

*for all $x, y \in A^+$. Then there exists a unique 3rd root mapping $T : A^+ \to B^+$ satisfying (3.155) and*

$$\|f(x) - T(x)\| \le \frac{1}{2} \tilde{\varphi}(x, x)$$

*for all $x \in A^+$.*

*Proof.* It follows from (3.167) that

$$\left\| f(x) - \frac{1}{2}f(8x) \right\| \le \frac{1}{2} \varphi(x, x)$$

for all $x \in A^+$. The rest of the proof is similar to the proof of Theorem 3.107.  □

**Corollary 3.115.** *Let* $0 < p < \frac{1}{3}$, $\theta_1$, $\theta_2$ *be non-negative real numbers and* $f : A^+ \to B^+$ *be a mapping satisfying* (3.168). *Then there exists a unique 3rd root mapping* $T : A^+ \to B^+$ *satisfying* (3.155) *and*

$$\|f(x) - T(x)\| \leq \frac{2\theta_1 + \theta_2}{2 - 8^p} \|x\|^p$$

*for all* $x \in A^+$.

*Proof.* Define $\varphi(x, y) = \theta_1(\|x\|^p + \|y\|^p) + \theta_2 \cdot \|x\|^{\frac{p}{2}} \cdot \|y\|^{\frac{p}{2}}$ and apply Theorem 3.114. Then we get the desired result.                                                                                                      □

## 3.9   Square Root and 3rd Root Functional Equations: The Fixed Point Method

In this section, we prove the Hyers-Ulam stability of the square root functional equation and the 3rd root functional equation in $C^*$-algebras via fixed point method [235].

### 3.9.1   Stability of the Square Root Functional Equation

In this section, we investigate the square root functional equation in $C^*$-algebras.

Now, we prove the Hyers-Ulam stability of the square root functional equation in $C^*$-algebras.

**Theorem 3.116.** *Let* $\varphi : A^+ \times A^+ \to [0, \infty)$ *be a function such that there exists* $L < 1$ *with*

$$\varphi(x, y) \leq \frac{L}{2}\varphi(4x, 4y) \tag{3.169}$$

*for all* $x, y \in A^+$. *Let* $f : A^+ \to B^+$ *be a mapping satisfying*

$$\left\| f\left(x + y + x^{\frac{1}{4}}y^{\frac{1}{2}}x^{\frac{1}{4}} + y^{\frac{1}{4}}x^{\frac{1}{2}}y^{\frac{1}{4}}\right) - f(x) - f(y) \right\|$$
$$\leq \varphi(x, y) \tag{3.170}$$

*for all* $x, y \in A^+$. *Then there exists a unique square root mapping* $S : A^+ \to A^+$ *satisfying* (3.154) *and*

$$\|f(x) - S(x)\| \leq \frac{L}{2 - 2L}\varphi(x, x) \tag{3.171}$$

*for all* $x \in A^+$.

*Proof.* Letting $y = x$ in (3.170), we get

$$\|f(4x) - 2f(x)\| \le \varphi(x, x) \tag{3.172}$$

for all $x \in A^+$. Consider the set

$$X := \{g : A^+ \to B^+\}$$

and introduce the generalized metric on $X$ defined by

$$d(g, h) = \inf\{\mu \in \mathbb{R}_+ : \|g(x) - h(x)\| \le \mu\varphi(x, x), \ \forall x \in A^+\},$$

where, as usual, $\inf \phi = +\infty$ which $(X, d)$ is complete.

Now, we consider the linear mapping $J : X \to X$ such that

$$Jg(x) := 2g\left(\frac{x}{4}\right)$$

for all $x \in A^+$. Let $g, h \in X$ be given such that $d(g, h) = \varepsilon$. Then

$$\|g(x) - h(x)\| \le \varphi(x, x)$$

for all $x \in A^+$. Hence we have

$$\|Jg(x) - Jh(x)\| = \left\|2g\left(\frac{x}{4}\right) - 2h\left(\frac{x}{4}\right)\right\| \le L\varphi(x, x)$$

for all $x \in A^+$. So, $d(g, h) = \varepsilon$ implies that $d(Jg, Jh) \le L\varepsilon$. This means that

$$d(Jg, Jh) \le Ld(g, h)$$

for all $g, h \in X$. It follows from (3.172) that

$$\left\|f(x) - 2f\left(\frac{x}{4}\right)\right\| \le \frac{L}{2}\varphi(x, x)$$

for all $x \in A^+$ and so $d(f, Jf) \le \frac{L}{2}$. By Theorem 1.3, there exists a mapping $S : A^+ \to B^+$ satisfying the following:

(1) $S$ is a fixed point of $J$, i.e.,

$$S\left(\frac{x}{4}\right) = \frac{1}{2}S(x) \tag{3.173}$$

for all $x \in A^+$. The mapping $S$ is a unique fixed point of $J$ in the set

$$M = \{g \in X : d(f, g) < \infty\}.$$

This implies that $S$ is a unique mapping satisfying (3.173) such that there exists a $\mu \in (0, \infty)$ satisfying

$$\|f(x) - S(x)\| \le \mu\varphi(x, x)$$

for all $x \in A^+$;

(2) $d(J^n f, S) \to 0$ as $n \to \infty$. This implies the equality

$$\lim_{n\to\infty} 2^n f\left(\frac{x}{4^n}\right) = S(x)$$

for all $x \in A^+$;

(3) $d(f, S) \le \frac{1}{1-L} d(f, Jf)$, which implies the inequality

$$d(f, S) \le \frac{L}{2 - 2L}.$$

This implies that the inequality (3.171) holds.

By (3.169) and (3.170), we have

$$2^n \left\| f\left(\frac{x + y + x^{\frac{1}{4}}y^{\frac{1}{2}}x^{\frac{1}{4}} + y^{\frac{1}{4}}x^{\frac{1}{2}}y^{\frac{1}{4}}}{4^n}\right) - f\left(\frac{x}{4^n}\right) - f\left(\frac{y}{4^n}\right) \right\|$$
$$\le 2^n \varphi\left(\frac{x}{4^n}, \frac{y}{4^n}\right)$$
$$\le L^n \varphi(x, y)$$

for all $x, y \in A^+$ and $n \in \mathbb{N}$. So, we have

$$\left\| S\left(x + y + x^{\frac{1}{4}}y^{\frac{1}{2}}x^{\frac{1}{4}} + y^{\frac{1}{4}}x^{\frac{1}{2}}y^{\frac{1}{4}}\right) - S(x) - S(y) \right\| = 0$$

for all $x, y \in A^+$. Thus the mapping $S : A^+ \to B^+$ is a square root mapping. This completes the proof. $\qquad\square$

**Corollary 3.117.** *Let $p > \frac{1}{2}$, $\theta_1$, $\theta_2$ be non-negative real numbers and $f : A^+ \to B^+$ be a mapping such that*

$$\left\| f\left(x + y + x^{\frac{1}{4}}y^{\frac{1}{2}}x^{\frac{1}{4}} + y^{\frac{1}{4}}x^{\frac{1}{2}}y^{\frac{1}{4}}\right) - f(x) - f(y) \right\|$$
$$\le \theta_1(\|x\|^p + \|y\|^p) + \theta_2 \cdot \|x\|^{\frac{p}{2}} \cdot \|y\|^{\frac{p}{2}} \qquad (3.174)$$

*for all $x, y \in A^+$. Then there exists a unique square root mapping $S : A^+ \to B^+$ satisfying (3.154) and*

$$\|f(x) - S(x)\| \leq \frac{2\theta_1 + \theta_2}{4^p - 2}\|x\|^p$$

*for all $x \in A^+$.*

*Proof.* The proof follows from Theorem 3.116 by taking
$\varphi(x, y) = \theta_1(\|x\|^p + \|y\|^p) + \theta_2 \cdot \|x\|^{\frac{p}{2}} \cdot \|y\|^{\frac{p}{2}}$ for all $x, y \in A^+$ and $L = 2^{1-2p}$.  $\square$

**Theorem 3.118.** *Let $\varphi : A^+ \times A^+ \to [0, \infty)$ be a function such that there exists $L < 1$ with*

$$\varphi(x, y) \leq 2L\varphi\left(\frac{x}{4}, \frac{y}{4}\right)$$

*for all $x, y \in A^+$. Let $f : A^+ \to B^+$ be a mapping satisfying (3.170). Then there exists a unique square root mapping $S : A^+ \to A^+$ satisfying (3.154) and*

$$\|f(x) - S(x)\| \leq \frac{1}{2 - 2L}\varphi(x, x)$$

*for all $x \in A^+$.*

*Proof.* Let $(X, d)$ be the generalized metric space defined in the proof of Theorem 3.116. Consider the linear mapping $J : X \to X$ such that

$$Jg(x) := \frac{1}{2}gt(4x)$$

for all $x \in A^+$. It follows from (3.172) that

$$\left\|f(x) - \frac{1}{2}f(4x)\right\| \leq \frac{1}{2}\varphi(x, x)$$

for all $x \in A^+$. So $d(f, Jf) \leq \frac{1}{2}$.

The rest of the proof is similar to the proof of Theorem 3.116. This completes the proof.  $\square$

**Corollary 3.119.** *Let $0 < p < \frac{1}{2}$, $\theta_1$, $\theta_2$ be non-negative real numbers and $f : A^+ \to B^+$ be a mapping satisfying (3.174). Then there exists a unique square root mapping $S : A^+ \to B^+$ satisfying (3.154) and*

$$\|f(x) - S(x)\| \leq \frac{2\theta_1 + \theta_2}{2 - 4^p}\|x\|^p$$

*for all $x \in A^+$.*

*Proof.* The proof follows from Theorem 3.118 by taking

$$\varphi(x, y) = \theta_1(\|x\|^p + \|y\|^p) + \theta_2 \cdot \|x\|^{\frac{p}{2}} \cdot \|y\|^{\frac{p}{2}}$$

for all $x, y \in A^+$ and $L = 2^{2p-1}$. $\qquad\square$

### 3.9.2 Stability of the 3rd Root Functional Equation

Now, we investigate the 3rd root functional equation in $C^*$-algebras.

Now, we prove the Hyers-Ulam stability of the 3rd root functional equation in $C^*$-algebras.

**Theorem 3.120.** *Let* $\varphi : A^+ \times A^+ \to [0, \infty)$ *be a function such that there exists an* $L < 1$ *with*

$$\varphi(x, y) \leq \frac{L}{2}\varphi(8x, 8y)$$

*for all* $x, y \in A^+$. *Let* $f : A^+ \to B^+$ *be a mapping satisfying*

$$\left\| f\left(x + y + 3x^{\frac{1}{3}}y^{\frac{1}{3}}x^{\frac{1}{3}} + 3y^{\frac{1}{3}}x^{\frac{1}{3}}y^{\frac{1}{3}}\right) - f(x) - f(y) \right\|$$
$$\leq \varphi(x, y) \tag{3.175}$$

*for all* $x, y \in A^+$. *Then there exists a unique 3rd root mapping* $T : A^+ \to A^+$ *satisfying* (3.155) *and*

$$\|f(x) - T(x)\| \leq \frac{L}{2 - 2L}\varphi(x, x)$$

*for all* $x \in A^+$.

*Proof.* Letting $y = x$ in (3.175), we get

$$\|f(8x) - 2f(x)\| \leq \varphi(x, x) \tag{3.176}$$

for all $x \in A^+$. Let $(X, d)$ be the generalized metric space defined in the proof of Theorem 3.116. Consider the linear mapping $J : X \to X$ such that

$$Jg(x) := 2g\left(\frac{x}{8}\right)$$

for all $x \in A^+$.

Now, we consider the linear mapping $J : X \to X$ such that

$$Jg(x) := 2g\left(\frac{x}{8}\right)$$

for all $x \in A^+$. It follows from (3.176) that

$$\left\| f(x) - 2f\left(\frac{x}{8}\right) \right\| \le \frac{L}{2}\varphi(x,x)$$

for all $x \in X$ and so $d(f, Jf) \le \frac{L}{2}$.

The rest of the proof is similar to the proof of Theorem 3.116. This completes the proof. $\qquad\square$

**Corollary 3.121.** *Let* $p > \frac{1}{3}$, $\theta_1$, $\theta_2$ *be non-negative real numbers and* $f : A^+ \to B^+$ *be a mapping such that*

$$\left\| f\left(x + y + 3x^{\frac{1}{3}}y^{\frac{1}{3}}x^{\frac{1}{3}} + 3y^{\frac{1}{3}}x^{\frac{1}{3}}y^{\frac{1}{3}}\right) - f(x) - f(y) \right\|$$
$$\le \theta_1(\|x\|^p + \|y\|^p) + \theta_2 \cdot \|x\|^{\frac{p}{2}} \cdot \|y\|^{\frac{p}{2}} \qquad (3.177)$$

*for all* $x, y \in A^+$. *Then there exists a unique 3rd root mapping* $T : A^+ \to B^+$ *satisfying (3.155) and*

$$\|f(x) - T(x)\| \le \frac{2\theta_1 + \theta_2}{8^p - 2}\|x\|^p$$

*for all* $x \in A^+$.

*Proof.* The proof follows from Theorem 3.120 by taking

$$\varphi(x,y) = \theta_1(\|x\|^p + \|y\|^p) + \theta_2 \cdot \|x\|^{\frac{p}{2}} \cdot \|y\|^{\frac{p}{2}}$$

for all $x, y \in A^+$ and $L = 2^{1-3p}$. $\qquad\square$

**Theorem 3.122.** *Let* $\varphi : A^+ \times A^+ \to [0, \infty)$ *be a function such that there exists* $L < 1$ *with*

$$\varphi(x,y) \le 2L\varphi\left(\frac{x}{8}, \frac{y}{8}\right)$$

*for all* $x, y \in A^+$. *Let* $f : A^+ \to B^+$ *be a mapping satisfying (3.175). Then there exists a unique 3rd root mapping* $T : A^+ \to A^+$ *satisfying (3.155) and*

$$\|f(x) - T(x)\| \le \frac{1}{2 - 2L}\varphi(x,x)$$

*for all* $x \in A^+$.

*Proof.* Let $(X, d)$ be the generalized metric space defined in the proof of Theorem 3.116. Consider the linear mapping $J : X \to X$ such that

$$Jg(x) := \frac{1}{2}g(8x)$$

for all $x \in A^+$. It follows from (3.176) that

$$\left\| f(x) - \frac{1}{2}f(8x) \right\| \leq \frac{1}{2}\varphi(x, x)$$

for all $x \in A^+$ and so $d(f, Jf) \leq \frac{1}{2}$.

The rest of the proof is similar to the proof of Theorem 3.116. This completes the proof. □

**Corollary 3.123.** *Let* $0 < p < \frac{1}{3}$, $\theta_1$, $\theta_2$ *be non-negative real numbers and* $f : A^+ \to B^+$ *be a mapping satisfying* (3.177)*. Then there exists a unique 3rd root mapping* $T : A^+ \to B^+$ *satisfying* (3.155) *and*

$$\|f(x) - T(x)\| \leq \frac{2\theta_1 + \theta_2}{2 - 8^p}\|x\|^p$$

*for all* $x \in A^+$.

*Proof.* The proof follows from Theorem 3.122 by taking

$$\varphi(x, y) = \theta_1(\|x\|^p + \|y\|^p) + \theta_2 \cdot \|x\|^{\frac{p}{2}} \cdot \|y\|^{\frac{p}{2}}$$

for all $x, y \in A^+$ and $L = 2^{3p-1}$. □

## 3.10 Positive-Additive Functional Equation

In this section, we consider a positive-additive functional equation in $C^*$-algebras [258]. Using fixed point and direct methods, we prove the stability of the positive-additive functional equation in $C^*$-algebras.

**Definition 3.124 ([96]).** Let $A$ be a $C^*$-algebra and $x \in A$ be a self-adjoint element, i.e., $x^* = x$. Then $x$ is said to be *positive* if it is of the form $yy^*$ for some $y \in A$. The set of positive elements of $A$ is denoted by $A^+$.

Note that $A^+$ is a closed convex cone (see [96]). It is well known that for a positive element $x$ and a positive integer $n$ there exists a unique positive element $y \in A^+$ such that $x = y^n$. We denote $y$ by $x^{\frac{1}{n}}$ (see [128]).

In this section, we introduce the following functional equation:

$$T\left(\left(x^{\frac{1}{m}} + y^{\frac{1}{m}}\right)^m\right) = \left(T(x)^{\frac{1}{m}} + T(y)^{\frac{1}{m}}\right)^m \tag{3.178}$$

for all $x, y \in A^+$ and a fixed integer $m$ greater than 1, which is called a *positive-additive functional equation*. Each solution of the positive-additive functional equation is called a *positive-additive mapping*.

Note that the function $f(x) = cx$ for any $c \geq 0$ in the set of non-negative real numbers is a solution of the functional equation (3.178).

Throughout this section, let $A^+$ and $B^+$ be the sets of positive elements in $C^*$-algebras $A$ and $B$, respectively. Assume that $m$ is a fixed integer greater than 1.

### 3.10.1  Stability of the Positive-Additive Functional Equations: The Fixed Point Method

Here we investigate the positive-additive functional equation (3.178) in $C^*$-algebras.

**Lemma 3.125.** *Let $T : A^+ \to B^+$ be a positive-additive mapping satisfying* (3.178). *Then $T$ satisfies*

$$T(2^{mn}x) = 2^{mn}T(x)$$

*for all $x \in A^+$ and $n \in \mathbb{Z}$.*

*Proof.* Putting $x = y$ in (3.178), we obtain $T(2^m x) = 2^m T(x)$ for all $x \in A^+$. So, one can show that

$$T(2^{mn}x) = 2^{mn}T(x)$$

for all $x \in A^+$ and $n \in \mathbb{Z}$.                                                                □

Using the fixed point method, we prove the Hyers-Ulam stability of the positive-additive functional equation (3.178) in $C^*$-algebras. Note that the fundamental ideas in the proofs of the main results in this section are contained in [62–64].

**Theorem 3.126.** *Let $\varphi : A^+ \times A^+ \to [0, \infty)$ be a function such that there exists $L < 1$ with*

$$\varphi(x, y) \leq \frac{L}{2^m}\varphi\left(2^m x, 2^m y\right) \tag{3.179}$$

*for all $x, y \in A^+$. Let $f : A^+ \to B^+$ be a mapping satisfying*

$$\left\| f\left(\left(x^{\frac{1}{m}} + y^{\frac{1}{m}}\right)^{m}\right) - \left(f(x)^{\frac{1}{m}} + f(y)^{\frac{1}{m}}\right)^{m}\right\| \leq \varphi(x, y) \tag{3.180}$$

*for all $x, y \in A^+$. Then there exists a unique positive-additive mapping $T : A^+ \to A^+$ satisfying (3.178) and*

$$\|f(x) - T(x)\| \leq \frac{L}{2^m - 2^m L} \varphi(x, x) \tag{3.181}$$

*for all $x \in A^+$.*

*Proof.* Letting $y = x$ in (3.180), we get

$$\|f(2^m x) - 2^m f(x)\| \leq \varphi(x, x) \tag{3.182}$$

for all $x \in A^+$. Consider the set

$$X := \{g : A^+ \to B^+\}$$

and introduce the generalized metric on $X$ defined by

$$d(g, h) = \inf\{\mu \in \mathbb{R}_+ : \|g(x) - h(x)\| \leq \mu \varphi(x, x), \ \forall x \in A^+\},$$

where, as usual, $\inf \phi = +\infty$ which $(X, d)$ is complete.

Now, we consider the linear mapping $J : X \to X$ such that

$$Jg(x) := 2^m g\left(\frac{x}{2^m}\right)$$

for all $x \in A^+$. Let $g, h \in X$ be given such that $d(g, h) = \varepsilon$. Then we have

$$\|g(x) - h(x)\| \leq \varphi(x, x)$$

for all $x \in A^+$ and so

$$\|Jg(x) - Jh(x)\| = \left\| 2^m g\left(\frac{x}{2^m}\right) - 2^m h\left(\frac{x}{2^m}\right) \right\| \leq L\varphi(x, x)$$

for all $x \in A^+$. So, $d(g, h) = \varepsilon$ implies that $d(Jg, Jh) \leq L\varepsilon$. This means that

$$d(Jg, Jh) \leq Ld(g, h)$$

for all $g, h \in X$. It follows from (3.182) that

$$\left\| f(x) - 2^m f\left(\frac{x}{2^m}\right) \right\| \leq \frac{L}{2^m} \varphi(x, x)$$

for all $x \in A^+$ and so $d(f, Jf) \leq \frac{L}{2^m}$. By Theorem 1.3, there exists a mapping $T : A^+ \to B^+$ satisfying the following:

(1) $T$ is a fixed point of $J$, i.e.,

$$T\left(\frac{x}{2^m}\right) = \frac{1}{2^m}T(x) \tag{3.183}$$

for all $x \in A^+$. The mapping $T$ is a unique fixed point of $J$ in the set

$$M = \{g \in X : d(f, g) < \infty\}.$$

This implies that $T$ is a unique mapping satisfying (3.183) such that there exists a $\mu \in (0, \infty)$ satisfying

$$\|f(x) - T(x)\| \leq \mu\varphi(x, x)$$

for all $x \in A^+$;

(2) $d(J^n f, T) \to 0$ as $n \to \infty$. This implies the equality

$$\lim_{n\to\infty} 2^{mn} f\left(\frac{x}{2^{mn}}\right) = T(x)$$

for all $x \in A^+$;

(3) $d(f, T) \leq \frac{1}{1-L}d(f, Jf)$, which implies the inequality

$$d(f, T) \leq \frac{L}{2^m - 2^m L}.$$

This implies that the inequality (3.181) holds.

By (3.179) and (3.180), we have

$$2^{mn}\left\| f\left(\frac{\left(x^{\frac{1}{m}} + y^{\frac{1}{m}}\right)^m}{2^{mn}}\right) - \left(\left(2^{mn}f\left(\frac{x}{2^{mn}}\right)\right)^{\frac{1}{m}} + \left(2^{mn}f\left(\frac{y}{2^{mn}}\right)\right)^{\frac{1}{m}}\right)^m \right\|$$
$$\leq 2^{mn}\varphi\left(\frac{x}{2^{mn}}, \frac{y}{2^{mn}}\right)$$
$$\leq L^{mn}\varphi(x, y)$$

for all $x, y \in A^+$ and $n \in \mathbb{N}$ and so

$$\left\| T\left(\left(x^{\frac{1}{m}} + y^{\frac{1}{m}}\right)^m\right) - \left(T(x)^{\frac{1}{m}} + T(y)^{\frac{1}{m}}\right)^m \right\| = 0$$

for all $x, y \in A^+$. Thus the mapping $T : A^+ \to B^+$ is positive-additive. This completes the proof.                                                                        $\square$

**Corollary 3.127.** *Let $p > 1$, $\theta_1$, $\theta_2$ be non-negative real numbers and $f : A^+ \to B^+$ be a mapping such that*

$$\left\| f\left(\left(x^{\frac{1}{m}} + y^{\frac{1}{m}}\right)^m\right) - \left(f(x)^{\frac{1}{m}} + f(y)^{\frac{1}{m}}\right)^m \right\|$$

$$\leq \theta_1(\|x\|^p + \|y\|^p) + \theta_2 \cdot \|x\|^{\frac{p}{2}} \cdot \|y\|^{\frac{p}{2}} \qquad (3.184)$$

*for all $x, y \in A^+$. Then there exists a unique positive-additive mapping $T : A^+ \to B^+$ satisfying (3.178) and*

$$\|f(x) - T(x)\| \leq \frac{2\theta_1 + \theta_2}{2^{mp} - 2^m}\|x\|^p$$

*for all $x \in A^+$.*

*Proof.* The proof follows from Theorem 3.126 by taking

$$\varphi(x, y) = \theta_1(\|x\|^p + \|y\|^p) + \theta_2 \cdot \|x\|^{\frac{p}{2}} \cdot \|y\|^{\frac{p}{2}}$$

for all $x, y \in A^+$ and $L = 2^{m-mp}$. $\qquad\square$

**Theorem 3.128.** *Let $\varphi : A^+ \times A^+ \to [0, \infty)$ be a function such that there exists an $L < 1$ with*

$$\varphi(x, y) \leq 2^m L \varphi\left(\frac{x}{2^m}, \frac{y}{2^m}\right)$$

*for all $x, y \in A^+$. Let $f : A^+ \to B^+$ be a mapping satisfying (3.180). Then there exists a unique positive-additive mapping $T : A^+ \to A^+$ satisfying (3.178) and*

$$\|f(x) - T(x)\| \leq \frac{1}{2^m - 2^m L}\varphi(x, x)$$

*for all $x \in A^+$.*

*Proof.* Let $(X, d)$ be the generalized metric space defined in the proof of Theorem 3.126. Consider the linear mapping $J : X \to X$ such that

$$Jg(x) := \frac{1}{2^m}g(2^m x)$$

for all $x \in A^+$. It follows from (3.182) that

$$\left\| f(x) - \frac{1}{2^m}f(2^m x) \right\| \leq \frac{1}{2^m}\varphi(x, x)$$

for all $x \in A^+$ and so $d(f, Jf) \leq \frac{1}{2^m}$.

The rest of the proof is similar to the proof of Theorem 3.126. This completes the proof. $\qquad\square$

**Corollary 3.129.** *Let* $0 < p < 1$, $\theta_1$, $\theta_2$ *be non-negative real numbers and* $f : A^+ \to B^+$ *be a mapping satisfying* (3.184). *Then there exists a unique positive-additive mapping* $T : A^+ \to B^+$ *satisfying* (3.178) *and*

$$\|f(x) - T(x)\| \leq \frac{2\theta_1 + \theta_2}{2^m - 2^{mp}} \|x\|^p$$

*for all* $x \in A^+$.

*Proof.* The proof follows from Theorem 3.128 by taking

$$\varphi(x, y) = \theta_1(\|x\|^p + \|y\|^p) + \theta_2 \cdot \|x\|^{\frac{p}{2}} \cdot \|y\|^{\frac{p}{2}}$$

for all $x, y \in A^+$ and $L = 2^{mp-m}$.                                                                    □

### 3.10.2   Stability of the Positive-Additive Functional Equations: The Direct Method

Now, using the direct method of Hyers, we prove the Hyers-Ulam stability of the positive-additive functional equation (3.178) in $C^*$-algebras.

**Theorem 3.130.** *Let* $f : A^+ \to B^+$ *be a mapping for which there exists a function* $\varphi : A^+ \times A^+ \to [0, \infty)$ *satisfying* (3.180) *and*

$$\tilde{\varphi}(x, y) := \sum_{j=1}^{\infty} 2^{mj} \varphi \left( \frac{x}{2^{mj}}, \frac{y}{2^{mj}} \right) < \infty \tag{3.185}$$

*for all* $x, y \in A^+$. *Then there exists a unique positive-additive mapping* $T : A^+ \to A^+$ *satisfying* (3.178) *and*

$$\|f(x) - T(x)\| \leq \frac{1}{2^m} \tilde{\varphi}(x, y) \tag{3.186}$$

*for all* $x \in A^+$.

*Proof.* It follows from (3.182) that

$$\left\| f(x) - 2^m f\left( \frac{x}{2^m} \right) \right\| \leq \varphi \left( \frac{x}{2^m}, \frac{x}{2^m} \right)$$

for all $x \in A^+$ and so

$$\left\| 2^{ml} f\left( \frac{x}{2^{ml}} \right) - 2^{mk} f\left( \frac{x}{2^{mk}} \right) \right\| \leq \frac{1}{2^m} \sum_{j=l+1}^{k} 2^{mj} \varphi \left( \frac{x}{2^{mj}}, \frac{x}{2^{mj}} \right) \tag{3.187}$$

for all $k, l \geq 1$ with $k > l$ and $x \in A^+$. It follows from (3.185) and (3.187) that the sequence $\{2^{mj}f\left(\frac{x}{2^{mj}}\right)\}$ is a Cauchy sequence for all $x \in A^+$. Since $B^+$ is complete, the sequence $\{2^{mj}f\left(\frac{x}{2^{mj}}\right)\}$ converges and so one can define the mapping $T : A^+ \to B^+$ by

$$T(x) := \lim_{j \to \infty} 2^{mj}f\left(\frac{x}{2^{mj}}\right)$$

for all $x \in A^+$. By (3.180) and (3.185), we have

$$\left\| T\left(\left(x^{\frac{1}{m}} + y^{\frac{1}{m}}\right)^m\right) - \left(T(x)^{\frac{1}{m}} + T(y)^{\frac{1}{m}}\right)^m \right\|$$

$$= \lim_{j \to \infty} 2^{mj} \left\| f\left(\frac{\left(x^{\frac{1}{m}} + y^{\frac{1}{m}}\right)^m}{2^{mj}}\right) \right.$$

$$\left. - \left(\left(2^{mj}f\left(\frac{x}{2^{mj}}\right)\right)^{\frac{1}{m}} + \left(2^{mj}f\left(\frac{y}{2^{mj}}\right)\right)^{\frac{1}{m}}\right)^m \right\|$$

$$\leq \lim_{j \to \infty} 2^{mj}\varphi\left(\frac{x}{2^{mj}}, \frac{y}{2^{mj}}\right)$$

$$= 0$$

for all $x, y \in A^+$ and so

$$T\left(\left(x^{\frac{1}{m}} + y^{\frac{1}{m}}\right)^m\right) - \left(T(x)^{\frac{1}{m}} + T(y)^{\frac{1}{m}}\right)^m = 0$$

for all $x, y \in A^+$. Hence the mapping $T : A^+ \to B^+$ is positive-additive. Moreover, letting $l = 0$ and passing the limit $k \to \infty$ in (3.187), we get (3.186). So there exists a positive-additive mapping $T : A^+ \to B^+$ satisfying (3.178) and (3.186).

Now, let $T' : A^+ \to B^+$ be another positive-additive mapping satisfying (3.178) and (3.186). Then we have

$$\|T(x) - T'(x)\|$$

$$= 2^{mq} \left\| T\left(\frac{x}{2^{mq}}\right) - T'\left(\frac{x}{2^{mq}}\right) \right\|$$

$$\leq 2^{mq} \left\| T\left(\frac{x}{2^{mq}}\right) - f\left(\frac{x}{2^{mq}}\right) \right\| + 2^{mq} \left\| T'\left(\frac{x}{2^{mq}}\right) - f\left(\frac{x}{2^{mq}}\right) \right\|$$

$$\leq \frac{2 \cdot 2^{mq}}{2^m} \tilde{\varphi}\left(\frac{x}{2^{mq}}, \frac{x}{2^{mq}}\right),$$

which tends to zero as $q \to \infty$ for all $x \in A^+$. So, we can conclude that $T(x) = T'(x)$ for all $x \in A^+$, which proves the uniqueness of $T$. This completes the proof. $\qquad \square$

**Corollary 3.131.** *Let $p > 1$, $\theta_1$, $\theta_2$ be non-negative real numbers and $f : A^+ \to B^+$ be a mapping satisfying (3.184). Then there exists a unique positive-additive mapping $T : A^+ \to B^+$ satisfying (3.178) and*

$$\|f(x) - T(x)\| \le \frac{2\theta_1 + \theta_2}{2^{mp} - 2^m} \|x\|^p$$

*for all $x \in A^+$.*

*Proof.* Define

$$\varphi(x, y) = \theta_1(\|x\|^p + \|y\|^p) + \theta_2 \cdot \|x\|^{\frac{p}{2}} \cdot \|y\|^{\frac{p}{2}}$$

and apply Theorem 3.130. Then we get the desired result.                    $\square$

**Theorem 3.132.** *Let $f : A^+ \to B^+$ be a mapping for which there exists a function $\varphi : A^+ \times A^+ \to [0, \infty)$ satisfying (3.180) such that*

$$\tilde{\varphi}(x, y) := \sum_{j=0}^{\infty} 2^{-mj} \varphi(2^{mj}x, 2^{mj}y) < \infty$$

*for all $x, y \in A^+$. Then there exists a unique positive-additive mapping $T : A^+ \to B^+$ satisfying (3.178) and*

$$\|f(x) - T(x)\| \le \frac{1}{2^m} \tilde{\varphi}(x, x)$$

*for all $x \in A^+$.*

*Proof.* It follows from (3.182) that

$$\left\| f(x) - \frac{1}{2^m} f(2^m x) \right\| \le \frac{1}{2^m} \varphi(x, x)$$

for all $x \in A^+$. The rest of the proof is similar to the proof of Theorem 3.130.    $\square$

**Corollary 3.133.** *Let $0 < p < 1$, $\theta_1$, $\theta_2$ be non-negative real numbers and $f : A^+ \to B^+$ be a mapping satisfying (3.184). Then there exists a unique positive-additive mapping $T : A^+ \to B^+$ satisfying (3.178) and*

$$\|f(x) - T(x)\| \le \frac{2\theta_1 + \theta_2}{2^m - 2^{mp}} \|x\|^p$$

*for all $x \in A^+$.*

*Proof.* Define

$$\varphi(x, y) = \theta_1(\|x\|^p + \|y\|^p) + \theta_2 \cdot \|x\|^{\frac{p}{2}} \cdot \|y\|^{\frac{p}{2}}$$

and apply Theorem 3.132. Then we get the desired result. □

## 3.11 Stability of ∗-Homomorphisms in $JC^*$-Algebras

It is shown that every almost unital almost linear mapping $f : \mathcal{A} \rightarrow \mathcal{B}$ of $JC^*$-algebra $\mathcal{A}$ to a $JC^*$-algebra $\mathcal{B}$ is a homomorphism when $f(2^n u \circ y) = f(2^n u) \circ f(y)$ for all unitaries $u \in \mathcal{A}$, $y \in \mathcal{A}$ and $n \geq 0$ and every almost unital almost linear continuous mapping $f : \mathcal{A} \rightarrow \mathcal{B}$ of a $JC^*$-algebra $\mathcal{A}$ of real rank zero to a $JC^*$-algebra $\mathcal{B}$ is a homomorphism when $f(2^n u \circ y) = f(2^n u) \circ f(y)$ for all $u \in \{v \in \mathcal{A} : v = v^*, \|v\| = 1, v \text{ is invertible}\}$, $y \in \mathcal{A}$ and $n \geq 0$.

Furthermore, we prove the Hyers-Ulam stability of ∗-homomorphisms in $JC^*$-algebras and $\mathbb{C}$-linear ∗-derivations on $JC^*$-algebras.

Our knowledge concerning the continuity properties of epimorphisms on Banach algebras, Jordan–Banach algebras, and, more generally, non-associative complete normed algebras, is now fairly complete and satisfactory (see [143] and [326]). A basic continuity problem consists in determining algebraic conditions on a Banach algebra $A$ which ensure that derivations on $A$ are continuous. In 1996, Villena [326] proved that derivations on semisimple Jordan–Banach algebras are continuous.

Let $E_1$ and $E_2$ be Banach spaces with the norms $\| \cdot \|$ and $\| \cdot \|$, respectively. Consider $f : E_1 \rightarrow E_2$ to be a mapping such that $f(tx)$ is continuous in $t \in \mathbb{R}$ for each fixed $x \in E_1$. Assume that there exist constants $\theta \geq 0$ and $p \in [0, 1)$ such that

$$\|f(x + y) - f(x) - f(y)\| \leq \theta(\|x\|^p + \|y\|^p)$$

for all $x, y \in E_1$. In [267], Th. M. Rassias showed that there exists a unique $\mathbb{R}$-linear mapping $T : E_1 \rightarrow E_2$ such that

$$\|f(x) - T(x)\| \leq \frac{2\theta}{2 - 2^p}\|x\|^p$$

for all $x \in E_1$. Găvruta [123] generalized the Rassias' result.

In [146], Jun et al. proved the following:

Let $X$ and $Y$ be Banach spaces. Denote by $\varphi : X \times X \rightarrow [0, \infty)$ a function such that

$$\varepsilon(x) := \sum_{j=1}^{\infty} 2^{-j}(\varphi(2^{j-1}x, 0) + \varphi(0, 2^{j-1}x) + \varphi(2^{j-1}x, 2^{j-1}x)) < \infty$$

for all $x \in X$. Suppose that $f, g, h : X \to Y$ are mappings satisfying

$$\|2f(\frac{x+y}{2}) - g(x) - h(y)\| \leq \varphi(x, y)$$

for all $x, y \in X$. Then there exists a unique additive mapping $T : X \to Y$ such that

$$\|2f(\frac{x}{2}) - T(x)\| \leq \|g(0)\| + \|h(0)\| + \varepsilon(x),$$

$$\|g(x) - T(x)\| \leq \|g(0)\| + 2\|h(0)\| + \varphi(x, 0) + \varepsilon(x),$$

$$\|h(x) - T(x)\| \leq 2\|g(0)\| + \|h(0)\| + \varphi(0, x) + \varepsilon(x)$$

for all $x \in X$.

In Theorem 7.2 of Johnson [143], Johnson also investigated almost algebra $*$-homomorphisms between Banach $*$-algebras:

Suppose that $\mathcal{U}$ and $\mathcal{B}$ are Banach $*$-algebras which satisfy the conditions of Theorem 3.1 in [143]. Then, for each positive $\epsilon$ and $K$, there exists a positive $\delta$ such that if $T \in L(\mathcal{U}, \mathcal{B})$ with $\|T\| < K$, $\|T^\vee\| < \delta$ and $\|T(x^*)^* - T(x)\| \leq \delta\|x\|$ for all $x \in \mathcal{U}$, then there exists a $*$-homomorphism $T' : \mathcal{U} \to \mathcal{B}$ with $\|T - T'\| < \epsilon$. Here $L(\mathcal{U}, \mathcal{B})$ is the space of bounded linear mappings from $\mathcal{U}$ into $\mathcal{B}$ and

$$T^\vee(x, y) = T(xy) - T(x)T(y)$$

for all $x, y \in \mathcal{U}$ (see [143] for details).

The original motivation to introduce the class of nonassociative algebras known as Jordan algebras came from quantum mechanics (see [323]).

Let $\mathcal{H}$ be a complex Hilbert space, regarded as the "state space" of a quantum mechanical system. Let $\mathcal{L}(\mathcal{H})$ be the real vector space of all bounded self-adjoint linear operators on $\mathcal{H}$, interpreted as the (bounded) *observables* of the system. In 1932, Jordan observed that $\mathcal{L}(\mathcal{H})$ is a (nonassociative) algebra via the *anticommutator product* $x \circ y := \frac{xy+yx}{2}$. A commutative algebra $X$ with product $x \circ y$ is called a *Jordan algebra* if

$$x^2 \circ (x \circ y) = x \circ (x^2 \circ y).$$

A complex Jordan algebra $\mathcal{C}$ with the product $x \circ y$ and involution $x \mapsto x^*$ is called a *JB$^*$-algebra* if $\mathcal{C}$ carries a Banach space norm $\| \cdot \|$ satisfying

$$\|x \circ y\| \leq \|x\| \cdot \|y\|, \quad \|\{xx^*x\}\| = \|x\|^3.$$

Here $\{xy^*z\} := x \circ (y^* \circ z) - y^* \circ (z \circ x) + z \circ (x \circ y^*)$ denotes the *Jordan triple product* of $x, y, z \in \mathcal{C}$. A unital Jordan $C^*$-subalgebra of a $C^*$-algebra endowed with the anticommutator product is called a *JC$^*$-algebra*.

Throughout this section, let $\mathcal{A}$ be a $JC^*$-algebra with the norm $\|\cdot\|$ and the unit $e$ and $\mathcal{B}$ be a $JC^*$-algebra with the norm $\|\cdot\|$ and the unit $e'$. Let

$$\mathcal{U}(\mathcal{A}) = \{u \in \mathcal{A} : u^*u = uu^* = e\}, \quad \mathcal{A}_{sa} = \{x \in \mathcal{A} : x = x^*\},$$

$$I_1(\mathcal{A}_{sa}) = \{v \in \mathcal{A}_{sa} : \|v\| = 1, v \text{ is invertible}\}.$$

In this section, we prove that every almost unital almost linear mapping $h : \mathcal{A} \to \mathcal{B}$ is a homomorphism when $h(3^n u \circ y) = h(3^n u) \circ h(y)$ for all $u \in \mathcal{U}(\mathcal{A})$, $y \in \mathcal{A}$ and $n \geq 0$ and, for a $JC^*$-algebra $\mathcal{A}$ of real rank zero, every almost unital almost linear continuous mapping $h : \mathcal{A} \to \mathcal{B}$ is a homomorphism when $h(3^n u \circ y) = h(3^n u) \circ h(y)$ for all $u \in I_1(\mathcal{A}_{sa})$, $y \in \mathcal{A}$ and $n \geq 0$.

Furthermore, we prove the Hyers-Ulam stability of ∗-homomorphisms between $JC^*$-algebras and $\mathbb{C}$-linear ∗-derivations on $JC^*$-algebras.

### 3.11.1  ∗-Homomorphisms in $JC^*$-Algebras

Now, we investigate ∗-homomorphisms in $JC^*$-algebras.

**Theorem 3.134.** *Let $f, g, h : \mathcal{A} \to \mathcal{B}$ be mappings satisfying $f(0) = 0$, $g(0) = 0$ and $h(0) = 0$ and let $f(2^n u \circ y) = f(2^n u) \circ f(y)$, $g(2^n u \circ y) = g(2^n u) \circ g(y)$ and $h(2^n u \circ y) = h(2^n u) \circ h(y)$ for $u \in \mathcal{U}(\mathcal{A})$, $y \in \mathcal{A}$ and $n \geq 0$ for which there exists a function $\varphi : \mathcal{A} \setminus \{0\} \times \mathcal{A} \setminus \{0\} \to [0, \infty)$ such that*

$$\tilde{\varphi}(x, y) := \sum_{j=0}^{\infty} 2^{-j} \varphi(2^{j-1}x, 2^{j-1}y) < \infty, \tag{3.188}$$

$$\left\| 2f\left(\frac{\mu x + \mu y}{2}\right) - \mu g(x) - \mu h(y) \right\| \leq \varphi(x, y), \tag{3.189}$$

$$\|f(2^n u^*) - f(2^n u)^*\| \leq \varphi(2^n u, 2^n u) \tag{3.190}$$

*for all $\mu \in \mathbb{T}^1 := \{\lambda \in \mathbb{C} \mid |\lambda| = 1\}$, $u \in \mathcal{U}(\mathcal{A})$, $x, y \in \mathcal{A}$ and $n \geq 0$. Assume that*

$$\lim_{n \to \infty} \frac{f(2^n e)}{2^n} = e'. \tag{3.191}$$

*Then the mappings $f, g, h : \mathcal{A} \to \mathcal{B}$ are ∗-homomorphisms.*

*Proof.* Put $\mu = 1 \in \mathbb{T}^1$. It follows from Corollary 2.5 of [146] that there exists a unique additive mapping $H : \mathcal{A} \to \mathcal{B}$ such that

$$\left\| 2f\left(\frac{x}{2}\right) - H(x) \right\| \leq \varepsilon(x), \quad \|g(x) - H(x)\| \leq \varphi(x, 0) + \varepsilon(x),$$

$$\|h(x) - H(x)\| \leq \varphi(0, x) + \varepsilon(x) \tag{3.192}$$

for all $x \in \mathcal{A} \setminus \{0\}$, where

$$\varepsilon(x) := \sum_{j=1}^{\infty} 2^{-j}(\varphi(2^{j-1}x, 0) + \varphi(0, 2^{j-1}x) + \varphi(2^{j-1}x, 2^{j-1}x)) < \infty$$

for all $x \in \mathcal{A} \setminus \{0\}$. The additive mapping $H : \mathcal{A} \to \mathcal{B}$ is given by

$$H(x) = \lim_{n \to \infty} \frac{1}{2^n} f(2^n x)$$

for all $x \in \mathcal{A}$ and

$$\lim_{n \to \infty} 2^{-n} f(2^n x) = \lim_{n \to \infty} 2^{-n} g(2^n x) = \lim_{n \to \infty} 2^{-n} h(2^n x)$$

for all $x \in \mathcal{A}$. Let $\tilde{f}(x) = 2f(\frac{x}{2})$ for all $x \in \mathcal{A}$. Then we have

$$\lim_{n \to \infty} \frac{1}{2^n} \tilde{f}(2^n x) = \lim_{n \to \infty} \frac{1}{2^n} f(2^n x)$$

for all $x \in \mathcal{A}$. By the assumption, we have

$$\|f(2^n \mu x) - \mu f(2^n x)\|$$
$$= \left\| f(2^n \mu x) - \frac{1}{2}\mu g(2^n x) - \frac{1}{2}\mu h(2^n x) + \frac{1}{2}\mu g(2^n x) + \frac{1}{2}\mu h(2^n x) - \mu f(2^n x) \right\|$$
$$\leq \frac{1}{2}\varphi(2^n x, 2^n x) + \frac{1}{2}|\mu|\varphi(2^n x, 2^n x)$$
$$= \varphi(2^n x, 2^n x)$$

for all $\mu \in \mathbb{T}^1$ and $x \in \mathcal{A} \setminus \{0\}$. Thus $2^{-n}\|f(2^n \mu x) - \mu f(2^n x)\| \to 0$ as $n \to \infty$ for all $\mu \in \mathbb{T}^1$ and $x \in \mathcal{A} \setminus \{0\}$. Hence we have

$$H(\mu x) = \lim_{n \to \infty} \frac{f(2^n \mu x)}{2^n} = \lim_{n \to \infty} \frac{\mu f(2^n x)}{2^n} = \mu H(x) \tag{3.193}$$

for all $\mu \in \mathbb{T}^1$ and $x \in \mathcal{A} \setminus \{0\}$.

Now, let $\lambda \in \mathbb{C}$ ($\lambda \neq 0$) and $M$ be an integer greater than $2|\lambda|$. Then we have $|\frac{\lambda}{M}| < \frac{1}{2} = 1 - \frac{2}{4}$. By Theorem 1 of Kadison and Pedersen [167], there exist four elements $\mu_1, \mu_2, \mu_3, \mu_4 \in \mathbb{T}^1$ such that $4\frac{\lambda}{M} = \mu_1 + \mu_2 + \mu_3 + \mu_4$. Note that $H(x) = H(2 \cdot \frac{1}{2}x) = 2H(\frac{1}{2}x)$ for all $x \in \mathcal{A}$ and so $H(\frac{1}{2}x) = \frac{1}{2}H(x)$ for all $x \in \mathcal{A}$. Thus, by (3.193), we have

$$H(\lambda x) = H\left(\frac{M}{4} \cdot 4\frac{\lambda}{M}x\right) = M \cdot H\left(\frac{1}{4} \cdot 4\frac{\lambda}{M}x\right)$$

$$= \frac{M}{4}H\left(4\frac{\lambda}{M}x\right) = \frac{M}{4}H(\mu_1 x + \mu_2 x + \mu_3 x + \mu_4 x)$$

$$= \frac{M}{4}(H(\mu_1 x) + H(\mu_2 x) + H(\mu_3 x) + H(\mu_4 x))$$

$$= \frac{M}{4}(\mu_1 + \mu_2 + \mu_3 + \mu_4)H(x)$$

$$= \frac{M}{4} \cdot 4\frac{\lambda}{M}H(x) = \lambda H(x)$$

for all $x \in \mathcal{A}$. Hence we have

$$H(\zeta x + \eta y) = H(\zeta x) + H(\eta y) = \zeta H(x) + \eta H(y)$$

for all $\zeta, \eta \in \mathbb{C} \setminus \{0\}$ and $x, y \in \mathcal{A}$ and $H(0x) = 0 = 0H(x)$ for all $x \in \mathcal{A}$. So the unique additive mapping $H : \mathcal{A} \to \mathcal{B}$ is a $\mathbb{C}$-linear mapping. By (3.188) and (3.190), we get

$$H(u^*) = \lim_{n\to\infty} \frac{f(2^n u^*)}{2^n} = \lim_{n\to\infty} \frac{f(2^n u)^*}{2^n} = \left(\lim_{n\to\infty} \frac{f(2^n u)}{2^n}\right)^* = H(u)^*$$

for all $u \in \mathcal{U}(\mathcal{A})$. Since $H$ is $\mathbb{C}$-linear and each $x \in \mathcal{A}$ is a finite linear combination of unitary elements (see [168]), say, $x = \sum_{j=1}^{m} \lambda_j u_j$ for all $\lambda_j \in \mathbb{C}$ and $u_j \in \mathcal{U}(\mathcal{A})$, it follows that

$$H(x^*) = H\left(\sum_{j=1}^{m} \overline{\lambda_j}u_j^*\right) = \sum_{j=1}^{m} \overline{\lambda_j}H(u_j^*) = \sum_{j=1}^{m} \overline{\lambda_j}H(u_j)^*$$

$$= \left(\sum_{j=1}^{m} \lambda_j H(u_j)\right)^* = H\left(\sum_{j=1}^{m} \lambda_j u_j\right)^* = H(x)^*$$

for all $x \in \mathcal{A}$. Since $f(2^n u \circ y) = f(2^n u) \circ f(y)$ for all $u \in \mathcal{U}(\mathcal{A})$, $y \in \mathcal{A}$ and $n \geq 0$, we have

$$H(u \circ y) = \lim_{n\to\infty} \frac{1}{2^n}f(2^n u \circ y) = \lim_{n\to\infty} \frac{1}{2^n}f(2^n u) \circ f(y)$$

$$= H(u) \circ f(y) \tag{3.194}$$

for all $u \in \mathcal{U}(\mathcal{A})$ and $y \in \mathcal{A}$. By the additivity of $H$ and (3.194), we have

$$2^n H(u \circ y) = H(2^n u \circ y) = H(u \circ (2^n y)) = H(u) \circ f(2^n y)$$

for all $u \in \mathcal{U}(\mathcal{A})$ and $y \in \mathcal{A}$ and so

$$H(u \circ y) = \frac{1}{2^n} H(u) \circ f(2^n y) = H(u) \circ \frac{1}{2^n} f(2^n y) \tag{3.195}$$

for all $u \in \mathcal{U}(\mathcal{A})$ and $y \in \mathcal{A}$. Taking $n \to \infty$ in (3.195), we obtain

$$H(u \circ y) = H(u) \circ H(y) \tag{3.196}$$

for all $u \in \mathcal{U}(\mathcal{A})$ and $y \in \mathcal{A}$. Since $H$ is $\mathbb{C}$-linear and each $x \in \mathcal{A}$ is a finite linear combination of unitary elements, i.e., $x = \sum_{j=1}^m \lambda_j u_j$ for all $\lambda_j \in \mathbb{C}$ and $u_j \in \mathcal{U}(\mathcal{A})$, it follows from (3.196) that

$$H(x \circ y) = H\left(\sum_{j=1}^m \lambda_j u_j \circ y\right) = \sum_{j=1}^m \lambda_j H(u_j \circ y)$$

$$= \sum_{j=1}^m \lambda_j H(u_j) \circ H(y) = H\left(\sum_{j=1}^m \lambda_j u_j\right) \circ H(y)$$

$$= H(x) \circ H(y)$$

for all $x, y \in \mathcal{A}$. By (3.194) and (3.196), we have

$$H(e) \circ H(y) = H(e \circ y) = H(e) \circ f(y)$$

for all $y \in \mathcal{A}$. Since $\lim_{n\to\infty} \frac{f(2^n e)}{2^n} = H(e) = e'$, it follows that

$$H(y) = f(y)$$

for all $y \in \mathcal{A}$. Similarly, $H(y) = g(y) = h(y)$ for all $y \in \mathcal{A}$. Therefore, the mapping $f, g, h : \mathcal{A} \to \mathcal{B}$ are $*$-homomorphisms. This completes the proof.    $\square$

**Corollary 3.135.** *Let $f, g, h : \mathcal{A} \to \mathcal{B}$ be mappings satisfying $f(0) = 0$, $g(0) = 0$ and $h(0) = 0$ and let $f(2^n u \circ y) = f(2^n u) \circ f(y)$, $g(2^n u \circ y) = g(2^n u) \circ g(y)$ and $h(2^n u \circ y) = h(2^n u) \circ h(y)$ for all $u \in \mathcal{U}(\mathcal{A})$, $y \in \mathcal{A}$ and $n \geq 0$ for which there exist constants $\theta \geq 0$ and $p \in [0, 1)$ such that*

$$\left\| 2f\left(\frac{\mu x + \mu y}{2}\right) - \mu g(x) - \mu h(y) \right\| \leq \theta(\|x\|^p + \|y\|^p)$$

*and*

$$\|f(2^n u^*) - f(2^n u)^*\| \leq 2^{np+1}\theta$$

*for all $\mu \in \mathbb{T}^1$, $u \in \mathcal{U}(\mathcal{A})$, $n \geq 0$ and $x, y \in \mathcal{A} \backslash \{0\}$. Assume that $\lim_{n\to\infty} \frac{f(2^n e)}{2^n} = e'$. Then the mappings $f, g$ and $h$ are $*$-homomorphisms.*

*Proof.* Define $\varphi(x, y) = \theta(\|x\|^p + \|y\|^p)$ for all $x, y \in \mathcal{A} \backslash \{0\}$ and apply Theorem 3.134. The we have the conclusion.    $\square$

**Theorem 3.136.** *Let $f, g, h : A \to B$ be mappings satisfying $f(0) = 0$, $g(0) = 0$ and $h(0) = 0$ and let $f(2^n u \circ y) = f(2^n u) \circ f(y)$, $g(2^n u \circ y) = g(2^n u) \circ g(y)$ and $h(2^n u \circ y) = h(2^n u) \circ h(y)$ for all $u \in \mathcal{U}(A)$, $y \in A$ and $n \geq 0$ for which there exists a function $\varphi : A \times A \to [0, \infty)$ satisfying (3.188), (3.190) and (3.191) such that*

$$\left\| 2f\left(\frac{\mu x + \mu y}{2}\right) - \mu g(x) - \mu h(y) \right\| \leq \varphi(x, y) \tag{3.197}$$

*for all $x, y \in A \setminus \{0\}$ and $\mu = 1, i$. If $f(tx)$ is continuous in $t \in \mathbb{R}$ for each fixed $x \in A$, then the mappings $f, g, h : A \to B$ are ∗-homomorphisms.*

*Proof.* Put $\mu = 1$ in (3.197). By the same reasoning as the proof of Theorem 3.134, there exists a unique additive mapping $H : A \to B$ satisfying the inequality (3.192). It is easy to show that, the additive mapping $H : A \to B$ is $\mathbb{R}$-linear.

Put $\mu = i$ in (3.197). By the same method as the proof of Theorem 3.134, one can obtain that

$$H(ix) = \lim_{n \to \infty} \frac{f(2^n i x)}{2^n} = \lim_{n \to \infty} \frac{i f(2^n x)}{2^n} = iH(x)$$

for all $x \in A$. For each element $\lambda \in \mathbb{C}$, $\lambda = s + it$ for all $s, t \in \mathbb{R}$. So, we have

$$H(\lambda x) = H(sx + itx) = sH(x) + tH(ix) = sH(x) + itH(x)$$
$$= (s + it)H(x) = \lambda H(x)$$

for all $\lambda \in \mathbb{C}$ and $x \in A$ and

$$H(\zeta x + \eta y) = H(\zeta x) + H(\eta y) = \zeta H(x) + \eta H(y)$$

for all $\zeta, \eta \in \mathbb{C}$ and $x, y \in A$. Hence the additive mapping $H : A \to B$ is $\mathbb{C}$-linear.

The rest of the proof is the same as the proof of Theorem 3.134. This completes the proof. □

From now on, assume that $A$ is a $JC^*$-algebra of real rank zero, where "real rank zero" means that the set of invertible self-adjoint elements is dense in the set of self-adjoint elements (see [54]).

Now, we investigate continuous ∗-homomorphisms between $JC^*$-algebras.

**Theorem 3.137.** *Let $f, g, h : A \to B$ be continuous mappings satisfying $f(0) = 0$, $g(0) = 0$ and $h(0) = 0$ and let $f(2^n u \circ y) = f(2^n u) \circ f(y)$, $g(2^n u \circ y) = g(2^n u) \circ g(y)$ and $h(2^n u \circ y) = h(2^n u) \circ h(y)$ for all $u \in I_1(A_{sa})$, $y \in A$ and all $n \geq 0$ for which there exists a function $\varphi : A \times A \to [0, \infty)$ satisfying (3.188), (3.189), (3.190) and (3.191). Then the mappings $f, g, h : A \to B$ are ∗-homomorphisms.*

*Proof.* By the same reasoning as the proof of Theorem 3.134, there exists a unique $\mathbb{C}$-linear involution mapping $H : \mathcal{A} \to \mathcal{B}$ satisfying the inequality (3.192). Since $f(2^n u \circ y) = f(2^n u) \circ f(y)$ for all $u \in I_1(\mathcal{A}_{sa})$, $y \in \mathcal{A}$ and $n \geq 0$, we have

$$H(u \circ y) = \lim_{n \to \infty} \frac{1}{2^n} f(2^n u \circ y) = \lim_{n \to \infty} \frac{1}{2^n} f(2^n u) \circ f(y)$$
$$= H(u) \circ f(y) \tag{3.198}$$

for all $u \in I_1(\mathcal{A}_{sa})$ and $y \in \mathcal{A}$. By the additivity of $H$ and (3.198), we have

$$2^n H(u \circ y) = H(2^n u \circ y) = H(u \circ (2^n y)) = H(u) \circ f(2^n y)$$

for all $u \in I_1(\mathcal{A}_{sa})$ and $y \in \mathcal{A}$. Hence it follows that

$$H(u \circ y) = \frac{1}{2^n} H(u) \circ f(2^n y) = H(u) \circ \frac{1}{2^n} f(2^n y) \tag{3.199}$$

for all $u \in I_1(\mathcal{A}_{sa})$ and $y \in \mathcal{A}$. Taking $n \to \infty$ in (3.199), we obtain

$$H(u \circ y) = H(u) \circ H(y) \tag{3.200}$$

for all $u \in I_1(\mathcal{A}_{sa})$ and $y \in \mathcal{A}$. By (3.198) and (3.200), we have

$$H(e) \circ H(y) = H(e \circ y) = H(e) \circ f(y)$$

for all $y \in \mathcal{A}$. Since $\lim_{n \to \infty} \frac{f(2^n e)}{2^n} = H(e) = e'$, we have

$$H(y) = f(y)$$

for all $y \in \mathcal{A}$. Similarly, $H(y) = g(y) = h(y)$ for all $y \in \mathcal{A}$. So $H : \mathcal{A} \to \mathcal{B}$ is continuous. But, by the assumption that $\mathcal{A}$ has real rank zero, it is easy to show that $I_1(\mathcal{A}_{sa})$ is dense in $\{x \in \mathcal{A}_{sa} : \|x\| = 1\}$. So, for each $w \in \{z \in \mathcal{A}_{sa} : \|z\| = 1\}$, there exists a sequence $\{\kappa_j\}$ such that $\kappa_j \to w$ as $j \to \infty$ and $\kappa_j \in I_1(\mathcal{A}_{sa})$. Since $H : \mathcal{A} \to \mathcal{B}$ is continuous, it follows from (3.200) that

$$H(w \circ y) = H\left( \lim_{j \to \infty} \kappa_j \circ y \right) = \lim_{j \to \infty} H(\kappa_j \circ y)$$
$$= \lim_{j \to \infty} H(\kappa_j) \circ H(y) = H\left( \lim_{j \to \infty} \kappa_j \right) \circ H(y)$$
$$= H(w) \circ H(y) \tag{3.201}$$

for all $w \in \{z \in \mathcal{A}_{sa} : \|z\| = 1\}$ and $y \in \mathcal{A}$.

For each $x \in \mathcal{A}$, let $x = \frac{x+x^*}{2} + i\frac{x-x^*}{2i}$, where $x_1 := \frac{x+x^*}{2}$ and $x_2 := \frac{x-x^*}{2i}$ are self-adjoint.

First, consider the case that $x_1 \neq 0, x_2 \neq 0$. Since $H : \mathcal{A} \to \mathcal{B}$ is $\mathbb{C}$-linear, it follows from (3.201) that

$$
\begin{aligned}
H(x \circ y) &= H(x_1 \circ y + ix_2 \circ y) = H\left( \|x_1\| \frac{x_1}{\|x_1\|} \circ y + i\|x_2\| \frac{x_2}{\|x_2\|} \circ y \right) \\
&= \|x_1\| H\left( \frac{x_1}{\|x_1\|} \circ y \right) + i\|x_2\| H\left( \frac{x_2}{\|x_2\|} \circ y \right) \\
&= \|x_1\| H\left( \frac{x_1}{\|x_1\|} \right) \circ H(y) + i\|x_2\| H\left( \frac{x_2}{\|x_2\|} \right) \circ H(y) \\
&= \left\{ H\left( \|x_1\| \frac{x_1}{\|x_1\|} \right) + iH\left( \|x_2\| \frac{x_2}{\|x_2\|} \right) \right\} \circ H(y) \\
&= H(x_1 + ix_2) \circ H(y) = H(x) \circ H(y)
\end{aligned}
$$

for all $y \in \mathcal{A}$.

Next, consider the case that $x_1 \neq 0$ and $x_2 = 0$. Since $H : \mathcal{A} \to \mathcal{B}$ is $\mathbb{C}$-linear, it follows from (3.201) that

$$
\begin{aligned}
H(x \circ y) &= H(x_1 \circ y) = H\left( \|x_1\| \frac{x_1}{\|x_1\|} \circ y \right) = \|x_1\| H\left( \frac{x_1}{\|x_1\|} \circ y \right) \\
&= \|x_1\| H\left( \frac{x_1}{\|x_1\|} \right) \circ H(y) = H\left( \|x_1\| \frac{x_1}{\|x_1\|} \right) \circ H(y) \\
&= H(x_1) \circ H(y) = H(x) \circ H(y)
\end{aligned}
$$

for all $y \in \mathcal{A}$.

Finally, consider the case that $x_1 = 0, x_2 \neq 0$. Since $H : \mathcal{A} \to \mathcal{B}$ is $\mathbb{C}$-linear, it follows from (3.201) that

$$
\begin{aligned}
H(x \circ y) &= H(ix_2 \circ y) = H\left( i\|x_2\| \frac{x_2}{\|x_2\|} \circ y \right) = i\|x_2\| H\left( \frac{x_2}{\|x_2\|} \circ y \right) \\
&= i\|x_2\| H\left( \frac{x_2}{\|x_2\|} \right) \circ H(y) = H\left( i\|x_2\| \frac{x_2}{\|x_2\|} \right) \circ H(y) \\
&= H(ix_2) \circ H(y) = H(x) \circ H(y)
\end{aligned}
$$

for all $y \in \mathcal{A}$. Hence we have

$$
H(x \circ y) = H(x) \circ H(y)
$$

for all $x, y \in \mathcal{A}$. Therefore, the mappings $f, g, h : \mathcal{A} \to \mathcal{B}$ are *-homomorphisms. This completes the proof. $\qquad\square$

**Corollary 3.138.** *Let $f, g, h : \mathcal{A} \to \mathcal{B}$ be continuous mappings satisfying $f(0) = 0$, $g(0) = 0$ and $h(0) = 0$ and let $f(2^n u \circ y) = f(2^n u) \circ f(y)$, $g(2^n u \circ y) = g(2^n u) \circ g(y)$ and $h(2^n u \circ y) = h(2^n u) \circ h(y)$ for all $u \in I_1(\mathcal{A}_{sa})$, $y \in \mathcal{A}$ and $n \geq 0$ for which there* *exist constants $\theta \geq 0$ and $p \in [0, 1)$ such that*

$$\left\| 2f\left(\frac{\mu x + \mu y}{2}\right) - \mu g(x) - \mu h(y) \right\| \leq \theta(\|x\|^p + \|y\|^p)$$

*and*

$$\|f(2^n u^*) - f(2^n u)^*\| \dagger \, 2^{np+1}\theta$$

*for all $\mu \in \mathbb{T}^1$, $u \in I_1(\mathcal{A}_{sa})$, $x, y \in \mathcal{A} \setminus \{0\}$ and $n \geq 0$. If $\lim_{n \to \infty} \frac{f(2^n e)}{2^n} = e'$, then the mappings $f, g, h : \mathcal{A} \to \mathcal{B}$ are $*$-homomorphisms.*

*Proof.* Define

$$\varphi(x, y) = \theta(\|x\|^p + \|y\|^p)$$

for all $x, y \in \mathcal{A} \setminus \{0\}$ and then apply Theorem 3.137. Then we get the desired result.                                                                                                $\square$

*Remark 3.139.* Let $f, g, h : \mathcal{A} \to \mathcal{B}$ be continuous mappings satisfying $f(0) = 0$, $g(0) = 0$ and $h(0) = 0$ and let $f(2^n u \circ y) = f(2^n u) \circ f(y)$, $g(2^n u \circ y) = g(2^n u) \circ g(y)$ and $h(2^n u \circ y) = h(2^n u) \circ h(y)$ for all $u \in I_1(\mathcal{A}_{sa})$, $y \in \mathcal{A}$ and $n \geq 0$ for which there exists a function $\varphi : \mathcal{A} \times \mathcal{A} \to [0, \infty)$ satisfying (3.188), (3.190), (3.191) and (3.197). Then the mappings $f, g, h : \mathcal{A} \to \mathcal{B}$ are $*$-homomorphisms.

Note that there exists a unique $\mathbb{C}$-linear mapping $H : \mathcal{A} \to \mathcal{B}$ satisfying the system of the inequalities (3.192).

### 3.11.2  Stability of *-Homomorphisms in JC*-Algebras

Now, we prove the Hyers-Ulam stability of $*$-homomorphisms in $JC^*$-algebras.

**Theorem 3.140.** *Let $f, g, h : \mathcal{A} \to \mathcal{B}$ be mappings with $f(0) = 0$, $g(0) = 0$ and $h(0) = 0$ for which there exists a function $\varphi : \mathcal{A}^4 \to [0, \infty)$ such that*

$$\tilde{\varphi}(x, y, z, w) = \sum_{j=0}^{\infty} 2^{-j} \varphi(2^j x, 2^j y, 2^j z, 2^j w) < \infty, \tag{3.202}$$

$$\left\| 2f\left(\frac{\mu x + \mu y + z \circ w}{2}\right) - \mu g(x) - \mu h(y) - f(z) \circ f(w) \right\|$$
$$\leq \varphi(x, y, z, w), \tag{3.203}$$

$$\|f(2^n u^*) - f(2^n u)^*\| \leq \varphi(2^n u, 2^n u, 0, 0) \tag{3.204}$$

*for all $\mu \in \mathbb{T}^1$, $u \in \mathcal{U}(\mathcal{A})$, $x, y, z, w \in \mathcal{A} \setminus \{0\}$ and $n \geq 0$. Then there exists a unique ∗-homomorphism $H : \mathcal{A} \rightarrow \mathcal{B}$ such that*

$$\left\| 2f\left(\tfrac{x}{2}\right) - H(x) \right\| \leq \varepsilon(x), \, \|g(x) - H(x)\| \leq \varphi(x, 0, 0, 0) + \varepsilon(x),$$

$$\|h(x) - H(x)\| \leq \varphi(0, x, 0, 0) + \varepsilon(x) \tag{3.205}$$

*for all $x \in \mathcal{A} \setminus \{0\}$, where*

$$\varepsilon(x) := \sum_{j=1}^{\infty} 2^{-j}(\varphi(2^{j-1}x, 0, 0, 0) + \varphi(0, 2^{j-1}x, 0, 0) + \varphi(2^{j-1}x, 2^{j-1}x, 0, 0))$$

$$< \infty.$$

*Proof.* Put $z = w = 0$ and $\mu = 1 \in \mathbb{T}^1$ in (3.203). By the same reasoning as in the proof of Theorem 3.134, there exists a unique $\mathbb{C}$-linear involutive mapping $H : \mathcal{A} \rightarrow \mathcal{B}$ satisfying the inequality (3.205). The $\mathbb{C}$-linear mapping $H : \mathcal{A} \rightarrow \mathcal{B}$ is given by

$$H(x) = \lim_{n \to \infty} \frac{1}{2^n} f(2^n x) \tag{3.206}$$

for all $x \in \mathcal{A}$. It follows from (3.206) that

$$H(x) = \lim_{n \to \infty} \frac{f(2^{2n}x)}{2^{2n}} \tag{3.207}$$

for all $x \in \mathcal{A}$. Let $x = y = 0$ in (3.203). Then we get

$$\left\| 2f\left(\frac{z \circ w}{2}\right) - f(z) \circ f(w) \right\| \leq \varphi(0, 0, z, w)$$

for all $z, w \in \mathcal{A}$. Since $\frac{1}{2^{2n}}\varphi(0, 0, 2^n z, 2^n w) \leq \frac{1}{2^n}\varphi(0, 0, 2^n z, 2^n w)$, it follows that

$$\frac{1}{2^{2n}} \left\| 2f\left(\frac{1}{2}2^n z \circ 2^n w\right) - f(2^n z) \circ f(2^n w) \right\| \tag{3.208}$$

$$\leq \frac{1}{2^{2n}}\varphi(0, 0, 2^n z, 2^n w) \tag{3.209}$$

$$\leq \frac{1}{2^n}\varphi(0, 0, 2^n z, 2^n w)$$

for all $z, w \in \mathcal{A}$. By (3.206), (3.207) and (3.208), we have

$$2H\left(\frac{z \circ w}{2}\right) = \lim_{n\to\infty} \frac{2f(\frac{1}{2}2^{2n}z \circ w)}{2^{2n}} = \lim_{n\to\infty} \frac{2f(\frac{1}{2}2^n z \circ 2^n w)}{2^n \cdot 2^n}$$

$$= \lim_{n\to\infty} \left(\frac{f(2^n z)}{2^n} \circ \frac{f(2^n w)}{2^n}\right) = \lim_{n\to\infty} \frac{f(2^n z)}{2^n} \circ \lim_{n\to\infty} \frac{f(2^n w)}{2^n}$$

$$= H(z) \circ H(w)$$

for all $z, w \in \mathcal{A}$. But, since $H$ is $\mathbb{C}$-linear,

$$H(z \circ w) = 2H\left(\frac{z \circ w}{2}\right) = H(z) \circ H(w)$$

for all $z, w \in \mathcal{A}$. Hence the $\mathbb{C}$-linear mapping $H : \mathcal{A} \to \mathcal{B}$ is a $*$-homomorphism satisfying the inequality (3.205). This completes the proof.                                                         $\square$

**Corollary 3.141.** *Let $f, g, h : \mathcal{A} \to \mathcal{B}$ be a mapping with $f(0) = 0$, $g(0) = 0$ and $h(0) = 0$ for which there exist constants $\theta \geq 0$ and $p \in [0, 1)$ such that*

$$\left\| 2f\left(\frac{\mu x + \mu y + z \circ w}{2}\right) - \mu g(x) - \mu h(y) - f(z) \circ f(w) \right\|$$
$$\leq \theta(\|x\|^p + \|y\|^p + \|z\|^p + \|w\|^p)$$

*and*

$$\|f(2^n u^*) - f(2^n u)^*\| \leq 2^{np+1}\theta$$

*for all $\mu \in \mathbb{T}^1$, $u \in \mathcal{U}(\mathcal{A})$, $x, y, z, w \in \mathcal{A} \setminus \{0\}$ and $n \geq 0$. Then there exists a unique $*$-homomorphism $H : \mathcal{A} \to \mathcal{B}$ such that*

$$\left\| 2f\left(\frac{x}{2}\right) - H(x) \right\| \leq \frac{1}{2 - 2^p}\theta\|x\|^p,$$

$$\|g(x) - H(x)\| \leq \frac{3 - 2^p}{2 - 2^p}\theta\|x\|^p,$$

$$\|h(x) - H(x)\| \leq \frac{3 - 2^p}{2 - 2^p}\theta\|x\|^p$$

*for all $x \in \mathcal{A} \setminus \{0\}$.*

*Proof.* Define

$$\varphi(x, y, z, w) = \theta(\|x\|^p + \|y\|^p + \|z\|^p + \|w\|^p)$$

and apply Theorem 3.140. Then we get the desired result.                                                         $\square$

*Remark 3.142.* Let $f, g, h : \mathcal{A} \to \mathcal{B}$ be a mapping with $f(0) = 0$, $g(0) = 0$ and $h(0) = 0$ for which there exists a function $\varphi : \mathcal{A}^4 \to [0, \infty)$ satisfying (3.202) and (3.203) such that

$$\left\| 2f\left(\frac{\mu x + \mu y + z \circ w}{2}\right) - \mu g(x) - \mu h(y) - f(z) \circ f(w) \right\| \leq \varphi(x, y, z, w)$$

for all $x, y, z, w \in \mathcal{A} \setminus \{0\}$ and $\mu = 1, i$. If $f(tx)$ is continuous in $t \in \mathbb{R}$ for each fixed $x \in \mathcal{A}$, then there exists a unique ∗-homomorphism $H : \mathcal{A} \to \mathcal{B}$ satisfying the inequality (3.205).

Note that there exists a unique $\mathbb{C}$-linear mapping $H : \mathcal{A} \to \mathcal{B}$ satisfying the inequality (3.205).

# Chapter 4
# Stability of Functional Inequalities in Banach Algebras

In this chapter, we study functional inequalities in Banach algebras via the direct and fixed point methods.

In Sect. 4.1, we consider the following additive functional inequality:

$$\|f(2x) + f(2y) + 2f(z)\| \leq \|2f(x + y + z)\|.$$

Next, we approximate the homomorphisms in proper $CQ^*$-algebras and derivations on proper $CQ^*$-algebras associated with the above additive functional inequality by direct method.

In Sect. 4.2, we consider the functional inequality

$$\|f(x) + f(y) + f(z) + f(w)\| \leq \|f(x) + f(y + z + w)\|.$$

Next, we prove that, if $f : A \to B$ is a multiplicative mapping such that

$$\|f(x) + f(y) + f(z) + \mu f(w)\| \leq \|f(x) + f(y + z + \mu w)\|$$

for all $x, y, z, w \in A$ and all $\mu \in \mathbb{T} := \{\lambda \in \mathbb{C} : |\lambda| = 1\}$. Then the mapping $f : A \to B$ is a $C^*$-algebra homomorphism. Moreover, by using fixed point method, we prove the Hyers-Ulam stability of the functional following inequality:

$$\|f(x) + f(y) + f(z) + f(w)\|$$
$$\leq \|f(x) + f(y + z + w)\|$$
$$+ \theta(\|x\|^p + \|y\|^p + \|z\|^p + \|w\|^p + \|x\|^{\frac{p}{4}} \cdot \|y\|^{\frac{p}{4}} \cdot \|z\|^{\frac{p}{4}} \cdot \|w\|^{\frac{p}{4}})$$

in real Banach spaces.

© Springer International Publishing Switzerland 2015
Y.J. Cho et al., *Stability of Functional Equations in Banach Algebras*, DOI 10.1007/978-3-319-18708-2_4

## 4.1   Stability of Additive Functional Inequalities in Banach Algebras

In this section, by the direct method, we prove the Hyers-Ulam stability of the following additive functional inequality:

$$\|f(2x) + f(2y) + 2f(z)\| \le \|2f(x + y + z)\|. \tag{4.1}$$

Then we consider homomorphisms in proper $CQ^*$-algebras and derivations on proper $CQ^*$-algebras associated with the additive functional inequality (4.1) (see [181, 182, 186, 254]).

### 4.1.1   Stability of $\mathbb{C}$-Linear Mappings in Banach Spaces

Now, we investigate the Hyers-Ulam stability of $\mathbb{C}$-linear mappings in Banach spaces associated with the additive functional inequality.

Now, we assume that $X, Y$ are Banach spaces.

**Lemma 4.1.** *Let $f : X \to Y$ be a mapping satisfying the following:*

$$\|f(2x) + f(2y) + 2f(z)\|_Y \le \|2f(x + y + z)\|_Y \tag{4.2}$$

*for all $x, y, z \in X$. Then $f$ is additive.*

*Proof.* Letting $x = y = z = 0$ in (4.2), we get $\|4f(0)\|_Y \le \|2f(0)\|_Y$ and so $f(0) = 0$. Letting $z = 0$ and replacing $y$ by $-x$ in (4.2), we get

$$\|f(2x) + f(-2x)\|_Y \le \|2f(0)\|_Y = 0$$

for all $x \in X$. Hence $f(-2x) = -f(2x)$ and so $f(-x) = -f(x)$ for all $x \in X$.

Letting $y = 0$ and replacing $z$ by $-x$ in (4.2), we get

$$\|f(2x) + 2f(-x)\|_Y \le \|2f(0)\|_Y = 0$$

for all $x \in X$. Thus we have $f(2x) = 2f(x)$ for all $x \in X$. Letting replacing $z$ by $-x - y$ in (4.2), we get

$$\|f(2x) + f(2y) - 2f(x + y)\|_Y \le \|2f(0)\|_Y = 0$$

for all $x, y \in X$. Thus we have

$$f(2x + 2y) = f(2x) + f(2y)$$

for all $x \in X$ and so

$$f(x + y) = f(x) + f(y)$$

for all $x, y \in X$. This completes the proof.                                          □

**Theorem 4.2.** *Let $f : X \to Y$ be a mapping with $f(0) = 0$. If there exists a function $\varphi : X^3 \to [0, \infty)$ satisfying the following;*

$$\|f(2x) + f(2y) + 2f(z)\|_Y \le \|2f(x + y + z)\|_Y + \varphi(x, y, z) \tag{4.3}$$

*and*

$$\tilde{\varphi}(x, y, z) := \sum_{j=0}^{\infty} \frac{1}{2^j} \varphi\left((-2)^j x, (-2)^j y, (-2)^j z\right) < \infty \tag{4.4}$$

*for all $x, y, z \in X$, then there exists a unique additive mapping $L : X \to Y$ such that*

$$\|f(x) - L(x)\|_Y \le \frac{1}{2} \tilde{\varphi}(0, -x, x) \tag{4.5}$$

*for all $x \in X$.*

*Proof.* Replacing $x, y, z$ by $0, -(-2)^n x, (-2)^n x$, respectively, and dividing by $2^{n+1}$ in (4.3). Since $f(0) = 0$, we get

$$\left\| \frac{f((-2)^{n+1} x)}{(-2)^{n+1}} - \frac{f((-2)^n x)}{(-2)^n} \right\|_Y \le \frac{1}{2^{n+1}} \varphi(0, -(-2)^n x, (-2)^n x)$$

for all $x \in X$. From the above inequality, we have

$$\left\| \frac{f((-2)^n x)}{(-2)^n} - \frac{f((-2)^q x)}{(-2)^q} \right\|_Y \le \sum_{j=q}^{n-1} \left\| \frac{f((-2)^{j+1} x)}{(-2)^{j+1}} - \frac{f((-2)^j x)}{(-2)^j} \right\|_Y$$

$$\le \sum_{j=q}^{n-1} \frac{1}{2^{j+1}} \varphi(0, -(-2)^j x, (-2)^j x)$$

for all $x \in X$ and $q, n \ge 1$ with $q < n$. From (4.4), the sequence $\{\frac{f((-2)^n x)}{(-2)^n}\}$ is a Cauchy sequence for all $x \in X$. Since $Y$ is complete, the sequence $\{\frac{f((-2)^n x)}{(-2)^n}\}$ converges for all $x \in X$ and so we can define a mapping $L : X \to Y$ by

$$L(x) := \lim_{n \to \infty} \frac{f((-2)^n x)}{(-2)^n}$$

for all $x \in X$.

In order to prove that $L$ satisfies (4.5), if we put $q = 0$ and let $n \to \infty$ in the above inequality, then we obtain

$$\|f(x) - L(x)\|_Y \le \sum_{j=0}^{\infty} \frac{1}{2^{j+1}} \varphi(0, -(-2)^j x, (-2)^j x) = \frac{1}{2} \tilde{\varphi}(0, -x, x)$$

for all $x \in X$. Replacing $x, y, z$ by $(-2)^n x, (-2)^n y, (-2)^n z$, respectively, and dividing by $2^n$ in (4.3), we get

$$\left\| \frac{f((-2)^n 2x)}{(-2)^n} + \frac{f((-2)^n 2y)}{(-2)^n} + \frac{2f((-2)^n z)}{(-2)^n} \right\|_Y$$

$$\le \left\| 2\frac{f((-2)^n (x + y + z))}{(-2)^n} \right\|_Y + \frac{1}{2^n} \varphi((-2)^n x, (-2)^n y, (-2)^n z)$$

for all $x, y, z \in X$. Since (4.4) gives

$$\lim_{n \to \infty} \frac{1}{2^n} \varphi((-2)^n x, (-2)^n y, (-2)^n z) = 0$$

for all $x, y, z \in X$, if we let $n \to \infty$ in the above inequality, then we get

$$\|L(2x) + L(2y) + 2L(z)\|_Y \le \|2L(x + y + z)\|_Y,$$

and so $L$ is additive by Lemma 4.1.

Now, to prove the uniqueness of $L$, let $L' : X \to Y$ be another additive mapping satisfying (4.5). Since $L$ and $L'$ are additive, we have

$$\|L(x) - L'(x)\|_Y = \frac{1}{2^n} \left\| L(2^n x) - L'(2^n x) \right\|_Y$$

$$\le \frac{1}{2^n} (\|L(2^n x) - f(2^n x)\|_Y + \|L'(2^n x) - f(2^n x)\|_Y)$$

$$\le \frac{1}{2^n} \cdot 2\tilde{\varphi}(0, -2^n x, 2^n x)$$

$$= 2 \sum_{j=0}^{\infty} \frac{1}{2^{n+j}} \varphi(0, (-2)^{j+n} x, (-2)^{j+n} x),$$

which goes to zero as $n \to \infty$ for all $x \in X$ by (4.4). Consequently, $L$ is a unique additive mapping satisfying (4.5). This completes the proof. $\square$

**Corollary 4.3.** *Let* $f : X \to Y$ *be a mapping. If there exists a function* $\varphi : X^3 \to [0, \infty)$ *satisfying* (4.3) *and*

$$\tilde{\varphi}(x, y, z) := \sum_{i=1}^{\infty} 2^i \varphi\left(\frac{x}{(-2)^i}, \frac{y}{(-2)^i}, \frac{z}{(-2)^i}\right) < \infty \qquad (4.6)$$

*for all* $x, y, z \in X$, *then there exists a unique additive mapping* $L : X \to Y$ *such that*

$$\|f(x) - L(x)\|_Y \leq \frac{1}{2}\tilde{\varphi}(0, -x, x) \qquad (4.7)$$

*for all* $x \in X$.

*Proof.* Since $\tilde{\varphi}(0, 0, 0) < \infty$ in (4.6), we have $\varphi(0, 0, 0) = 0$ and so $f(0) = 0$. Replacing $x, y, z$ by $0, -\frac{x}{(-2)^n}, \frac{x}{(-2)^n}$, respectively, and multiplying by $2^{n-1}$ in (4.3), we get

$$\left\|(-2)^{n-1}f\left(\frac{x}{(-2)^{n-1}}\right) - (-2)^n f\left(\frac{x}{(-2)^n}\right)\right\|_Y$$

$$\leq 2^{n-1}\varphi\left(0, -\frac{x}{(-2)^n}, \frac{x}{(-2)^n}\right)$$

for all $x \in X$. From the above inequality, we have

$$\left\|(-2)^n f\left(\frac{x}{(-2)^n}\right) - (-2)^q f\left(\frac{x}{(-2)^q}\right)\right\|_Y$$

$$\leq \sum_{i=q+1}^{n} \left\|(-2)^i f\left(\frac{x}{(-2)^i}\right) - (-2)^{i-1} f\left(\frac{x}{(-2)^{i-1}}\right)\right\|_Y$$

$$\leq \sum_{i=q+1}^{n} 2^{i-1}\varphi\left(0, -\frac{x}{(-2)^i}, \frac{x}{(-2)^i}\right)$$

for all $x \in X$ and $q, n \geq 1$ with $q < n$. From (4.6), the sequence $\left\{(-2)^n f\left(\frac{x}{(-2)^n}\right)\right\}$ is a Cauchy sequence for all $x \in X$. Since $Y$ is complete, the sequence $\left\{(-2)^n f\left(\frac{x}{(-2)^n}\right)\right\}$ converges for all $x \in X$ and so we can define a mapping $L : X \to Y$ by

$$L(x) := \lim_{n \to \infty} (-2)^n f\left(\frac{x}{(-2)^n}\right)$$

for all $x \in X$.

In order to prove that $L$ satisfies (4.7), if we put $q = 0$ and let $n \to \infty$ in the above inequality, then we obtain

$$\|f(x) - L(x)\|_Y \leq \sum_{i=1}^{\infty} 2^{i-1}\varphi\left(0, -\frac{x}{(-2)^i}, \frac{x}{(-2)^i}\right) = \frac{1}{2}\tilde{\varphi}(0, -x, x)$$

for all $x \in X$. Replacing $x, y, z$ by $\frac{x}{(-2)^n}, \frac{y}{(-2)^n}, \frac{z}{(-2)^n}$, respectively, and multiplying by $2^n$ in (4.3), we get

$$\left\| (-2)^n f\left(\frac{2x}{(-2)^n}\right) + (-2)^n f\left(\frac{2y}{(-2)^n}\right) + 2 \cdot (-2)^n f\left(\frac{z}{(-2)^n}\right) \right\|_Y$$
$$\leq \left\| 2 \cdot (-2)^n f\left(\frac{x+y+z}{(-2)^n}\right) \right\|_Y + 2^n \varphi\left(\frac{x}{(-2)^n}, \frac{y}{(-2)^n}, \frac{z}{(-2)^n}\right)$$

for all $x, y, z \in X$. Since (4.6) gives

$$\lim_{n\to\infty} 2^n \varphi\left(\frac{x}{(-2)^n}, \frac{y}{(-2)^n}, \frac{z}{(-2)^n}\right) = 0$$

for all $x, y, z \in X$, if we let $n \to \infty$ in the above inequality, then we get

$$\|L(2x) + L(2y) + 2L(z)\|_Y \leq \|2L(x+y+z)\|_Y$$

and so $L$ is additive by Lemma 4.1.

The rest of the proof is the same as in the corresponding part of the proof of Theorem 4.2. This completes the proof. □

**Lemma 4.4.** *Let $f : X \to Y$ be a mapping satisfying*

$$\|f(2x) + \mu f(2y) + 2f(z)\|_Y \leq \|2f(x + \mu y + z)\|_Y \tag{4.8}$$

*for all $\mu \in \mathbb{T}^1$ and all $x, y, z \in X$. Then $f$ is $\mathbb{C}$-linear.*

*Proof.* If we put $\mu = 1$ in (4.8), then $f$ is additive by Lemma 4.1. Replacing $x, y, z$ by $\mu x, -x, 0$ in (4.8), respectively, we get $f(2\mu x) + \mu f(-2x) = 0$ and so $f(\mu x) = \mu f(x)$ for all $\mu \in \mathbb{T}^1$ and $x \in X$. Thus we have

$$f(\mu x + \bar{\mu} x) = f(\mu x) + f(\bar{\mu} x) = \mu f(x) + \bar{\mu} f(x)$$

for all $\mu \in \mathbb{T}^1$ and $x \in X$ and so $f(tx) = tf(x)$ for any real number $t$ with $|t| \leq 1$ and all $x \in X$.

On the other hand, since $f(2x) = 2f(x)$, we get $f(2^n x) = 2^n f(x)$ for all $n \geq 1$. So, for any real number $t$, there exists an integer $n \geq 1$ with $|t| \leq 2^n$. Thus we have

$$f(tx) = f\left(2^n \cdot \frac{t}{2^n} x\right) = 2^n f\left(\frac{t}{2^n} x\right) = 2^n \cdot \frac{t}{2^n} f(x) = tf(x).$$

Now, we consider any $\alpha \in \mathbb{C}$ with $\alpha = t + si$ for some real numbers $t, s$. Since $f(ix) = if(x)$ all $x \in X$, we have

$$f(\alpha x) = f(tx) + f(six) = tf(x) + sf(ix) = tf(x) + sif(x) = \alpha f(x)$$

and so $f$ is $\mathbb{C}$-linear. This completes the proof. □

**Theorem 4.5.** *Let* $f : X \to Y$ *be a mapping with* $f(0) = 0$. *If there exists a function* $\varphi : X^3 \to [0, \infty)$ *satisfying* (4.4) *and*

$$\|f(2x) + \mu f(2y) + 2f(z)\|_Y \leq \|2f(x + \mu y + z)\|_Y + \varphi(x, y, z) \tag{4.9}$$

*for all* $\mu \in \mathbb{T}^1$ *and* $x, y, z \in X$, *then there exists a unique* $\mathbb{C}$-*linear mapping* $L : X \to Y$ *satisfying* (4.5).

*Proof.* If we put $\mu = 1$ in (4.9), then, by Theorem 4.2, there exists a unique additive mapping $L : X \to Y$ defined by

$$L(x) := \lim_{n \to \infty} \frac{f((-2)^n x)}{(-2)^n}$$

for all $x \in X$, which satisfies (4.5). By the similar method to the corresponding part in the proof of Theorem 4.2, $L$ satisfies

$$\|L(2x) + \mu L(2y) + 2L(z)\|_Y \leq \|2L(x + \mu y + z)\|_Y$$

for all $\mu \in \mathbb{T}^1$ and $x, y, z \in X$. Thus Lemma 4.4 gives that $L$ is $\mathbb{C}$-linear. This completes the proof.                                                                  □

**Corollary 4.6.** *Let* $f : X \to Y$ *be a mapping. If there exists a function* $\varphi : X^3 \to [0, \infty)$ *satisfying* (4.6) *and* (4.9), *then there exists a unique* $\mathbb{C}$-*linear mapping* $L : X \to Y$ *satisfying* (4.7).

*Proof.* If we put $\mu = 1$ in (4.9), then, by Corollary 4.3, there exists a unique additive mapping $L : X \to Y$ defined by

$$L(x) := \lim_{n \to \infty} (-2)^n f\left(\frac{x}{(-2)^n}\right)$$

for all $x \in X$ which satisfies (4.7).

The rest of the proof is the same as in the corresponding part of the proof of Theorem 4.5.                                                                                    □

## 4.1.2   Stability of Homomorphisms in Proper CQ*-Algebras

Now, we investigate the Hyers-Ulam stability of isomorphisms in proper $CQ^*$-algebras associated with the additive functional inequality.

**Theorem 4.7.** *Let* $f : A \to B$ *be a mapping with* $f(0) = 0$. *If there exists a function* $\varphi : A^3 \to [0, \infty)$ *satisfying*

$$\|f(2x) + \mu f(2y) + 2f(z)\|_B \leq \|2f(x + \mu y + z)\|_B + \varphi(x, y, z) \tag{4.10}$$

*and*

$$\tilde{\varphi}(x, y, z) := \sum_{j=0}^{\infty} \frac{1}{2^j} \varphi\left((-2)^j x, (-2)^j y, (-2)^j z\right) < \infty, \tag{4.11}$$

*for all $\mu \in \mathbb{T}^1$ and $x, y, z \in A$. If, in addition, there exists a function $\phi : A^2 \to [0, \infty)$ satisfying*

$$\|f(xy) - f(x)f(y)\|_B \leq \phi(x, y) \tag{4.12}$$

*and*

$$\lim_{n \to \infty} \frac{1}{4^n} \phi((-2)^n x, (-2)^n y) = 0 \tag{4.13}$$

*for all $x, y \in A$ whenever the multiplication is defined, then there exists a unique proper $CQ^*$-algebra homomorphism $h : A \to B$ such that*

$$\|f(x) - h(x)\|_B \leq \frac{1}{2}\tilde{\varphi}(0, -x, x) \tag{4.14}$$

*for all $x \in A$.*

*Proof.* By Theorem 4.5, we have a unique $\mathbb{C}$-linear mapping $h : A \to B$ defined by

$$h(x) := \lim_{n \to \infty} \frac{f((-2)^n x)}{(-2)^n}$$

for all $x \in A$ which satisfies (4.14).

Now, we show that $h(xy) = h(x)h(y)$ for all $x, y \in A$ whenever the multiplication is defined. Replacing $x, y$ by $(-2)^n x, (-2)^n y$, respectively, and dividing by $4^n$ in (4.12), we get

$$\left\| \frac{1}{4^n}[f((-2)^n x (-2)^n y) - f((-2)^n x)f((-2)^n y)] \right\|_B$$

$$\leq \frac{1}{4^n} \phi((-2)^n x, (-2)^n y)$$

for all $x, y \in A$ whenever the multiplication is defined. Also, we have

$$\lim_{n \to \infty} \frac{1}{4^n} f((-2)^n x (-2)^n y) = \lim_{n \to \infty} \frac{f((-2)^{2n} xy)}{(-2)^{2n}} = h(xy)$$

and

$$\lim_{n \to \infty} \frac{1}{4^n} f((-2)^n x) f((-2)^n y) = \lim_{n \to \infty} \frac{f((-2)^n x)}{(-2)^n} \cdot \lim_{n \to \infty} \frac{f((-2)^n x)}{(-2)^n}$$

$$= h(x) h(y)$$

for all $x, y \in A$ whenever the multiplication is defined. If we let $n \to \infty$ in the above inequality, then (4.13) gives $h(xy) = h(x) h(y)$ for all $x, y \in A$ whenever the multiplication is defined. This completes the proof.  □

**Corollary 4.8.** *Let $\theta, p$ be nonnegative real numbers with $p < 1$ and $f : A \to B$ be a mapping satisfying*

$$\|f(2x) + \mu f(2y) + 2f(z)\|_B$$
$$\leq \|2f(x + \mu y + z)\|_B + \theta(\|x\|_A^p + \|y\|_A^p + \|z\|_A^p)$$

*and*

$$\|f(xy) - f(x)f(y)\|_B \leq \theta(\|x\|_A^{2p} + \|y\|_A^{2p})$$

*for all $\mu \in \mathbb{T}^1$ and $x, y, z \in A$ whenever the multiplication is defined. Then there exists a unique proper $CQ^*$-algebra homomorphism $h : A \to B$ such that*

$$\|f(x) - h(x)\|_B \leq \frac{2\theta}{2 - 2^p} \|x\|_A^p$$

*for all $x \in A$*

*Proof.* Let $\varphi : A^3 \to [0, \infty)$ be a mapping defined by

$$\varphi(x, y, z) = \theta(\|x\|_A^p + \|y\|_A^p + \|z\|_A^p).$$

for all $x, y, z \in A$. For any $p < 1$, we have

$$\tilde{\varphi}(x, y, z) := \sum_{j=0}^{\infty} \frac{1}{2^j} \varphi\left((-2)^j x, (-2)^j y, (-2)^j z\right)$$

$$= \sum_{j=0}^{\infty} \frac{2^{pj}}{2^j} \theta(\|x\|_A^p + \|y\|_A^p + \|z\|_A^p)$$

$$= \frac{2\theta}{2 - 2^p} (\|x\|_A^p + \|y\|_A^p + \|z\|_A^p).$$

In addition, let $\phi : A^2 \to [0, \infty)$ be $\phi(x, y) = \theta(\|x\|_A^{2p} + \|y\|_A^{2p})$. For any $p < 1$, we have

$$\lim_{n \to \infty} \frac{1}{4^n} \phi((-2)^n x, (-2)^n y) = \lim_{n \to \infty} \frac{2^{2pn}}{4^n} \theta(\|x\|_A^{2p} + \|y\|_A^{2p}) = 0$$

for all $x, y \in A$. By applying Theorem 4.7, there exists a unique proper $CQ^*$-algebra homomorphism $h : A \to B$ such that

$$\|f(x) - h(x)\|_B \le \frac{1}{2} \tilde\varphi(0, -x, x) = \frac{2\theta}{2 - 2^p} \|x\|_A^p$$

for all $x \in A$. This completes the proof.    □

**Corollary 4.9.** *Let $\theta, p$ be nonnegative real numbers with $p < 1$ and $f : A \to B$ be a mapping satisfying*

$$\|f(2x) + \mu f(2y) + 2f(z)\|_B$$
$$\le \|2f(x + \mu y + z)\|_B + \theta(\|x\|_A^p + \|y\|_A^p + \|z\|_A^p)$$

*and*

$$\|f(xy) - f(x)f(y)\|_B \le \theta \cdot \|x\|_A^p \cdot \|y\|_A^p$$

*for all $x, y \in A$ whenever the multiplication is defined. Then there exists a unique proper $CQ^*$-algebra homomorphism $h : A \to B$ such that*

$$\|f(x) - h(x)\|_B \le \frac{2\theta}{2 - 2^p} \|x\|_A^p$$

*for all $x \in A$*

*Proof.* Let a mapping $\varphi : A^3 \to [0, \infty)$ be defined by

$$\varphi(x, y, z) = \theta(\|x\|_A^p + \|y\|_A^p + \|z\|_A^p)$$

for all $x, y, z \in A$ and $\phi : A^2 \to [0, \infty)$ be a mapping defined by

$$\phi(x, y) = \theta \cdot \|x\|_A^p \cdot \|y\|_A^p$$

for all $x, y \in A$. When $p < 1$, we have $\tilde\varphi(x, y, z) < \infty$ and

$$\lim_{n \to \infty} \frac{1}{4^n} \phi((-2)^n x, (-2)^n y) = \lim_{n \to \infty} \frac{2^{2pn}}{4^n} \cdot \theta \cdot \|x\|_A^p \cdot \|y\|_A^p = 0$$

for all $x, y, z \in A$. By applying Theorem 4.7, there exists a unique proper $CQ^*$-algebra homomorphism $h : A \to B$ such that

$$\|f(x) - h(x)\|_B \leq \frac{1}{2}\tilde{\varphi}(0, -x, x) = \frac{2\theta}{2 - 2^p}\|x\|_A^p$$

for all $x \in A$. This completes the proof.                                                                        □

**Theorem 4.10.** *Let $f : A \to B$ be a mapping. Suppose that there exists a function $\varphi : A^3 \to [0, \infty)$ satisfying (4.10) and*

$$\tilde{\varphi}(x, y, z) := \sum_{i=1}^{\infty} 2^i \varphi\left(\frac{x}{(-2)^i}, \frac{y}{(-2)^i}, \frac{z}{(-2)^i}\right) < \infty \tag{4.15}$$

*for all $x, y, z \in A$ If, in addition, there exists a function $\phi : A^2 \to [0, \infty)$ satisfying (4.12) and*

$$\lim_{n \to \infty} 4^n \phi\left(\frac{x}{(-2)^n}, \frac{y}{(-2)^n}\right) = 0 \tag{4.16}$$

*for all $x, y \in A$ whenever the multiplication is defined, then there exists a unique proper $CQ^*$-algebra homomorphism $h : A \to B$ such that*

$$\|f(x) - h(x)\|_B \leq \frac{1}{2}\tilde{\varphi}(0, -x, x) \tag{4.17}$$

*for all $x \in A$.*

*Proof.* By Corollary 4.12, we have a unique $\mathbb{C}$-linear mapping $h : A \to B$ defined by

$$h(x) := \lim_{n \to \infty} (-2)^n f\left(\frac{x}{(-2)^n}\right)$$

for all $x \in A$, which satisfies (4.17). Now, replacing $x, y$ by $\frac{x}{(-2)^n}, \frac{y}{(-2)^n}$, respectively, and multiplying by $4^n$ in (4.12), we get

$$\left\|4^n\left[f\left(\frac{x}{(-2)^n} \cdot \frac{y}{(-2)^n}\right) - f\left(\frac{x}{(-2)^n}\right)f\left(\frac{y}{(-2)^n}\right)\right]\right\|_B$$
$$\leq 4^n \phi\left(\frac{x}{(-2)^n}, \frac{y}{(-2)^n}\right)$$

for all $x, y \in A$ whenever the multiplication is defined. Since

$$\lim_{n \to \infty} 4^n f\left(\frac{x}{(-2)^n} \cdot \frac{y}{(-2)^n}\right) = \lim_{n \to \infty} (-2)^{2n} f\left(\frac{xy}{(-2)^{2n}}\right) = h(xy)$$

and

$$\lim_{n \to \infty} 4^n f\left(\frac{x}{(-2)^n}\right) f\left(\frac{y}{(-2)^n}\right)$$

$$= \lim_{n \to \infty} (-2)^n f\left(\frac{x}{(-2)^n}\right) \cdot \lim_{n \to \infty} (-2)^n f\left(\frac{y}{(-2)^n}\right)$$

$$= h(x)h(y)$$

for all $x, y \in A$ whenever the multiplication is defined. If we let $n \to \infty$ in the above inequality then (4.16) gives $h(xy) = h(x)h(y)$ for all $x, y \in A$ whenever the multiplication is defined. This completes the proof.     $\square$

**Corollary 4.11.** *Let $\theta, p$ be nonnegative real numbers with $p > 1$ and $f : A \to B$ be a mapping satisfying*

$$\|f(2x) + \mu f(2y) + 2f(z)\|_B$$

$$\leq \|2f(x + \mu y + z)\|_B + \theta(\|x\|_A^p + \|y\|_A^p + \|z\|_A^p)$$

*and*

$$\|f(xy) - f(x)f(y)\|_B \leq \theta(\|x\|_A^{2p} + \|y\|_A^{2p})$$

*for all $\mu \in \mathbb{T}^1$ and $x, y, z \in A$ whenever the multiplication is defined. Then there exists a unique proper $CQ^*$-algebra homomorphism $h : A \to B$ such that*

$$\|f(x) - h(x)\|_B \leq \frac{2\theta}{2^p - 2}\|x\|_A^p$$

*for all $x \in A$*

*Proof.* Let $\varphi : A^3 \to [0, \infty)$ be a mapping defined by

$$\varphi(x, y, z) = \theta(\|x\|_A^p + \|y\|_A^p + \|z\|_A^p)$$

for all $x, y, z \in A$. When $p > 1$, we get

$$\tilde{\varphi}(x, y, z) := \sum_{i=1}^{\infty} 2^i \varphi\left(\frac{x}{(-2)^i}, \frac{y}{(-2)^i}, \frac{z}{(-2)^i}\right)$$

$$= \sum_{i=1}^{\infty} \frac{2^i}{2^{pi}} \theta(\|x\|_A^p + \|y\|_A^p + \|z\|_A^p)$$

$$= \frac{2\theta}{2^p - 2}(\|x\|_A^p + \|y\|_A^p + \|z\|_A^p)$$

for all $x, y, z \in A$. In addition, let $\phi : A^2 \to [0, \infty)$ be a mapping defined by

$$\phi(x, y) = \theta(\|x\|_A^{2p} + \|y\|_A^{2p})$$

for all $x, y \in A$. When $p > 1$, we get

$$\lim_{n \to \infty} 4^n \phi \left( \frac{x}{(-2)^n}, \frac{y}{(-2)^n} \right) = \lim_{n \to \infty} \frac{4^n}{2^{2pn}} \theta(\|x\|_A^{2p} + \|y\|_A^{2p}) = 0$$

for all $x, y \in A$. By applying Theorem 4.10, there exists a unique proper $CQ^*$-algebra homomorphism $h : A \to B$ such that

$$\|f(x) - h(x)\|_B \leq \frac{1}{2} \tilde{\phi}(0, -x, x) = \frac{2\theta}{2^p - 2} \|x\|_A^p$$

for all $x \in A$. This completes the proof.                                    $\square$

**Corollary 4.12.** *Let $\theta, p$ be nonnegative real numbers with $p > 1$ and $f : A \to B$ be a mapping satisfying*

$$\|f(2x) + \mu f(2y) + 2f(z)\|_B$$
$$\leq \|2f(x + \mu y + z)\|_B + \theta(\|x\|_A^p + \|y\|_A^p + \|z\|_A^p)$$

*and*

$$\|f(xy) - f(x)f(y)\|_B \leq \theta \cdot \|x\|_A^p \cdot \|y\|_A^p$$

*for all $\mu \in \mathbb{T}^1$ and $x, y, z \in A$ whenever the multiplication is defined. Then there exists a unique proper $CQ^*$-algebra homomorphism $h : A \to B$ such that*

$$\|f(x) - h(x)\|_B \leq \frac{2\theta}{2^p - 2} \|x\|_A^p$$

*for all $x \in A$.*

*Proof.* Let $\varphi : A^3 \to [0, \infty)$ be a mapping defined by

$$\varphi(x, y, z) = \theta(\|x\|_A^p + \|y\|_A^p + \|z\|_A^p)$$

for all $x, y, z \in A$ and $\phi : A^2 \to [0, \infty)$ be a mapping defined by

$$\phi(x, y) = \theta \cdot \|x\|_A^p \cdot \|y\|_A^p$$

for all $x, y \in A$. For any $p > 1$, we have $\tilde{\varphi}(x, y, z) < \infty$ and

$$\lim_{n \to \infty} 4^n \phi \left( \frac{x}{(-2)^n}, \frac{y}{(-2)^n} \right) = \lim_{n \to \infty} \frac{4^n}{2^{2pn}} \cdot \theta \cdot \|x\|_A^p \cdot \|y\|_A^p = 0$$

for all $x, y, z \in A$. By applying Theorem 4.10, there exists a unique proper $CQ^*$-algebra homomorphism $h : A \rightarrow B$ such that

$$\|f(x) - h(x)\|_B \leq \frac{1}{2}\tilde{\varphi}(0, -x, x) = \frac{2\theta}{2^p - 2}\|x\|_A^p$$

for all $x \in A$. This completes the proof.                                    $\square$

### 4.1.3   Stability of Derivations in Proper CQ*-Algebras

Now, we consider the Hyers-Ulam stability of derivations on proper $CQ^*$-algebras associated with the additive functional inequality.

**Theorem 4.13.** *Let $f : A \rightarrow A$ be a mapping with $f(0) = 0$. Suppose that there exists a function $\varphi : A^3 \rightarrow [0, \infty)$ such that*

$$\|f(2x) + \mu f(2y) + 2f(z)\|_A \leq \|2f(x + \mu y + z)\|_A + \varphi(x, y, z) \qquad (4.18)$$

*and*

$$\tilde{\varphi}(x, y, z) := \sum_{j=0}^{\infty} \frac{1}{2^j}\varphi\left((-2)^j x, (-2)^j y, (-2)^j z\right) < \infty \qquad (4.19)$$

*for all $\mu \in \mathbb{T}^1$ and $x, y, z \in A$. If, in addition, there exists a function $\psi : A^2 \rightarrow [0, \infty)$ such that*

$$\|f(xy) - f(x)y - xf(y)\|_A \leq \psi(x, y) \qquad (4.20)$$

*and*

$$\lim_{n \to \infty} \frac{1}{4^n}\psi((-2)^n x, (-2)^n y) = 0 \qquad (4.21)$$

*for all $x, y \in A$ whenever the multiplication is defined, then there exists a unique derivation $\delta$ on $A$ such that*

$$\|f(x) - \delta(x)\|_A \leq \frac{1}{2}\tilde{\varphi}(0, -x, x) \qquad (4.22)$$

*for all $x \in A$.*

*Proof.* By Theorem 4.5, we have a unique $\mathbb{C}$-linear mapping $\delta : A \rightarrow A$ defined by

$$\delta(x) := \lim_{n \to \infty} \frac{f((-2)^n x)}{(-2)^n}$$

for all $x \in A$ which satisfies (4.22).

Now, we show that $\delta(xy) = \delta(x)\delta(y)$ for all $x, y \in A$ whenever the multiplication is defined. Replacing $x, y$ by $(-2)^n x, (-2)^n y$, respectively, and dividing by $4^n$ in (4.3), we have

$$\left\| \frac{1}{4^n} [f((-2)^n x(-2)^n y) - f((-2)^n x)(-2)^n y - (-2)^n x f((-2)^n y)] \right\|_A$$

$$\leq \frac{1}{4^n} \psi((-2)^n x, (-2)^n y)$$

for all $x, y \in A$ whenever the multiplication is defined. Also, we have

$$\lim_{n \to \infty} \frac{1}{4^n} f((-2)^n x(-2)^n y) = \lim_{n \to \infty} \frac{f((-2)^{2n} xy)}{(-2)^{2n}} = \delta(xy),$$

$$\lim_{n \to \infty} \frac{1}{4^n} f((-2)^n x) \cdot (-2)^n y = \lim_{n \to \infty} \frac{f((-2)^n x)}{(-2)^n} \cdot \frac{(-2)^n y}{(-2)^n} = \delta(x)y$$

and

$$\lim_{n \to \infty} \frac{1}{4^n} (-2)^n x \cdot f((-2)^n y) = \lim_{n \to \infty} \frac{(-2)^n x}{(-2)^n} \cdot \frac{(-2)^n y}{(-2)^n} = x\delta(y)$$

for all $x, y \in A$ whenever the multiplication is defined. If we let $n \to \infty$ in the above inequality, then (4.21) gives $\delta(xy) = \delta(x)y - x\delta(y)$ for all $x, y \in A$ whenever the multiplication is defined. This completes the proof.   $\square$

**Corollary 4.14.** *Let $\theta, p$ be nonnegative real numbers with $p < 1$ and $f : A \to A$ be a mapping satisfying*

$$\|f(2x) + \mu f(2y) + 2f(z)\|_A$$

$$\leq \|2f(x + \mu y + z)\|_A + \theta(\|x\|_A^p + \|y\|_A^p + \|z\|_A^p)$$

*and*

$$\|f(xy) - f(x)y - xf(y)\|_A \leq \theta(\|x\|_A^{2p} + \|y\|_A^{2p})$$

*for all $\mu \in \mathbb{T}^1$ and $x, y, z \in A$ whenever the multiplication is defined. Then there exists a unique derivation $\delta$ on $A$ such that*

$$\|f(x) - \delta(x)\|_A \leq \frac{2\theta}{2 - 2^p} \|x\|_A^p$$

*for all $x \in A$.*

*Proof.* Let $\varphi : A^3 \to [0, \infty)$ be a mapping defined by

$$\varphi(x, y, z) = \theta(\|x\|_A^p + \|y\|_A^p + \|z\|_A^p)$$

for all $x, y, z \in A$. For any $p < 1$, we have

$$\tilde{\varphi}(x, y, z) := \sum_{j=0}^{\infty} \frac{1}{2^j} \varphi\left((-2)^j x, (-2)^j y, (-2)^j z\right)$$

$$= \sum_{j=0}^{\infty} \frac{2^{pj}}{2^j} \theta(\|x\|_A^p + \|y\|_A^p + \|z\|_A^p)$$

$$= \frac{2\theta}{2 - 2^p}(\|x\|_A^p + \|y\|_A^p + \|z\|_A^p).$$

In addition, let $\psi : A^2 \to [0, \infty)$ be a mapping defined by

$$\psi(x, y) = \theta(\|x\|_A^{2p} + \|y\|_A^{2p})$$

for all $x, y \in A$. For any $p < 1$, we have

$$\lim_{n \to \infty} \frac{1}{4^n} \psi((-2)^n x, (-2)^n y) = \lim_{n \to \infty} \frac{2^{2pn}}{4^n} \theta(\|x\|_A^{2p} + \|y\|_A^{2p}) = 0$$

for all $x, y \in A$. By applying Theorem 4.13, there exists a unique proper $CQ^*$-algebra homomorphism $h : A \to B$ such that

$$\|f(x) - \delta(x)\|_A \le \frac{1}{2}\tilde{\varphi}(0, -x, x) = \frac{2\theta}{2 - 2^p}\|x\|_A^p$$

for all $x \in A$. This completes the proof.                                  □

**Corollary 4.15.** *Let $\theta, p$ be nonnegative real numbers with $p < 1$ and $f : A \to B$ be a mapping satisfying*

$$\|f(2x) + \mu f(2y) + 2f(z)\|_A \le \|2f(x + \mu y + z)\|_A + \theta(\|x\|_A^p + \|y\|_A^p + \|z\|_A^p)$$

*and*

$$\|f(xy) - f(x)y - xf(y)\|_A \le \theta \cdot \|x\|_A^p \cdot \|y\|_A^p$$

*for all $x, y, z \in A$ whenever the multiplication is defined. Then there exists a unique derivation $\delta$ on $A$ such that*

$$\|f(x) - \delta(x)\|_A \le \frac{2\theta}{2 - 2^p}\|x\|_A^p$$

*for all $x \in A$.*

*Proof.* Let $\varphi : A^3 \to [0, \infty)$ be a mapping defined by

$$\varphi(x, y, z) = \theta(\|x\|_A^p + \|y\|_A^p + \|z\|_A^p)$$

for all $x, y, z \in A$ and $\psi : A^2 \to [0, \infty)$ be a mapping by

$$\psi(x, y) = \theta \cdot \|x\|_A^p \cdot \|y\|_A^p$$

for all $x, y \in A$. For any $p < 1$, we have $\tilde{\varphi}(x, y, z) < \infty$ and

$$\lim_{n \to \infty} \frac{1}{4^n} \phi((-2)^n x, (-2)^n y) = \lim_{n \to \infty} \frac{2^{2pn}}{4^n} \cdot \theta \cdot \|x\|_A^p \cdot \|y\|_A^p = 0$$

for all $x, y, z \in A$. By applying Theorem 4.13, there exists a unique proper $CQ^*$-algebra homomorphism $\delta : A \to A$ such that

$$\|f(x) - \delta(x)\|_A \le \frac{1}{2}\tilde{\varphi}(0, -x, x) = \frac{2\theta}{2 - 2^p}\|x\|_A^p$$

for all $x \in A$. This completes the proof. $\square$

**Theorem 4.16.** *Let $f : A \to A$ be a mapping. Suppose that there exists a function $\varphi : A^3 \to [0, \infty)$ satisfying (4.18) and*

$$\tilde{\varphi}(x, y, z) := \sum_{i=1}^{\infty} 2^i \varphi\left(\frac{x}{(-2)^i}, \frac{y}{(-2)^i}, \frac{z}{(-2)^i}\right) < \infty \tag{4.23}$$

*for all $x, y, z \in A$. If, in addition, there exists a function $\psi : A^2 \to [0, \infty)$ satisfying (4.20) and*

$$\lim_{n \to \infty} 4^n \psi\left(\frac{x}{(-2)^n}, \frac{y}{(-2)^n}\right) = 0 \tag{4.24}$$

*for all $x, y \in A$, then there exists a unique derivation $\delta$ on $A$ satisfying*

$$\|f(x) - \delta(x)\|_A \le \frac{1}{2}\tilde{\varphi}(0, -x, x) \tag{4.25}$$

*for all $x \in A$.*

*Proof.* By Corollary 4.18, we have a unique $\mathbb{C}$-linear mapping $\delta : A \to A$ defined by

$$\delta(x) := \lim_{n \to \infty} (-2)^n f\left(\frac{x}{(-2)^n}\right)$$

for all $x \in A$, which satisfies (4.25). Now, replacing $x, y$ by $\frac{x}{(-2)^n}, \frac{y}{(-2)^n}$, respectively, and multiplying by $4^n$ in (4.20), we get

$$\left\| 4^n \left[ f\left( \frac{x}{(-2)^n} \cdot \frac{y}{(-2)^n} \right) - f\left( \frac{x}{(-2)^n} \right) f\left( \frac{y}{(-2)^n} \right) \right] \right\|_A$$
$$\leq 4^n \psi \left( \frac{x}{(-2)^n}, \frac{y}{(-2)^n} \right)$$

for all $x, y \in A$ whenever the multiplication is defined. Also, we have

$$\lim_{n \to \infty} 4^n f\left( \frac{x}{(-2)^n} \cdot \frac{y}{(-2)^n} \right) = \lim_{n \to \infty} (-2)^{2n} f\left( \frac{xy}{(-2)^{2n}} \right) = \delta(xy),$$

$$\lim_{n \to \infty} 4^n f\left( \frac{x}{(-2)^n} \right) \cdot \frac{y}{(-2)^n}$$
$$= \lim_{n \to \infty} (-2)^n f\left( \frac{x}{(-2)^n} \right) \cdot \lim_{n \to \infty} \frac{(-2)^n y}{(-2)^n} = \delta(x)y$$

and

$$\lim_{n \to \infty} 4^n \frac{x}{(-2)^n} \cdot f\left( \frac{y}{(-2)^n} \right)$$
$$= \lim_{n \to \infty} \frac{(-2)^n x}{(-2)^n} \cdot \lim_{n \to \infty} (-2)^n f\left( \frac{y}{(-2)^n} \right) = x\delta(y)$$

for all $x, y \in A$ whenever the multiplication is defined. If we let $n \to \infty$ in the above inequality, then (4.24) gives $\delta(xy) = \delta(x)y - x\delta(y)$ for all $x, y \in A$ whenever the multiplication is defined. This completes the proof. $\qquad \square$

**Corollary 4.17.** *Let $\theta, p$ be nonnegative real numbers with $p > 1$ and $f : A \to A$ be a mapping satisfying*

$$\| f(2x) + \mu f(2y) + 2f(z) \|_A$$
$$\leq \| 2f(x + \mu y + z) \|_A + \theta(\|x\|_A^p + \|y\|_A^p + \|z\|_A^p)$$

*and*

$$\| f(xy) - f(x)y - xf(y) \|_A \leq \theta(\|x\|_A^{2p} + \|y\|_A^{2p})$$

*for all $\mu \in \mathbb{T}^1$ and $x, y, z \in A$ whenever the multiplication is defined. Then there exists a unique derivation $\delta$ on $A$ such that*

$$\|f(x) - \delta(x)\|_A \leq \frac{2\theta}{2^p - 2} \|x\|_A^p$$

*for all $x \in A$.*

*Proof.* Let $\varphi : A^3 \to [0, \infty)$ be a mapping by

$$\varphi(x, y, z) = \theta(\|x\|_A^p + \|y\|_A^p + \|z\|_A^p)$$

for all $x, y, z \in A$. For any $p > 1$, we have

$$\tilde{\varphi}(x, y, z) := \sum_{i=1}^{\infty} 2^i \varphi\left(\frac{x}{(-2)^i}, \frac{y}{(-2)^i}, \frac{z}{(-2)^i}\right)$$

$$= \sum_{i=1}^{\infty} \frac{2^i}{2^{pi}} \theta(\|x\|_A^p + \|y\|_A^p + \|z\|_A^p)$$

$$= \frac{2\theta}{2^p - 2}(\|x\|_A^p + \|y\|_A^p + \|z\|_A^p).$$

In addition, let $\psi : A^2 \to [0, \infty)$ be a mapping defined by

$$\phi(x, y) = \theta(\|x\|_A^{2p} + \|y\|_A^{2p})$$

for all $x, y \in A$. For any $p > 1$, we get

$$\lim_{n \to \infty} 4^n \psi\left(\frac{x}{(-2)^n}, \frac{y}{(-2)^n}\right) = \lim_{n \to \infty} \frac{4^n}{2^{2pn}} \theta(\|x\|_A^{2p} + \|y\|_A^{2p}) = 0$$

for all $x, y \in A$. By applying Theorem 4.16, there exists a unique proper $CQ^*$-algebra homomorphism $\delta : A \to A$ such that

$$\|f(x) - \delta(x)\|_A \leq \frac{1}{2}\tilde{\varphi}(0, -x, x) = \frac{2\theta}{2^p - 2} \|x\|_A^p$$

for all $x \in A$. This completes the proof. $\qquad\square$

**Corollary 4.18.** *Let $\theta, p$ be nonnegative real numbers with $p > 1$ and $f : A \to B$ be a mapping satisfying*

$$\|f(2x) + \mu f(2y) + 2f(z)\|_A$$

$$\leq \|2f(x + \mu y + z)\|_A + \theta(\|x\|_A^p + \|y\|_A^p + \|z\|_A^p)$$

*and*

$$\|f(xy) - f(x)y - xf(y)\|_A \leq \theta \cdot \|x\|_A^p \cdot \|y\|_A^p$$

*for all $\mu \in \mathbb{T}^1$ and $x, y, z \in A$ whenever the multiplication is defined. Then there exists a unique derivation $\delta$ on $A$ such that*

$$\|f(x) - \delta(x)\|_A \leq \frac{2\theta}{2^p - 2}\|x\|_A^p$$

*for all $x \in A$.*

*Proof.* Let $\varphi : A^3 \to [0, \infty)$ be a mapping defined by

$$\varphi(x, y, z) = \theta(\|x\|_A^p + \|y\|_A^p + \|z\|_A^p)$$

for all $x, y, z \in A$ and $\psi : A^2 \to [0, \infty)$ be a mapping defined by

$$\psi(x, y) = \theta \cdot \|x\|_A^p \cdot \|y\|_A^p$$

for all $x, y \in A$. When $p > 1$, we have $\tilde{\varphi}(x, y, z) < \infty$ and

$$\lim_{n \to \infty} 4^n \phi \left( \frac{x}{(-2)^n}, \frac{y}{(-2)^n} \right) = \lim_{n \to \infty} \frac{4^n}{2^{2pn}} \cdot \theta \cdot \|x\|_A^p \cdot \|y\|_A^p = 0$$

for all $x, y, z \in A$. By applying Theorem 4.10, there exists a unique proper $CQ^*$-algebra homomorphism $\delta : A \to A$ such that

$$\|f(x) - \delta(x)\|_A \leq \frac{1}{2}\tilde{\varphi}(0, -x, x) = \frac{2\theta}{2^p - 2}\|x\|_A^p$$

for all $x \in A$. This completes the proof.                                    □

## 4.2   Stability of Functional Inequalities over $C^*$-Algebras

In this section, we investigate a module linear mapping associated with the following functional inequality.

$$\|f(x) + f(y) + f(z) + f(w)\| \leq \|f(x) + f(y + z + w)\|. \tag{4.26}$$

This is applied to understand homomorphisms in $C^*$-algebras. Moreover, we prove the Hyers-Ulam stability of the functional following inequality:

$$\|f(x) + f(y) + f(z) + f(w)\|$$
$$\leq \|f(x) + f(y + z + w)\|$$
$$+ \theta(\|x\|^p + \|y\|^p + \|z\|^p + \|w\|^p + \|x\|^{\frac{p}{4}} \cdot \|y\|^{\frac{p}{4}} \cdot \|z\|^{\frac{p}{4}} \cdot \|w\|^{\frac{p}{4}}) \quad (4.27)$$

in real Banach spaces. Using the fixed point method, we prove the Hyers-Ulam stability of the functional inequality (4.27) in real Banach spaces.

### 4.2.1 Functional Inequalities in Normed Modules over $C^*$-Algebras

Throughout this section, let $A$ be a unital $C^*$-algebra with the unitary group $U(A)$ and the unit $e$ and let $B$ be a $C^*$-algebra. Assume that $X$ is a normed $A$-module with the norm $\| \cdot \|$ and $Y$ is a normed $A$-module with the norm $\| \cdot \|$.

Now, we investigate an $A$-linear mapping associated with the functional inequality (4.26).

**Theorem 4.19.** *Let* $f : X \rightarrow Y$ *be a mapping such that*

$$\|f(x) + f(y) + f(z) + uf(w)\| \leq \|f(x) + f(y + z + uw)\| \quad (4.28)$$

*for all* $x, y, z, w \in X$ *and* $u \in U(A)$. *Then the mapping* $f : X \rightarrow Y$ *is* $A$-*linear.*

*Proof.* Letting $x = y = z = w = 0$ and $u = e \in U(A)$ in (4.28), we have

$$\|4f(0)\| \leq \|2f(0)\|.$$

So $f(0) = 0$. Letting $x = w = 0$ in (4.28), we have

$$\|f(y) + f(z)\| \leq \|f(y + z)\| \quad (4.29)$$

for all $y, z \in X$. Replacing $y$ and $z$ by $x$ and $y + z + w$ in (4.29), respectively, we get

$$\|f(x) + f(y + z + w)\| \leq \|f(x + y + z + w)\|$$

for all $x, y, z, w \in X$ and so

$$\|f(x) + f(y) + f(z) + f(w)\| \leq \|f(x + y + z + w)\| \quad (4.30)$$

for all $x, y, z, w \in X$. Letting $z = w = 0$ and $y = -x$ in (4.30), we have

$$\|f(x) + f(-x)\| \leq \|f(0)\| = 0$$

for all $x \in X$. Hence $f(-x) = -f(x)$ for all $x \in X$. Letting $z = -x - y$ and $w = 0$ in (4.30), we have

$$\|f(x) + f(y) - f(x + y)\| = \|f(x) + f(y) + f(-x - y)\| \le \|f(0)\| = 0$$

for all $x, y \in X$. Thus

$$f(x + y) = f(x) + f(y)$$

for all $x, y \in X$. Letting $z = -uw$ and $x = y = 0$ in (4.28), we have

$$\| -f(uw) + uf(w)\| = \|f(-uw) + uf(w)\| \le \|2f(0)\| = 0$$

for all $w \in X$ and all $u \in U(A)$. Thus we have

$$f(uw) = uf(w) \qquad (4.31)$$

for all $u \in U(A)$ and all $w \in X$.

Now, let $a \in A$ $(a \ne 0)$ and $M$ be an integer greater than $4|a|$. Then we have

$$\left|\frac{a}{M}\right| < \frac{1}{4} < 1 - \frac{2}{3} = \frac{1}{3}.$$

By Theorem 1 of Kadison and Pedersen [167], there exist three elements $u_1, u_2, u_3 \in U(A)$ such that $3\frac{a}{M} = u_1 + u_2 + u_3$. So, by (4.31), we have

$$
\begin{aligned}
f(ax) &= f\left(\frac{M}{3} \cdot 3\frac{a}{M}x\right) = M \cdot f\left(\frac{1}{3} \cdot 3\frac{a}{M}x\right) = \frac{M}{3}f\left(3\frac{a}{M}x\right) \\
&= \frac{M}{3}h(u_1x + u_2x + u_3x) = \frac{M}{3}(f(u_1x) + f(u_2x) + f(u_3x)) \\
&= \frac{M}{3}(u_1 + u_2 + u_3)f(x) = \frac{M}{3} \cdot 3\frac{a}{M}f(x) = af(x)
\end{aligned}
$$

for all $x \in X$. Therefore, $f : X \to Y$ is $A$-linear. This completes the proof.   $\square$

**Corollary 4.20.** *Let $f : A \to B$ be a multiplicative mapping such that*

$$\|f(x) + f(y) + f(z) + \mu f(w)\| \le \|f(x) + f(y + z + \mu w)\| \qquad (4.32)$$

*for all $x, y, z, w \in A$ and $\mu \in \mathbb{T} := \{\lambda \in \mathbb{C} : |\lambda| = 1\}$. Then the mapping $f : A \to B$ is a $C^*$-algebra homomorphism.*

*Proof.* By Theorem 4.19, the multiplicative mapping $f : A \to B$ is $\mathbb{C}$-linear since $C^*$-algebras are normed modules over $\mathbb{C}$. So, the multiplicative mapping $f : A \to B$ is a $C^*$-algebra homomorphism.   $\square$

Assume that $X$ is a real normed linear space and $Y$ is a real Banach space. Now, we prove the Hyers-Ulam stability of the functional inequality (4.27) in real Banach spaces.

**Theorem 4.21.** *Assume that $f : X \to Y$ is an odd mapping for which there exist the constants $\theta \geq 0$ and $p \in \mathbb{R}$ such that $p \neq 1$ and $f : X \to Y$ satisfies the following functional inequality:*

$$\|f(x) + f(y) + f(z) + f(w)\|$$
$$\leq \|f(x) + f(y + z + w)\|$$
$$+ \theta(\|x\|^p + \|y\|^p + \|z\|^p + \|w\|^p + \|x\|^{\frac{p}{4}} \cdot \|y\|^{\frac{p}{4}} \cdot \|z\|^{\frac{p}{4}} \cdot \|w\|^{\frac{p}{4}}) \quad (4.33)$$

*for all $x, y, z, w \in X$. Then there exists a unique Cauchy additive mapping $A : X \to Y$ such that*

$$\|f(x) - A(x)\| \leq \frac{2^p + 2}{|2^p - 2|} \theta \|x\|^p \tag{4.34}$$

*for all $x \in X$. If, in addition, $f : X \to Y$ is a mapping such that the transformation $t \to f(tx)$ is continuous in $t \in \mathbb{R}$ for each fixed $x \in X$, then $A$ is an $\mathbb{R}$-linear mapping.*

*Proof.* Since $f$ is odd, $f(0) = 0$ and $f(-x) = -f(x)$ for all $x \in X$. Letting $x = 0$, $z = y$ and $w = -2y$ in (4.33), we have

$$\|2f(y) - f(2y)\| \leq (2 + 2^p)\theta\|y\|^p \tag{4.35}$$

for all $y \in X$ and so

$$\left\| f(x) - 2f\left(\frac{x}{2}\right) \right\| \leq \frac{2 + 2^p}{2^p} \theta \|x\|^p$$

for all $x \in X$. Hence we have

$$\left\| 2^l f\left(\frac{x}{2^l}\right) - 2^m f\left(\frac{x}{2^m}\right) \right\| \leq \frac{2 + 2^p}{2^p} \sum_{j=l}^{m-1} \frac{2^j}{2^{pj}} \theta \|x\|^p \tag{4.36}$$

for all $m, l \geq 1$ with $m > l$ and $x \in X$.

Assume that $p > 1$. It follows from (4.36) that the sequence $\{2^k f(\frac{x}{2^k})\}$ is a Cauchy sequence for all $x \in X$. Since $Y$ is complete, the sequence $\{2^k f(\frac{x}{2^k})\}$ converges and so one can define the mapping $A : X \to Y$ by

$$A(x) := \lim_{k \to \infty} 2^k f\left(\frac{x}{2^k}\right)$$

for all $x \in X$. Letting $l = 0$ and $m \to \infty$ in (4.36), we get

$$\|f(x) - A(x)\| \leq \frac{2^p + 2}{2^p - 2} \theta \|x\|^p$$

for all $x \in X$. It follows from (4.33) that

$$\left\| 2^k f \left( \frac{x}{2^k} \right) + 2^k f \left( \frac{y}{2^k} \right) + 2^k f \left( \frac{z}{2^k} \right) + 2^k f \left( \frac{w}{2^k} \right) \right\|$$

$$\leq \left\| 2^k f \left( \frac{x}{2^k} \right) + 2^k f \left( \frac{y + z + w}{2^k} \right) \right\|$$

$$+ \frac{2^k \theta}{2^{pk}} (\|x\|^p + \|y\|^p + \|z\|^p + \|w\|^p + \|x\|^{\frac{\ell}{4}} \cdot \|y\|^{\frac{\ell}{4}} \cdot \|z\|^{\frac{\ell}{4}} \cdot \|w\|^{\frac{\ell}{4}}) \quad (4.37)$$

for all $x, y, z, w \in X$. Letting $k \to \infty$ in (4.37), we get

$$\|A(x) + A(y) + A(z) + A(w)\| \leq \|A(x) + A(y + z + w)\| \quad (4.38)$$

for all $x, y, z, w \in X$. It is easy to show that $A : X \to Y$ is odd. Letting $w = -y - z$ and $x = 0$ in (4.38), we get $A(y + z) = A(y) + A(z)$ for all $y, z \in X$. So, there exists a Cauchy additive mapping $A : X \to Y$ satisfying (4.34) for the case $p > 1$.

Now, let $T : X \to Y$ be another Cauchy additive mapping satisfying (4.33). Then we have

$$\|A(x) - T(x)\| = 2^q \left\| A \left( \frac{x}{2^q} \right) - T \left( \frac{x}{2^q} \right) \right\|$$

$$\leq 2^q \left( \left\| L \left( \frac{x}{2^q} \right) - f \left( \frac{x}{2^q} \right) \right\| + \left\| T \left( \frac{x}{2^q} \right) - f \left( \frac{x}{2^q} \right) \right\| \right)$$

$$\leq \frac{2^p + 2}{2^p - 2} \cdot \frac{2 \cdot 2^q}{2^{pq}} \theta \|x\|^p,$$

which tends to zero as $q \to \infty$ for all $x \in X$. So, we can conclude that $A(x) = T(x)$ for all $x \in X$. This proves the uniqueness of $A$.

Assume that $p < 1$. It follows from (4.35) that

$$\left\| f(x) - \frac{1}{2} f(2x) \right\| \leq \frac{2^p + 2}{2} \theta \|x\|^p$$

for all $x \in X$. Hence we have

$$\left\| \frac{1}{2^l} f(2^l x) - \frac{1}{2^m} f(2^m x) \right\| \leq \frac{2^p + 2}{2} \sum_{j=l}^{m-1} \frac{2^{pj}}{2^j} \theta \|x\|^p \quad (4.39)$$

for all $m, l \geq 1$ with $m > l$ and $x \in X$. It follows from (4.39) that the sequence $\{\frac{1}{2^k} f(2^k x)\}$ is a Cauchy sequence for all $x \in X$. Since $Y$ is complete, the sequence $\{\frac{1}{2^k} f(2^k x)\}$ converges and so one can define the mapping $A : X \to Y$ by

$$A(x) := \lim_{k \to \infty} \frac{1}{2^k} f \left( 2^k x \right)$$

for all $x \in X$. Letting $l = 0$ and $m \to \infty$ in (4.39), we get

$$\|f(x) - A(x)\| \le \frac{2^p + 2}{2 - 2^p} \theta \|x\|^p$$

for all $x \in X$.

The rest of the proof is similar to the above proof. So, there exists a unique Cauchy additive mapping $A : X \to Y$ such that

$$\|f(x) - A(x)\| \le \frac{2^p + 2}{|2^p - 2|} \theta \|x\|^p \tag{4.40}$$

for all $x \in X$. Assume that $f : X \to Y$ is a mapping such that the transformation $t \to f(tx)$ is continuous in $t \in \mathbb{R}$ for each fixed $x \in X$. By the same reasoning as in the proof of Theorem 4.19, one can prove that $A$ is an $\mathbb{R}$-linear mapping. This completes the proof. □

Using the fixed point method, we prove the Hyers-Ulam stability of the functional inequality (4.27) in Banach spaces.

**Theorem 4.22.** *Let $f : X \to Y$ be an odd mapping for which there exists a function $\varphi : X^4 \to [0, \infty)$ such that there exists $L < 1$ such that*

$$\varphi(x, y, z, w) \le \frac{1}{2} L \varphi(2x, 2y, 2z, 2w)$$

*for all $x, y, z, w \in X$ and*

$$\|f(x) + f(y) + f(z) + f(w)\|$$
$$\le \|f(x) + f(y + z + w)\| + \varphi(x, y, z, w) \tag{4.41}$$

*for all $x, y, z, w \in X$. Then there exists a unique Cauchy additive mapping $A : X \to Y$ such that*

$$\|f(x) - A(x)\| \le \frac{L}{2 - 2L} \varphi(0, x, x, -2x) \tag{4.42}$$

*for all $x \in X$.*

*Proof.* Consider the set $S := \{g : X \to Y\}$ and introduce the generalized metric on $S$ as follows:

$$d(g, h) = \inf\{K \in \mathbb{R}_+ : \|g(x) - h(x)\| \le K\varphi(0, x, x, -2x), \ \forall x \in X\}.$$

which $(S, d)$ is complete.

Now, we consider the linear mapping $J : S \to S$ defined by

$$Jg(x) := 2g\left(\frac{x}{2}\right)$$

for all $x \in X$. Now, we have

$$d(Jg, Jh) \leq Ld(g, h)$$

for all $g, h \in S$. Since $f : X \to Y$ is odd, $f(0) = 0$ and $f(-x) = -f(x)$ for all $x \in X$. Letting $z = y = x$ and $w = -2x$ in (4.41), we have

$$\|2f(x) - f(2x)\| = \|2f(x) + f(-2x)\| \leq \varphi(0, x, x, -2x) \qquad (4.43)$$

for all $x \in X$. It follows from (4.43) that

$$\left\| f(x) - 2f\left(\frac{x}{2}\right) \right\| \leq \varphi\left(0, \frac{x}{2}, \frac{x}{2}, -x\right) \leq \frac{L}{2}\varphi(0, x, x, -2x)$$

for all $x \in X$. Hence $d(f, Jf) \leq \frac{L}{2}$. By Theorem 1.3, there exists a mapping $A : X \to Y$ satisfying the following:

(1) $A$ is a fixed point of $J$, i.e.,

$$A\left(\frac{x}{2}\right) = \frac{1}{2}A(x) \qquad (4.44)$$

for all $x \in X$. Then $A : X \to Y$ is an odd mapping. The mapping $A$ is a unique fixed point of $J$ in the set

$$M = \{g \in S : d(f, g) < \infty\}.$$

This implies that $A$ is a unique mapping satisfying (4.44) such that there exists a $K \in (0, \infty)$ satisfying

$$\|f(x) - A(x)\| \leq K\varphi(0, x, x, -2x)$$

for all $x \in X$;

(2) $d(J^n f, A) \to 0$ as $n \to \infty$. This implies the equality

$$\lim_{n \to \infty} 2^n f\left(\frac{x}{2^n}\right) = A(x) \qquad (4.45)$$

for all $x \in X$;

(3) $d(f, A) \leq \frac{1}{1-L}d(f, Jf)$, which implies the inequality

$$d(f, A) \leq \frac{L}{2 - 2L}.$$

This implies that the inequality (4.42) holds.

It follows from (4.41) and (4.45) that

$$\|A(x) + A(y) + A(z) + A(w)\| \leq \|A(x) + A(y + z + w)\|$$

for all $x, y, z, w \in X$. By Theorem 4.19, the mapping $A : X \to Y$ is a Cauchy additive mapping. Therefore, there exists a unique Cauchy additive mapping $A : X \to Y$ satisfying (4.43). This completes the proof.                                                                          $\square$

**Corollary 4.23.** *Let $r > 1$ and $\theta$ be nonnegative real numbers and $f : X \to Y$ be an odd mapping such that*

$$\|f(x) + f(y) + f(z) + f(w)\|$$
$$\leq \|f(x) + f(y + z + w)\|$$
$$+ \theta(\|x\|^r + \|y\|^r + \|z\|^r + \|w\|^r + \|x\|^{\frac{r}{4}} \cdot \|y\|^{\frac{r}{4}} \cdot \|z\|^{\frac{r}{4}} \cdot \|w\|^{\frac{r}{4}}) \quad (4.46)$$

*for all $x, y, z, w \in X$. Then there exists a unique Cauchy additive mapping $A : X \to Y$ such that*

$$\|f(x) - A(x)\| \leq \frac{2^r + 2}{2^r - 2} \theta \|x\|^r$$

*for all $x \in X$.*

*Proof.* The proof follows from Theorem 4.22 by taking

$$\varphi(x, y, z, w)$$
$$:= \theta(\|x\|^r + \|y\|^r + \|z\|^r + \|w\|^r + \|x\|^{\frac{r}{4}} \cdot \|y\|^{\frac{r}{4}} \cdot \|z\|^{\frac{r}{4}} \cdot \|w\|^{\frac{r}{4}})$$

for all $x, y, z, w \in X$ and $L = 2^{1-r}$ and so we get the desired result.        $\square$

*Remark 4.24.* Let $f : X \to Y$ be an odd mapping for which there exists a function $\varphi : X^4 \to [0, \infty)$ satisfying (4.41). By the similar method to the proof of Theorem 4.22, one can show that, if there exists $L < 1$ such that

$$\varphi(x, y, z, w) \leq 2L\varphi\left(\frac{x}{2}, \frac{y}{2}, \frac{z}{2}, \frac{w}{2}\right)$$

for all $x, y, z, w \in X$, then there exists a unique Cauchy additive mapping $A : X \to Y$ such that

$$\|f(x) - A(x)\| \leq \frac{1}{2 - 2L} \varphi(0, x, x, -2x)$$

for all $x \in X$.

For the case $0 < r < 1$, one can obtain the similar result to Corollary 4.23:

Let $0 < r < 1$, $\theta \geq 0$ be real numbers and $f : X \to Y$ be an odd mapping satisfying (4.46). Then there exists a unique Cauchy additive mapping $A : X \to Y$ such that

$$\|f(x) - A(x)\| \le \frac{2 + 2^r}{2 - 2^r} \theta \|x\|^r$$

for all $x \in X$.

## 4.2.2   On Additive Functional Inequalities in Normed Modules over $C^*$-Algebras

In [12], An investigated the following additive functional inequality:

$$\|f(x) + f(y) + f(z) + f(w)\| \le \|f(x + y) + f(z + w)\| \qquad (4.47)$$

in normed modules over a $C^*$-algebra. This is applied to understand homomorphisms in $C^*$-algebras. Moreover, he proved the Hyers-Ulam stability of the functional following inequality:

$$\|f(x) + f(y) + f(z) + f(w)\|$$
$$\le \|f(x + y + z + w)\| + \theta \|x\|^p \|y\|^p \|z\|^p \|w\|^p \qquad (4.48)$$

in real Banach spaces, where $\theta$ and $p$ are positive real numbers with $4p \ne 1$.

Gilányi [126] showed that, if $f$ satisfies the functional following inequality:

$$\|2f(x) + 2f(y) - f(x - y)\| \le \|f(x + y)\|, \qquad (4.49)$$

then $f$ satisfies the Jordan-von Neumann functional equation:

$$2f(x) + 2f(y) = f(x + y) + f(x - y).$$

Fechner [113] and Gilányi [127] proved the Hyers-Ulam stability of the functional inequality (4.49). Park et al. [253] investigated the functional following inequality:

$$\|f(x) + f(y) + f(z)\| \le \|f(x + y + z)\| \qquad (4.50)$$

in Banach spaces and proved the Hyers-Ulam stability of the functional inequality (4.50) in Banach spaces.

Now, let $A$ be a unital $C^*$-algebra with the unitary group $U(A)$ and the unit $e$ and let $B$ be a $C^*$-algebra. Assume that $X$ is a normed $A$-module with the norm $\| \cdot \|_X$ and $Y$ is a normed $A$-module with the norm $\| \cdot \|_Y$.

**Theorem 4.25.** *Let $f : X \to Y$ be a mapping such that*

$$\|uf(x) + f(y) + f(z) + f(w)\|_Y \le \|f(ux + y) + f(z + w)\|_Y \qquad (4.51)$$

*for all $x, y, z, w \in X$ and $u \in U(A)$. Then the mapping $f : X \to Y$ is A-linear.*

*Proof.* Letting $x = y = z = w = 0$ and $u = e \in U(A)$ in (4.51), we have

$$\|4f(0)\|_Y \leq \|2f(0)\|_Y$$

and so $f(0) = 0$. Letting $z = w = 0$ in (4.51), we have

$$\|f(x) + f(y)\|_Y \leq \|f(x + y)\|_Y \tag{4.52}$$

for all $x, y \in X$. Replacing $x$ and $y$ by $x + y$ and $z + w$ in (4.52), respectively, we have

$$\|f(x + y) + f(z + w)\|_Y \leq \|f(x + y + z + w)\|_Y$$

for all $x, y, z, w \in X$ and so

$$\|f(x) + f(y) + f(z) + f(w)\|_Y \leq \|f(x + y + z + w)\|_Y \tag{4.53}$$

for all $x, y, z, w \in X$. Letting $z = w = 0$ and $y = -x$ in (4.53), we have

$$\|f(x) + f(-x)\|_Y \leq \|f(0)\|_Y = 0$$

for all $x \in X$. Hence $f(-x) = -f(x)$ for all $x \in X$. Letting $z = -x - y$ and $w = 0$ in (4.53), we have

$$\|f(x) + f(y) - f(x + y)\|_Y = \|f(x) + f(y) + f(-x - y)\|_Y \leq \|f(0)\|_Y = 0$$

for all $x, y \in X$. Thus we have

$$f(x + y) = f(x) + f(y)$$

for all $x, y \in X$. Letting $y = -ux$ and $y = w = 0$ in (4.51), we have

$$\|f(ux) - f(ux)\|_Y = \|f(ux) + f(-ux)\|_Y \leq \|2f(0)\|_Y = 0$$

for all $x \in X$ and $u \in U(A)$. Thus we have

$$f(ux) = uf(x) \tag{4.54}$$

for all $u \in U(A)$ and $x \in X$.

Now, let $a \in A$ with $a \neq 0$ and $M$ be an integer greater than $4|a|$. Then we have

$$\left|\frac{a}{M}\right| < \frac{1}{4} < 1 - \frac{2}{3} = \frac{1}{3}.$$

By Theorem 1 of Kadison and Pedersen [167], there exist three elements $u_1, u_2, u_3 \in U(A)$ such that $3\frac{a}{M} = u_1 + u_2 + u_3$ and so, by (4.54), we have

$$f(ax) = f\left(\frac{M}{3} \cdot 3\frac{a}{M}x\right) = M \cdot f\left(\frac{1}{3} \cdot 3\frac{a}{M}x\right) = \frac{M}{3}f\left(3\frac{a}{M}x\right)$$
$$= \frac{M}{3}h(u_1x + u_2x + u_3x) = \frac{M}{3}(f(u_1x) + f(u_2x) + f(u_3x))$$
$$= \frac{M}{3}(u_1 + u_2 + u_3)f(x) = \frac{M}{3} \cdot 3\frac{a}{M}f(x) = af(x)$$

for all $x \in X$. Therefore, $f : X \to Y$ is $A$-linear. This completes the proof. □

**Corollary 4.26.** *Let* $f : A \to B$ *be a multiplicative mapping such that*

$$\|\mu f(x) + f(y) + f(z) + f(w)\| \leq \|f(\mu x + y) + f(z + w)\| \qquad (4.55)$$

*for all* $x, y, z, w \in A$ *and* $\mu \in \mathbb{T} := \{\lambda \in \mathbb{C} : |\lambda| = 1\}$. *Then the mapping* $f : A \to B$ *is a $C^*$-algebra homomorphism.*

*Proof.* By Theorem 4.25, the multiplicative mapping $f : A \to B$ is $\mathbb{C}$-linear since $C^*$-algebras are normed modules over $\mathbb{C}$. So, the multiplicative mapping $f : A \to B$ is a $C^*$-algebra homomorphism. □

**Theorem 4.27.** *Let $X$ be a real normed linear space and $Y$ be a real Banach space. Assume that $f : X \to Y$ is an approximately additive odd mapping for which there exist the constants $\theta \geq 0$ and $p \in \mathbb{R}$ such that $4p \neq 1$ and $f$ satisfies the general Cauchy-Rassias inequality:*

$$\|f(x) + f(y) + f(z) + f(w)\|$$
$$\leq \|f(x + y + z + w)\| + \theta \|x\|^p \|y\|^p \|z\|^p \|w\|^p \qquad (4.56)$$

*for all $x, y, z, w \in X$. Then there exists a unique additive mapping $L : X \to Y$ such that*

$$\|f(x) - L(x)\| \leq \frac{3^p \theta}{|81^p - 3|} \|x\|^{4p} \qquad (4.57)$$

*for all $x \in X$. If, in addition, $f : X \to Y$ is a mapping such that the transformation $t \to f(tx)$ is continuous in $t \in \mathbb{R}$ for each fixed $x \in X$, then $L$ is an $\mathbb{R}$-linear mapping.*

*Proof.* Since $f$ is odd, $f(0) = 0$ and $f(-x) = -f(x)$ for all $x \in X$. Letting $y = z = x$ and $w = -3x$ in (4.56), we have

$$\|3f(x) - f(3x)\| \leq 3^p \theta \|x\|^{4p} \qquad (4.58)$$

for all $x \in X$ and so

$$\left\|f(x) - 3f\left(\frac{x}{3}\right)\right\| \leq \frac{\theta}{27^p} \|x\|^{4p}$$

for all $x \in X$. Hence we have

$$\left\| 3^l f\left(\frac{x}{3^l}\right) - 3^m f\left(\frac{x}{3^m}\right) \right\| \leq \frac{\theta}{27^p} \sum_{j=l}^{m-1} \frac{3^j}{81^{pj}} ||x||^{4p} \tag{4.59}$$

for all $m, l \geq 1$ with $m > l$ and $x \in X$.

Assume that $p > \frac{1}{4}$. It follows from (4.59) that the sequence $\{3^k f(\frac{x}{3^k})\}$ is a Cauchy sequence for all $x \in X$. Since $Y$ is complete, the sequence $\{3^k f(\frac{x}{3^k})\}$ converges and so one can define the mapping $L : X \to Y$ by

$$L(x) := \lim_{k \to \infty} 3^k f\left(\frac{x}{3^k}\right)$$

for all $x \in X$. Letting $l = 0$ and $m \to \infty$ in (4.59), we have

$$\|f(x) - L(x)\| \leq \frac{3^p \theta}{81^p - 3} ||x||^{4p}$$

for all $x \in X$. It follows from (4.56) that

$$\left\| 3^k f\left(\frac{x}{3^k}\right) + 3^k f\left(\frac{y}{3^k}\right) + 3^k f\left(\frac{z}{3^k}\right) + 3^k f\left(\frac{w}{3^k}\right) \right\|$$
$$\leq \left\| 3^k f\left(\frac{x+y+z+w}{3^k}\right) \right\| + \frac{3^k \theta}{81^{pk}} ||x||^p ||y||^p ||z||^p ||w||^p \tag{4.60}$$

for all $x, y, z, w \in X$. Letting $k \to \infty$ in (4.60), we get

$$\|L(x) + L(y) + L(z) + L(w)\| \leq \|L(x + y + z + w)\| \tag{4.61}$$

for all $x, y, z, w \in X$. It is easy to show that $L : X \to Y$ is odd. Letting $z = -x - y$ and $w = 0$ in (4.61), we get $L(x + y) = L(x) + L(y)$ for all $x, y \in X$. So, there exists an additive mapping $L : X \to Y$ satisfying (4.57) for the case $p > \frac{1}{4}$.

Now, let $T : X \to Y$ be another additive mapping satisfying (4.57). Then we have

$$\|L(x) - T(x)\| = 3^q \left\| L\left(\frac{x}{3^q}\right) - T\left(\frac{x}{3^q}\right) \right\|$$
$$\leq 3^q \left( \left\| L\left(\frac{x}{3^q}\right) - f\left(\frac{x}{3^q}\right) \right\| + \left\| T\left(\frac{x}{3^q}\right) - f\left(\frac{x}{3^q}\right) \right\| \right)$$
$$\leq \frac{3^p \theta}{81^p - 3} \cdot \frac{2 \cdot 3^q}{81^q} ||x||^{4p},$$

which tends to zero as $q \to \infty$ for all $x \in X$. So we can conclude that $L(x) = T(x)$ for all $x \in X$. This proves the uniqueness of $L$.

Assume that $p < \frac{1}{4}$. It follows from (4.58) that

$$\left\| f(x) - \frac{1}{3}f(3x) \right\| \le 3^{p-1}\theta \|x\|^{4p}$$

for all $x \in X$. Hence we have

$$\left\| \frac{1}{3^l}f(3^l x) - \frac{1}{3^m}f(3^m x) \right\| \le 3^{p-1}\theta \sum_{j=l}^{m-1} \frac{81^{pj}}{3^j}||x||^{4p} \qquad (4.62)$$

for all $m, l \ge 1$ with $m > l$ and $x \in X$. It follows from (4.62) that the sequence $\{\frac{1}{3^k}f(3^k x)\}$ is a Cauchy sequence for all $x \in X$. Since $Y$ is complete, the sequence $\{\frac{1}{3^k}f(3^k x)\}$ converges and so one can define the mapping $L : X \to Y$ by

$$L(x) := \lim_{k\to\infty} \frac{1}{3^k}f\left(3^k x\right)$$

for all $x \in X$. Letting $l = 0$ and $m \to \infty$ in (4.62), we have

$$\|f(x) - L(x)\| \le \frac{3^p\theta}{3 - 81^p}||x||^{4p}$$

for all $x \in X$.

The rest of the proof is similar to the above proof. So, there exists a unique additive mapping $L : X \to Y$ satisfying

$$\|f(x) - L(x)\| \le \frac{3^p\theta}{|81^p - 3|}\|x\|^{4p} \qquad (4.63)$$

for all $x \in X$. Assume that $f : X \to Y$ is a mapping such that the transformation $t \to f(tx)$ is continuous in $t \in \mathbb{R}$ for each fixed $x \in X$. By the same reasoning as in the proof of Theorem 4.25, one can prove that $L$ is an $\mathbb{R}$-linear mapping This completes the proof.                                                                              $\square$

### 4.2.3  Generalization of Cauchy-Rassias Inequalities via the Fixed Point Method

Now, we improve the results of An's results [12], which presented at last pages. In fact, we get a better error estimation of main result of An by applying the fixed point alternative theorem.

**Theorem 4.28.** *Let X be a real normed linear space and Y be a real Banach space. Assume that $f : X \to Y$ is an approximately additive odd mapping satisfying the general Cauchy-Rassias inequality:*

$$\|f(x) + f(y) + f(z) + f(w)\|$$
$$\leq \|f(x + y + z + w)\| + \varphi(x, y, z, w) \qquad (4.64)$$

*for all $x, y, z, w \in X$, where $\varphi : X^4 \to [0, \infty)$ is a given function. If there exists $0 < L < 1$ such that*

$$\varphi(x, y, z, w) \leq \frac{1}{3} L \varphi(3x, 3y, 3z, 3w) \qquad (4.65)$$

*for all $x, y, z, w \in X$. Then there exists a unique additive mapping $A : X \to Y$ such that*

$$\|f(x) - A(x)\| \leq \frac{L}{3 - 3L} \varphi(x, x, x, -3x) \qquad (4.66)$$

*for all $x \in X$. If, in addition, $f : X \to Y$ is a mapping such that the transformation $t \to f(tx)$ is continuous in $t \in \mathbb{R}$ for each fixed $x \in X$, then A is an $\mathbb{R}$-linear mapping.*

*Proof.* Since $f$ is odd, $f(0) = 0$ and $f(-x) = -f(x)$ for all $x \in X$. Consider the set $S := \{g : X \to Y\}$ and introduce the generalized metric on $S$ as follows:

$$d(g, h) = \inf\{C \in \mathbb{R}_+ : \|g(x) - h(x)\| \leq C \varphi(x, x, x, -3x), \ \forall x \in X\}.$$

Now, we show that $(S, d)$ is complete. Let $\{h_n\}$ be a Cauchy sequence in $(S, d)$. Then, for any $\varepsilon > 0$, there exists an integer $N_\varepsilon > 0$ such that $d(h_m, h_n) < \varepsilon$ for all $m, n \geq N_\varepsilon$. Then there exists $C \in (0, \varepsilon)$ such that

$$\|h_m(x) - h_n(x)\| \leq C \varphi(x, x, x, -3x) \leq \varepsilon \varphi(x, x, x, -3x) \qquad (4.67)$$

for all $m, n \geq N_\varepsilon$ and $x \in X$. Since $Y$ is complete, $\{h_n(x)\}$ converges for each $x \in X$. Thus a mapping $h : X \to Y$ can be defined by

$$h(x) := \lim_{n \to \infty} h_n(x) \qquad (4.68)$$

for all $x \in X$. Letting $n \to \infty$ in (4.67), we have

$$m \geq N_\varepsilon \implies \|h_m(x) - h_n(x)\| \leq \varepsilon \varphi(x, x, x, -3x)$$
$$\implies \varepsilon \in \{C \in \mathbb{R}_+ : \|g(x) - h(x)\| \leq C \varphi(x, x, x, -3x), \ \forall x \in X\}$$
$$\implies d(h_m, h) \leq \varepsilon$$

for all $x \in X$. This means that the Cauchy sequence $\{h_n\}$ converges to $h$ in $(S, d)$. Hence $(S, d)$ is complete.

Now, we consider the linear mapping $\Lambda : S \to S$ defined by

$$\Lambda g(x) := 3g\left(\frac{x}{3}\right) \tag{4.69}$$

for all $x \in X$. We show that $\Lambda$ is a strictly contractive on $S$. For any $g, h \in S$, let $C_{g,h} \geq 0$ be an arbitrary constant with $d(g, h) \leq C_{g,h}$. Then we have

$$d(g, h) \leq C_{g,h} \implies \|g(x) - h(x)\| \leq C_{g,h}\varphi(x, x, x, -3x)$$

$$\implies \left\|3g\left(\frac{x}{3}\right) - 3h\left(\frac{x}{3}\right)\right\| \leq 3C_{g,h}\varphi\left(\frac{x}{3}, \frac{x}{3}, \frac{x}{3}, -x\right)$$

$$\implies \left\|3g\left(\frac{x}{3}\right) - 3h\left(\frac{x}{3}\right)\right\| \leq LC_{g,h}\varphi(x, x, x, -3x)$$

for all $x \in X$. This means $d(\Lambda g, \Lambda h) \leq LC_{g,h}$. Hence we see that $d(\Lambda g, \Lambda h) \leq Ld(g, h)$ for any $g, h \in S$. Therefore, $\Lambda$ is a strictly contractive on $S$ with the Lipschitz constant $0 < L < 1$.

Letting $y = z = x$ and $w = -3x$ in (4.64), we have

$$\|3f(x) - f(3x)\| \leq \varphi(x, x, x, -3x) \tag{4.70}$$

for all $x \in X$ and so

$$\left\|f(x) - 3f\left(\frac{x}{3}\right)\right\| \leq \varphi\left(\frac{x}{3}, \frac{x}{3}, \frac{x}{3}, -x\right) \leq \frac{1}{3}L\varphi(x, x, x, -3x)$$

for all $x \in X$. Thus we have

$$d(f, \Lambda f) \leq \frac{L}{3}.$$

Therefore, it follows of Theorem 1.3 that the sequence $\{\Lambda^n f\}$ converges to the unique fixed point $A$ of $\Lambda$, i.e.,

$$A(x) = (\Lambda f)(x) =:= \lim_{k \to \infty} 3^k f\left(\frac{x}{3^k}\right)$$

and $A(3x) = 3A(x)$ for all $x \in X$. Also, we have

$$d(A, f) \leq \frac{1}{1-L}d(\Lambda f, f) \leq \frac{L}{3 - 3L},$$

which means that (4.66) holds.

Assume that $f : X \to Y$ is a mapping such that the transformation $t \to f(tx)$ is continuous in $t \in \mathbb{R}$ for each fixed $x \in X$. By the same reasoning as in the proof of Theorem 4.25, one can prove that $A$ is an $\mathbb{R}$-linear mapping. This completes the proof. $\qquad\square$

In the following, we get a better error estimation of the main result of An [12].

**Corollary 4.29.** *Let $X$ be a real normed linear space and $Y$ be a real Banach space. Assume that $f : X \to Y$ is an approximately additive odd mapping for which there exist the constants $\theta \geq 0$ and $p \in \mathbb{R}$ such that $4p \neq 1$ and $f$ satisfies the general Cauchy-Rassias inequality:*

$$\|f(x) + f(y) + f(z) + f(w)\|$$
$$\leq \|f(x + y + z + w)\| + \theta \|x\|^p \|y\|^p \|z\|^p \|w\|^p$$

*for all $x, y, z, w \in X$. Then there exists a unique additive mapping $A : X \to Y$ such that*

$$\|f(x) - A(x)\| \leq \frac{3^{p-1}\theta}{|81^p - 3|} \|x\|^{4p}$$

*for all $x \in X$. If, in addition, $f : X \to Y$ is a mapping such that the transformation $t \to f(tx)$ is continuous in $t \in \mathbb{R}$ for each fixed $x \in X$, then $A$ is an $\mathbb{R}$-linear mapping.*

*Proof.* In Theorem 4.28, take

$$\varphi(x, y, z, w) := \theta \|x\|^p \|y\|^p \|z\|^p \|w\|^p$$

for all $x, y, z, w \in X$ and $L = 81^{1-p}$. Then we have desired result. $\square$

**Corollary 4.30.** *Let $X$ be a real normed linear space and $Y$ be a real Banach space. Assume that $f : X \to Y$ is an approximately additive odd mapping for which there exist the constants $\theta \geq 0$ and $p \in \mathbb{R}$ such that $4p \neq 1$ and $f$ satisfies the general Cauchy-Rassias inequality:*

$$\|f(x) + f(y) + f(z) + f(w)\|$$
$$\leq \|f(x + y + z + w)\| + \theta(\|x\|^p + \|y\|^p + \|z\|^p + \|w\|^p)$$

*for all $x, y, z, w \in X$. Then there exists a unique additive mapping $A : X \to Y$ such that*

$$\|f(x) - A(x)\| \leq \frac{(3^p + 3)\theta}{|3^p - 3|} \|x\|^p$$

*for all $x \in X$. If, in addition, $f : X \to Y$ is a mapping such that the transformation $t \to f(tx)$ is continuous in $t \in \mathbb{R}$ for each fixed $x \in X$, then $A$ is an $\mathbb{R}$-linear mapping.*

*Proof.* In Theorem 4.28, take

$$\varphi(x, y, z, w) := \theta \|x\|^p + \|y\|^p + \|z\|^p + \|w\|^p$$

for all $x, y, z, w \in X$ and $L = 3^{1-p}$. Then we have desired result. $\square$

# Chapter 5
# Stability of Functional Equations in $C^*$-Ternary Algebras

Ternary algebraic operations were considered in the nineteenth century by several mathematicians such as Cayley [68] who introduced the notion of cubic matrix which, in turn, was generalized by Kapranov et al. [169]. The simplest example of such non-trivial ternary operation is given by the following composition rule:

$$\{a, b, c\}_{ijk} = \sum_{1 \le l, m, n \le N} a_{nil} b_{ljm} c_{mkn}$$

for each $i, j, k = 1, 2, \cdots, N$.

Ternary structures and their generalization, the so-called $n$-ary structures, raise certain hopes in view of their applications in physics. Some significant applications are as follows (see [171] and [172]):

(1) The algebra of nonions generated by two matrices

$$\begin{pmatrix} 0\ 1\ 0 \\ 0\ 0\ 1 \\ 1\ 0\ 0 \end{pmatrix}, \quad \begin{pmatrix} 0\ \ 1\ 0 \\ 0\ \ 0\ \omega \\ \omega^2\ 0\ 0, \end{pmatrix}$$

where $\omega = e^{\frac{2\pi i}{3}}$, was introduced by Sylvester as a ternary analog of Hamiltons quaternions (see [2]).

(2) The quark model inspired a particular brand of ternary algebraic systems. The so-called Nambu mechanics is based on such structures (see [93]).

There are also some applications, although still hypothetical, in the fractional quantum Hall effect, the nonstandard statistics, supersymmetric theory and the Yang-Baxter equation (see [2, 171] and [325]).

In Sect. 5.1, we prove the Hyers-Ulam stability of $C^*$-ternary 3-derivations and of $C^*$-ternary 3-homomorphisms for the following functional equation:

© Springer International Publishing Switzerland 2015
Y.J. Cho et al., *Stability of Functional Equations in Banach Algebras*, DOI 10.1007/978-3-319-18708-2_5

$$f(x_1 + x_2, y_1 + y_2, z_1 + z_2) = \sum_{1 \leq i,j,k \leq 2} f(x_i, y_j, z_k)$$

in $C^*$-ternary algebras.

In Sect. 5.2, we consider the following Apollonius type additive functional equation:

$$f(z - x) + f(z - y) = -\frac{1}{2}f(x + y) + 2f\left(z - \frac{x + y}{4}\right)$$

and prove the superstability of $C^*$-ternary homomorphisms, $C^*$-ternary derivations, $JB^*$-triple homomorphisms and $JB^*$-triple derivations by using the fixed point method.

In Sect. 5.3, we prove the Hyers-Ulam stability of bi-$\theta$-derivations on $JB^*$-triples.


## 5.1  $C^*$-Ternary 3-Homomorphism and $C^*$-Ternary 3-Derivations

In this section, we prove the Hyers-Ulam stability of $C^*$-ternary 3-derivations and of $C^*$-ternary 3-homomorphisms for the following functional equation:

$$f(x_1 + x_2, y_1 + y_2, z_1 + z_2) = \sum_{1 \leq i,j,k \leq 2} f(x_i, y_j, z_k)$$

in $C^*$-ternary algebras (see [94]).

Let $X$ and $Y$ be complex vector spaces. A mapping $f : X \times X \times X \to Y$ is called a 3-*additive mapping* if $f$ is additive for each variable and a mapping $f : X \times X \times X \to Y$ is called a 3-$\mathbb{C}$-*linear mapping* if $f$ is $\mathbb{C}$-linear for each variable. A 3-$\mathbb{C}$-linear mapping $H : A \times A \times A \to B$ is called a $C^*$-*ternary 3-homomorphism* if it satisfies

$$H([x_1, y_1, z_1], [x_2, y_2, z_2], [x_3, y_3, z_3])$$
$$= [H(x_1, x_2, x_3), H(y_1, y_2, y_3), H(z_1, z_2, z_3)]$$

for all $x_1, y_1, z_1, x_2, y_2, z_2, x_3, y_3, z_3 \in A$. For a given mapping $f : A^3 \to B$, we define

$$D_{\lambda,\mu,\nu}f(x_1, x_2, y_1, y_2, z_1, z_2)$$
$$:= f(\lambda x_1 + \lambda x_2, \mu y_1 + \mu y_2, \nu z_1 + \nu z_2) - \lambda\mu\nu \sum_{1 \leq i,j,k \leq 2} f(x_i, y_j, z_k)$$

for all $\lambda, \mu, \nu \in S^1 := \{\lambda \in \mathbb{C} : |\lambda| = 1\}$ and $x_1, x_2, y_1, y_2, z_1, z_2 \in A$.

Bae and Park [21] proved the Hyers–Ulam stability of 3-homomorphisms in $C^*$-ternary algebras for the following functional equation:

$$D_{\lambda,\mu,\nu}f(x_1, x_2, y_1, y_2, z_1, z_2) = 0.$$

**Lemma 5.1 ([21]).** *Let X and Y be complex vector spaces and $f : X \times X \times X \to Y$ be a 3-additive mapping such that*

$$f(\lambda x, \mu y, \nu z) = \lambda\mu\nu f(x, y, z)$$

*for all $\lambda, \mu, \nu \in S^1$ and $x, y, z \in X$. Then f is 3-$\mathbb{C}$-linear.*

**Theorem 5.2 ([21]).** *Let $p, q, r \in (0, \infty)$ with $p + q + r < 3$ and $\theta \in (0, \infty)$ and $f : A^3 \to B$ be a mapping such that*

$$\|D_{\lambda,\mu,\nu}f(x_1, x_2, y_1, y_2, z_1, z_2)\| \tag{5.1}$$

$$\leq \theta \cdot \max\{\|x_1\|, \|x_2\|\}^p \cdot \max\{\|y_1\|, \|y_2\|\}^q \cdot \max\{\|z_1\|, \|z_2\|\}^r$$

*and*

$$\|f([x_1, y_1, z_1], [x_2, y_2, z_2], [x_3, y_3, z_3])$$

$$-[f(x_1, x_2, x_3), f(y_1, y_2, y_3), f(z_1, z_2, z_3)]\| \tag{5.2}$$

$$\leq \theta \sum_{i=1}^{3} \|x_i\|^p \cdot \|y_i\|^q \cdot \|z_i\|^r$$

*for all $\lambda, \mu, \nu \in S^1$ and $x_1, x_2, x_3, y_1, y_2, y_3, z_1, z_2, z_3 \in A$. Then there exists a unique $C^*$-ternary 3-homomorphism $H : A^3 \to B$ such that*

$$\|f(x, y, z) - H(x, y, z)\| \leq \frac{\theta}{2^3 - 2^{p+q+r}} \|x\|^p \cdot \|y\|^q \cdot \|z\|^r \tag{5.3}$$

*for all $x, y, z \in A$.*

### 5.1.1   *C\*-Ternary 3-Homomorphisms in C\*-Ternary Algebras*

Now, we investigate $C^*$-ternary 3-homomorphisms in $C^*$-ternary algebras.

**Theorem 5.3.** *Let $p, q, r \in (0, \infty)$ with $p + q + r < 3$ and $\theta \in (0, \infty)$ and $f : A^3 \to B$ be a mapping satisfying (5.1) and (5.2). If there exists $(x_0, y_0, z_0) \in A^3$ such that*

$$\lim_{n \to \infty} \frac{1}{8^n} f(2^n x_0, 2^n y_0, 2^n z_0) = e',$$

*then the mapping f is a $C^*$-ternary 3-homomorphism.*

*Proof.* By Theorem 5.2, there exists a unique $C^*$-ternary 3-homomorphism $H : A^3 \to B$ satisfying (5.3). Note that

$$H(x, y, z) := \lim_{n \to \infty} \frac{1}{8^n} f(2^n x, 2^n y, 2^n z)$$

for all $x, y, z \in A$. By the assumption, we have

$$H(x_0, y_0, z_0) = \lim_{n \to \infty} \frac{1}{8^n} f(2^n x_0, 2^n y_0, 2^n z_0) = e'.$$

It follows from (5.2) that

$$\|[H(x_1, x_2, x_3), H(y_1, y_2, y_3), H(z_1, z_2, z_3)]$$
$$\quad -[H(x_1, x_2, x_3), H(y_1, y_2, y_3), f(z_1, z_2, z_3)]\|$$
$$= \|H([x_1, y_1, z_1], [x_2, y_2, z_2], [x_3, y_3, z_3])$$
$$\quad -[H(x_1, x_2, x_3), H(y_1, y_2, y_3), f(z_1, z_2, z_3)]\|$$
$$= \lim_{n \to \infty} \frac{1}{8^{2n}} \|f([2^n x_1, 2^n y_1, z_1], [2^n x_2, 2^n y_2, z_2], [2^n x_3, 2^n y_3, z_3])$$
$$\quad -[f(2^n x_1, 2^n x_2, 2^n x_3), f(2^n y_1, 2^n y_2, 2^n y_3), f(z_1, z_2, z_3)]\|$$
$$\leq \lim_{n \to \infty} \frac{\theta 2^{n(p+q)}}{8^{2n}} \sum_{i=1}^{3} \|x_i\|^p \cdot \|y_i\|^q \cdot \|z_i\|^r$$
$$= 0$$

for all $x_1, x_2, x_3, y_1, y_2, y_3, z_1, z_2, z_3 \in A$ and so

$$[H(x_1, x_2, x_3), H(y_1, y_2, y_3), H(z_1, z_2, z_3)]$$
$$= [H(x_1, x_2, x_3), H(y_1, y_2, y_3), f(z_1, z_2, z_3)]$$

for all $x_1, x_2, x_3, y_1, y_2, y_3, z_1, z_2, z_3 \in A$. Letting $x_1 = y_1 = x_0, x_2 = y_2 = y_0$ and $x_3 = y_3 = z_0$ in the last equality, we get $f(z_1, z_2, z_3) = H(z_1, z_2, z_3)$ for all $z_1, z_2, z_3 \in A$. Therefore, the mapping $f$ is a $C^*$-ternary 3-homomorphism. This completes the proof. $\qquad\square$

**Theorem 5.4.** *Let* $p_i, q_i, r_i \in (0, \infty)$ $(i = 1, 2, 3)$ *such that* $p_i \neq 1$ *or* $q_i \neq 1$ *or* $r_i \neq 1$ *for some* $1 \leq i \leq 3$, $\theta, \eta \in (0, \infty)$ *and* $f : A^3 \to B$ *be a mapping such that*

$$\|D_{\lambda, \mu, \nu} f(x_1, x_2, y_1, y_2, z_1, z_2)\|$$
$$\leq \theta(\|x_1\|^{p_1} \cdot \|x_2\|^{p_2} \cdot \|y_1\|^{q_1} \cdot \|y_2\|^{q_2} \qquad\qquad (5.4)$$
$$\quad + \|y_1\|^{q_1} \cdot \|y_2\|^{q_2} \cdot \|z_1\|^{r_1} \cdot \|z_2\|^{r_2} + \|x_1\|^{p_1} \cdot \|x_2\|^{p_2} \cdot \|z_1\|^{r_1} \cdot \|z_2\|^{r_2})$$

*and*

$$\|f([x_1, y_1, z_1], [x_2, y_2, z_2], [x_3, y_3, z_3])$$
$$-[f(x_1, x_2, x_3), f(y_1, y_2, y_3), f(z_1, z_2, z_3)]\|$$
$$\leq \eta \|x_1\|^{p_1} \cdot \|x_2\|^{p_2} \cdot \|x_3\|^{p_3} \tag{5.5}$$
$$\cdot \|y_1\|^{q_1} \cdot \|y_2\|^{q_2} \cdot \|y_3\|^{q_3} \cdot \|z_1\|^{r_1} \cdot \|z_2\|^{r_2} \cdot \|z_3\|^{r_3}$$

*for all* $\lambda, \mu, \nu \in S^1$ *and* $x_1, x_2, x_3, y_1, y_2, y_3, z_1, z_2, z_3 \in A$. *Then the mapping* $f : A^3 \to B$ *is a* $C^*$-*ternary 3-homomorphism.*

*Proof.* Letting $x_i = y_j = z_k = 0$ $(i, j, k = 1, 2)$ in (5.4), we get $f(0, 0, 0) = 0$. Putting $\lambda = \mu = \nu = 1, x_2 = 0$ and $y_j = z_k = 0$ $(j, k = 1, 2)$ in (5.4), we have $f(x_1, 0, 0) = 0$ for all $x_1 \in A$. Similarly, we get $f(0, y_1, 0) = f(0, 0, z_1) = 0$ for all $y_1, z_1 \in A$. Setting $\lambda = \mu = \nu = 1, x_2 = 0, y_2 = 0$ and $z_1 = z_2 = 0$, we have $f(x_1, y_1, 0) = 0$ for all $x_1, y_1 \in A$. Similarly, we get $f(x_1, 0, z_1) = f(0, y_1, z_1) = 0$ for all $x_1, y_1, z_1 \in A$.

Now, letting $\lambda = \mu = \nu = 1$ and $y_2 = z_2 = 0$ in (5.4), we have

$$f(x_1 + x_2, y_1, z_1) = f(x_1, y_1, z_1) + f(x_2, y_1, z_1)$$

for all $x_1, x_2, y_1, z_1 \in A$. Similarly, one can show that the other equations hold and so $f$ is 3-additive. Letting $x_2 = y_2 = z_2 = 0$ in (5.4), we get

$$f(\lambda x_1, \mu y_1, \nu z_1) = \lambda \mu \nu f(x_1, y_1, z_1)$$

for all $\lambda, \mu, \nu \in S^1$ and $x_1, y_1, z_1 \in A$. So, by Lemma 5.1, the mapping $f$ is 3-$\mathbb{C}$-linear.

Without any loss of generality, we may suppose that $p_1 \neq 1$. Let $p_1 < 1$. It follows from (5.5) that

$$\|f([x_1, y_1, z_1], [x_2, y_2, z_2], [x_3, y_3, z_3])$$
$$-[f(x_1, x_2, x_3), f(y_1, y_2, y_3), f(z_1, z_2, z_3)]\|$$
$$= \lim_{n \to \infty} \frac{1}{3^n} \|f([3^n x_1, y_1, z_1], [x_2, y_2, z_2], [x_3, y_3, z_3])$$
$$-[f(3^n x_1, x_2, x_3), f(y_1, y_2, y_3), f(z_1, z_2, z_3)]\|$$
$$\leq \eta \lim_{n \to \infty} \frac{3^{n p_1}}{3^n} (\|x_1\|^{p_1} \cdot \|x_2\|^{p_2} \cdot \|x_3\|^{p_3} \cdot \|y_1\|^{q_1} \cdot \|y_2\|^{q_2}$$
$$\cdot \|y_3\|^{q_3} \cdot \|z_1\|^{r_1} \cdot \|z_2\|^{r_2} \cdot \|z_3\|^{r_3})$$
$$= 0$$

for all $x_1, x_2, x_3, y_1, y_2, y_3, z_1, z_2, z_3 \in A$. Let $p_1 > 1$. It follows from (5.5) that

$$\|f([x_1, y_1, z_1], [x_2, y_2, z_2], [x_3, y_3, z_3])$$
$$-[f(x_1, x_2, x_3), f(y_1, y_2, y_3), f(z_1, z_2, z_3)]\|$$
$$= \lim_{n \to \infty} 3^n \left\| f\left( \left[ \frac{1}{3^n} x_1, y_1, z_1 \right], [x_2, y_2, z_2], [x_3, y_3, z_3] \right) \right.$$
$$\left. - \left[ f\left( \frac{1}{3^n} x_1, x_2, x_3 \right), f(y_1, y_2, y_3), f(z_1, z_2, z_3) \right] \right\|$$
$$\leq \eta \lim_{n \to \infty} \frac{3^n}{3^{np_1}} (\|x_1\|^{p_1} \cdot \|x_2\|^{p_2} \cdot \|x_3\|^{p_3} \cdot \|y_1\|^{q_1} \cdot \|y_2\|^{q_2}$$
$$\cdot \|y_3\|^{q_3} \cdot \|z_1\|^{r_1} \cdot \|z_2\|^{r_2} \cdot \|z_3\|^{r_3})$$
$$= 0$$

for all $x_1, x_2, x_3, y_1, y_2, y_3, z_1, z_2, z_3 \in A$. Therefore, we have

$$f([x_1, y_1, z_1], [x_2, y_2, z_2], [x_3, y_3, z_3])$$
$$= [f(x_1, x_2, x_3), f(y_1, y_2, y_3), f(z_1, z_2, z_3)]$$

for all $x_1, x_2, x_3, y_1, y_2, y_3, z_1, z_2, z_3 \in A$. Therefore, the mapping $f : A^3 \to B$ is a $C^*$-ternary 3-homomorphism. This completes the proof.                    $\square$

*Remark 5.5.* Let $\varphi : A^6 \to [0, \infty)$ and $\psi : A^9 \to [0, \infty)$ be the functions such that

$$\varphi(x_1, \cdots, x_6) = 0$$

if $x_i = 0$ for some $1 \leq i \leq 6$ and

$$\frac{1}{3^n} \psi(x_1, \cdots, 3^n x_i, \cdots, x_9) = 0$$

or

$$3^n \psi \left( x_1, \cdots, \frac{1}{3^n} x_i, \cdots, x_9 \right) = 0.$$

Suppose that $f : A^3 \to B$ is a mapping satisfying

$$\|D_{\lambda, \mu, \nu} f(x_1, x_2, y_1, y_2, z_1, z_2)\| \leq \varphi(x_1, x_2, y_1, y_2, z_1, z_2)$$

and

$$\|f([x_1, y_1, z_1], [x_2, y_2, z_2], [x_3, y_3, z_3])$$
$$-[f(x_1, x_2, x_3), f(y_1, y_2, y_3), f(z_1, z_2, z_3)]\|$$
$$\leq \psi(x_1, x_2, x_3, y_1, y_2, y_3, z_1, z_2, z_3)$$

for all $\lambda, \mu, \nu \in S^1$ and $x_1, x_2, x_3, y_1, y_2, y_3, z_1, z_2, z_3 \in A$. Then the mapping $f$ is a $C^*$-ternary 3-homomorphism.

**Corollary 5.6.** *Let $p_i, q_i, r_i \in (0, \infty)$ ($i = 1, 2, 3$) such that $p_i \neq 1$ or $q_i \neq 1$ or $r_i \neq 1$ for some $1 \leq i \leq 3$, $\theta, \eta \in (0, \infty)$ and $f : A^3 \to B$ be a mapping such that*

$$\|D_{\lambda, \mu, \nu} f(x_1, x_2, y_1, y_2, z_1, z_2)\|$$
$$\leq \theta \|x_1\|^{p_1} \cdot \|x_2\|^{p_2} \cdot \|y_1\|^{q_1} \cdot \|y_2\|^{q_2} \cdot \|z_1\|^{r_1} \cdot \|z_2\|^{r_2}$$

*and*

$$\|f([x_1, y_1, z_1], [x_2, y_2, z_2], [x_3, y_3, z_3])$$
$$-[f(x_1, x_2, x_3), f(y_1, y_2, y_3), f(z_1, z_2, z_3)]\|$$
$$\leq \eta \|x_1\|^{p_1} \cdot \|x_2\|^{p_2} \cdot \|x_3\|^{p_3} \cdot \|y_1\|^{q_1} \cdot \|y_2\|^{q_2} \cdot \|y_3\|^{q_3} \cdot \|z_1\|^{r_1} \cdot \|z_2\|^{r_2} \cdot \|z_3\|^{r_3}$$

*for all $\lambda, \mu, \nu \in S^1$ and $x_1, x_2, x_3, y_1, y_2, y_3, z_1, z_2, z_3 \in A$. Then the mapping $f : A^3 \to B$ is a $C^*$-ternary 3-homomorphism.*

### 5.1.2 $C^*$-Ternary 3-Derivations in $C^*$-Ternary Algebras

Now, we investigate $C^*$-ternary 3-derivations in $C^*$-ternary algebras.

**Definition 5.7.** A 3-$\mathbb{C}$-linear mapping $D : A^3 \to A$ is called a $C^*$-*ternary 3-derivation* if it satisfies the following:

$$D([x_1, y_1, z_1], [x_2, y_2, z_2], [x_3, y_3, z_3])$$
$$= [D(x_1, x_2, x_3), [y_1, y_2^*, y_3], [z_1, z_2^*, z_3]]$$
$$+ [[x_1, x_2^*, x_3], D(y_1, y_2, y_3), [z_1, z_2^*, z_3]]$$
$$+ [[x_1, x_2^*, x_3], [y_1, y_2^*, y_3], D(z_1, z_2, z_3)]$$

for all $x_1, x_2, x_3, y_1, y_2, y_3, z_1, z_2, z_3 \in A$.

**Theorem 5.8.** *Let $p, q, r \in (0, \infty)$ with $p + q + r < 3$ and $\theta \in (0, \infty)$, and let $f : A^3 \to A$ be a mapping such that*

$$\|D_{\lambda, \mu, \nu} f(x_1, x_2, y_1, y_2, z_1, z_2)\| \tag{5.6}$$
$$\leq \theta \cdot \max\{\|x_1\|, \|x_2\|\}^p \cdot \max\{\|y_1\|, \|y_2\|\}^q \cdot \max\{\|z_1\|, \|z_2\|\}^r$$

*and*

$$\|f([x_1, y_1, z_1], [x_2, y_2, z_2], [x_3, y_3, z_3])$$
$$-[f(x_1, x_2, x_3), [y_1, y_2^*, y_3], [z_1, z_2^*, z_3]]$$

$$-[[x_1, x_2^*, x_3], f(y_1, y_2, y_3), [z_1, z_2^*, z_3]]$$
$$-[[x_1, x_2^*, x_3], [y_1, y_2^*, y_3], f(z_1, z_2, z_3)]\|$$

$$\leq \theta \sum_{i=1}^{3} \|x_i\|^p \cdot \|y_i\|^q \cdot \|z_i\|^r \tag{5.7}$$

for all $\lambda, \mu, \nu \in S^1$ and $x_1, x_2, x_3, y_1, y_2, y_3, z_1, z_2, z_3 \in A$. Then there exists a unique $C^*$-ternary 3-derivation $\delta : A^3 \to A$ such that

$$\|f(x, y, z) - \delta(x, y, z)\| \leq \frac{\theta}{2^3 - 2^{p+q+r}} \|x\|^p \cdot \|y\|^q \cdot \|z\|^r \tag{5.8}$$

for all $x, y, z \in A$.

*Proof.* By the same method as in the proof of [21, Theorem 1.2], we obtain a 3-$\mathbb{C}$-linear mapping $\delta : A^3 \to A$ satisfying (5.8) and the mapping

$$\delta(x, y, z) := \lim_{j \to \infty} \frac{1}{8^j} f(2^j x, 2^j y, 2^j z)$$

for all $x, y, z \in A$. It follows from (5.7) that

$$\|\delta([x_1, y_1, z_1], [x_2, y_2, z_2], [x_3, y_3, z_3])$$
$$-[\delta(x_1, x_2, x_3), [y_1, y_2^*, y_3], [z_1, z_2^*, z_3]]$$
$$-[[x_1, x_2^*, x_3], \delta(y_1, y_2, y_3), [z_1, z_2^*, z_3]]$$
$$-[[x_1, x_2^*, x_3], [y_1, y_2^*, y_3], \delta(z_1, z_2, z_3)]\|$$

$$= \lim_{n \to \infty} \frac{1}{8^{3n}} \|f(2^{3n}[x_1, y_1, z_1], 2^{3n}[x_2, y_2, z_2], 2^{3n}[x_3, y_3, z_3])$$

$$-[f(2^n x_1, 2^n x_2, 2^n x_3), [2^n y_1, 2^n y_2^*, 2^n y_3], [2^n z_1, 2^n z_2^*, 2^n z_3]]$$
$$-[[2^n x_1, 2^n x_2^*, 2^n x_3], f(2^n y_1, 2^n y_2, 2^n y_3), [2^n z_1, 2^n z_2^*, 2^n z_3]]$$
$$-[[2^n x_1, 2^n x_2^*, 2^n x_3], [2^n y_1, 2^n y_2^*, 2^n y_3], f(2^n z_1, 2^n z_2, 2^n z_3)]\|$$

$$\leq \lim_{n \to \infty} \frac{\theta 2^{n(p+q+r)}}{8^{3n}} \sum_{i=1}^{3} \|x_i\|^p \cdot \|y_i\|^q \cdot \|z_i\|^r = 0$$

for all $x_1, x_2, x_3, y_1, y_2, y_3, z_1, z_2, z_3 \in A$.

Now, let $T : A^3 \to A$ be another 3-derivation satisfying (5.8). Then we have

$$\|\delta(x, y, z) - T(x, y, z)\| = \frac{1}{8^n} \|\delta(2^n x, 2^n y, 2^n z) - T(2^n x, 2^n y, 2^n z)\|$$

$$\leq \frac{1}{8^n} \|\delta(2^n x, 2^n y, 2^n z) - f(2^n x, 2^n y, 2^n z)\|$$

$$+\frac{1}{8^n}\|f(2^n x, 2^n y, 2^n z) - T(2^n x, 2^n y, 2^n z)\|$$

$$\leq \frac{\theta 2^{(p+q+r-3)n+1}}{2^3 - 2^{p+q+r}}\|x\|^p \cdot \|y\|^q \cdot \|z\|^r,$$

which tends to zero as $n \to \infty$ for all $x, y, z \in A$. So we can conclude that $\delta(x, y, z) = T(x, y, z)$ for all $x, y, z \in A$. This proves the uniqueness of $\delta$. Therefore, the mapping $\delta : A^3 \to A$ is a unique $C^*$-ternary 3-derivation satisfying (5.8). This completes the proof.                                                                             □

**Corollary 5.9.** *Let* $\epsilon \in (0, \infty)$ *and* $f : A^3 \to A$ *be a mapping satisfying*

$$\|D_{\lambda,\mu,\nu}f(x_1, x_2, y_1, y_2, z_1, z_2)\| \leq \epsilon$$

*and*

$$\|f([x_1, y_1, z_1], [x_2, y_2, z_2], [x_3, y_3, z_3])$$
$$-[f(x_1, x_2, x_3), [y_1, y_2^*, y_3], [z_1, z_2^*, z_3]]$$
$$-[[x_1, x_2^*, x_3], f(y_1, y_2, y_3), [z_1, z_2^*, z_3]]$$
$$-[[x_1, x_2^*, x_3], [y_1, y_2^*, y_3], f(z_1, z_2, z_3)]\|$$
$$\leq 3\epsilon$$

*for all* $\lambda, \mu, \nu \in S^1$ *and* $x_1, x_2, x_3, y_1, y_2, y_3, z_1, z_2, z_3 \in A$. *Then there exists a unique* $C^*$-*ternary 3-derivation* $\delta : A^3 \to A$ *such that*

$$\|f(x, y, z) - \delta(x, y, z)\| \leq \frac{\epsilon}{7}$$

*for all* $x, y, z \in A$.

## 5.2   Apollonius Type Additive Functional Equations

C. Park and Th. M. Rassias proved the superstability of $C^*$-ternary homomorphisms, $C^*$-ternary derivations, $JB^*$-triple homomorphisms and $JB^*$-triple derivations associated with the following Apollonius type additive functional equation:

$$f(z - x) + f(z - y) = -\frac{1}{2}f(x + y) + 2f\left(z - \frac{x + y}{4}\right)$$

by using the direct method (see [244–248]).

In this section, under the conditions of the theorems, we can show that the mappings $f$ must be zero and we correct some conditions. Furthermore, we prove the superstability of $C^*$-ternary homomorphisms, $C^*$-ternary derivations, $JB^*$-triple homomorphisms and $JB^*$-triple derivations by using fixed point method.

In an inner product space, the following equality holds:

$$\|z - x\|^2 + \|z - y\|^2 = \frac{1}{2}\|x - y\|^2 + 2\left\|z - \frac{x+y}{2}\right\|^2,$$

which is called the *Apollonius' identity*. The following functional equation, which was motivated by this equation:

$$Q(z - x) + Q(z - y) = \frac{1}{2}Q(x - y) + 2Q\left(z - \frac{x+y}{2}\right), \tag{5.9}$$

is quadratic. For this reason, the function equation (5.9) is called a *quadratic functional equation of Apollonius type* and each solution of the functional equation (5.9) is said to be a *quadratic mapping of Apollonius type*. Jun and Kim [144] investigated the quadratic functional equation of Apollonius type.

Now, employing the above equality (5.9), we introduce a new functional equation, which is called the *Apollonius type additive functional equation* and whose solution of the functional equation is said to be the *Apollonius type additive mapping*:

$$L(z - x) + L(z - y) = -\frac{1}{2}L(x + y) + 2L\left(z - \frac{x+y}{4}\right).$$

### 5.2.1  Homomorphisms in $C^*$-Ternary Algebras

Assume that $A$ is a $C^*$-ternary algebra with the norm $\|\cdot\|_A$ and that $B$ is a $C^*$-ternary algebra with the norm $\|\cdot\|_B$.

Now, we investigate homomorphisms in $C^*$-ternary algebras.

**Lemma 5.10.** *Let $f : A \to B$ be a mapping such that*

$$\left\|f(z - x) + f(z - y) + \frac{1}{2}f(x + y)\right\|_B \le \left\|2f\left(z - \frac{x+y}{4}\right)\right\|_B \tag{5.10}$$

*for all $x, y, z \in A$. Then $f$ is additive.*

*Proof.* Letting $x = y = z = 0$ in (5.10), we get $\left\|\frac{5}{2}f(0)\right\|_B \le \|2f(0)\|_B$ and so $f(0) = 0$. Letting $z = 0$ and $y = -x$ in (5.10), we have

$$\|f(-x) + f(x)\|_B \le \|2f(0)\|_B = 0$$

for all $x \in A$. Hence $f(-x) = -f(x)$ for all $x \in A$. Letting $x = y = 2z$ in (5.10), we have

$$\left\| 2f(-z) + \frac{1}{2}f(4z) \right\|_B \leq \| 2f(0) \|_B = 0$$

for all $z \in A$ and so

$$f(4z) = -4f(-z) = 4f(z)$$

for all $z \in A$. Letting $z = \frac{x+y}{4}$ in (5.10), we have

$$\left\| f\left( \frac{-3x+y}{4} \right) + f\left( \frac{x-3y}{4} \right) + \frac{1}{2}f(x+y) \right\|_B \leq \| 2f(0) \|_B = 0$$

for all $x, y \in A$ and so

$$f\left( \frac{-3x+y}{4} \right) + f\left( \frac{x-3y}{4} \right) + \frac{1}{2}f(x+y) = 0 \qquad (5.11)$$

for all $x, y \in A$. Let $w_1 = \frac{-3x+y}{4}$ and $w_2 = \frac{x-3y}{4}$ in (5.11). Then we have

$$f(w_1) + f(w_2) = -\frac{1}{2}f(-2w_1 - 2w_2) = \frac{1}{2}f(2w_1 + 2w_2) = 2f\left( \frac{w_1 + w_2}{2} \right)$$

for all $w_1, w_2 \in A$ and so $f$ is additive. This completes the proof. $\qquad \square$

**Theorem 5.11.** *Let $r \neq 1$, $\theta$ be a nonnegative real number and $f : A \to B$ be a mapping such that*

$$\left\| f(\mu z - \mu x) + \mu f(z - y) + \frac{\mu}{2}f(x+y) \right\|_B$$
$$\leq \left\| 2f\left( z - \frac{x+y}{4} \right) \right\|_B \qquad (5.12)$$

*and*

$$\| f([x, y, z]) - [f(x), f(y), f(z)] \|_B$$
$$\leq \theta (\|x\|_A^{3r} + \|y\|_A^{3r} + \|z\|_A^{3r}) \qquad (5.13)$$

*for all $\mu \in \mathbb{T}^1 := \{ \lambda \in \mathbb{C} : |\lambda| = 1 \}$ and all $x, y, z \in A$. Then the mapping $f : A \to B$ is a $C^*$-ternary algebra homomorphism.*

*Proof.* Assume $r > 1$. Let $\mu = 1$ in (5.12). By Lemma 5.10, the mapping $f : A \to B$ is additive. Letting $y = -x$ and $z = 0$, we get

$$\| f(-\mu x) + \mu f(x) \|_B \leq \| 2f(0) \|_B = 0$$

for all $x \in A$ and $\mu \in \mathbb{T}^1$ and so

$$-f(\mu x) + \mu f(x) = f(-\mu x) + \mu f(x) = 0$$

for all $x \in A$ and $\mu \in \mathbb{T}^1$. Hence $f(\mu x) = \mu f(x)$ for all $x \in A$ and $\mu \in \mathbb{T}^1$. By the same reasoning as in the proof of Theorem 2.1 of Park [227], the mapping $f : A \to B$ is $\mathbb{C}$-linear. It follows from (5.13) that

$$\|f([x, y, z]) - [f(x), f(y), f(z)]\|_B$$
$$= \lim_{n \to \infty} 8^n \left\| f\left(\frac{[x, y, z]}{2^n \cdot 2^n \cdot 2^n}\right) - \left[f\left(\frac{x}{2^n}\right), f\left(\frac{y}{2^n}\right), f\left(\frac{z}{2^n}\right)\right]\right\|_B$$
$$\leq \lim_{n \to \infty} \frac{8^n \theta}{8^{nr}} (\|x\|_A^{3r} + \|y\|_A^{3r} + \|z\|_A^{3r})$$
$$= 0$$

for all $x, y, z \in A$ and so

$$f([x, y, z]) = [f(x), f(y), f(z)]$$

for all $x, y, z \in A$. Hence the mapping $f : A \to B$ is a $C^*$-ternary algebra homomorphism.

Similarly, one obtains the result for the case $r < 1$. This completes the proof. $\square$

### 5.2.2   Derivations in $C^*$-Ternary Algebras

Assume that $A$ is a $C^*$-ternary algebra with the norm $\| \cdot \|_A$. Now, we investigate derivations on $C^*$-ternary algebras.

**Theorem 5.12.** *Let $r \neq 1$ and $\theta$ be nonnegative real numbers, and let $f : A \to A$ be a mapping such that*

$$\left\|f(\mu z - \mu x) + \mu f(z - y) + \frac{\mu}{2}f(x + y)\right\|_A$$
$$\leq \left\|2f\left(z - \frac{x + y}{4}\right)\right\|_A \tag{5.14}$$

*and*

$$\|f([x, y, z]) - [f(x), y, z] - [x, f(y), z] - [x, y, f(z)]\|_A$$
$$\leq \theta(\|x\|_A^{3r} + \|y\|_A^{3r} + \|z\|_A^{3r})$$

*for all $x, y, z \in A$. Then the mapping $f : A \to A$ is a $C^*$-ternary derivation.*

*Proof.* Assume $r > 1$. By the same reasoning as in the proof of Theorem 5.11, the mapping $f : A \to A$ is $\mathbb{C}$-linear. It follows from (5.15) that

$$\|f([x, y, z]) - [f(x), y, z] - [x, f(y), z] - [x, y, f(z)]\|_A$$

$$= \lim_{n \to \infty} 8^n \left\| f\left(\frac{[x, y, z]}{8^n}\right) - \left[f\left(\frac{x}{2^n}\right), \frac{y}{2^n}, \frac{z}{2^n}\right]\right.$$

$$\left. - \left[\frac{x}{2^n}, f\left(\frac{y}{2^n}\right), \frac{z}{2^n}\right] - \left[\frac{x}{2^n}, \frac{y}{2^n}, f\left(\frac{z}{2^n}\right)\right]\right\|_A$$

$$\leq \lim_{n \to \infty} \frac{8^n \theta}{8^{nr}} (\|x\|_A^{3r} + \|y\|_A^{3r} + \|z\|_A^{3r})$$

$$= 0$$

for all $x, y, z \in A$ and so

$$f([x, y, z]) = [f(x), y, z] + [x, f(y), z] + [x, y, f(z)]$$

for all $x, y, z \in A$. Thus the mapping $f : A \to A$ is a $C^*$-ternary derivation.

Similarly, one obtains the result for the case $r < 1$. This completes the proof. □

### 5.2.3  Homomorphisms in JB*-Triples

Assume that $\mathcal{J}$ is a $JB^*$-triple with the norm $\|\cdot\|_{\mathcal{J}}$ and that $\mathcal{L}$ is a $JB^*$-triple with the norm $\|\cdot\|_{\mathcal{L}}$.

Now, we investigate homomorphisms in $JB^*$-triples.

**Theorem 5.13.** *Let $r \neq 1$, $\theta$ be a nonnegative real number and $f : \mathcal{J} \to \mathcal{L}$ be a mapping such that*

$$\left\|f(\mu z - \mu x) + \mu f(z - y) + \frac{\mu}{2} f(x + y)\right\|_{\mathcal{L}}$$

$$\leq \left\|2f\left(z - \frac{x + y}{4}\right)\right\|_{\mathcal{L}} \tag{5.15}$$

*and*

$$\|f(\{xyz\}) - \{f(x)f(y)f(z)\}\|_{\mathcal{L}}$$

$$\leq \theta(\|x\|_{\mathcal{J}}^{3r} + \|y\|_{\mathcal{J}}^{3r} + \|z\|_{\mathcal{J}}^{3r}) \tag{5.16}$$

*for all $\mu \in \mathbb{T}^1$ and all $x, y, z \in \mathcal{J}$. Then the mapping $f : \mathcal{J} \to \mathcal{L}$ is a $JB^*$-triple homomorphism.*

*Proof.* Assume $r > 1$. By the same reasoning as in the proof of Theorem 5.11, the mapping $f : \mathcal{J} \to \mathcal{L}$ is $\mathbb{C}$-linear. It follows from (5.16) that

$$\|f(\{xy\}]) - \{f(x)f(y)f(z)\}\|_{\mathcal{L}}$$

$$= \lim_{n \to \infty} 8^n \left\| f\left(\frac{\{xyz\}}{2^n \cdot 2^n \cdot 2^n}\right) - \left\{ f\left(\frac{x}{2^n}\right) f\left(\frac{y}{2^n}\right) f\left(\frac{z}{2^n}\right) \right\} \right\|_{\mathcal{L}}$$

$$\leq \lim_{n \to \infty} \frac{8^n \theta}{8^{nr}} (\|x\|_{\mathcal{J}}^{3r} + \|y\|_{\mathcal{J}}^{3r} + \|z\|_{\mathcal{J}}^{3r})$$

$$= 0$$

for all $x, y, z \in \mathcal{J}$ and so

$$f(\{xyz\}) = \{f(x)f(y)f(z)\}$$

for all $x, y, z \in \mathcal{J}$. Hence the mapping $f : \mathcal{J} \to \mathcal{L}$ is a $JB^*$-triple homomorphism. Similarly, one obtains the result for the case $r < 1$. This completes the proof. $\square$

### 5.2.4   Derivations in $JB^*$-Triples

Assume that $\mathcal{J}$ is a $JB^*$-triple with the norm $\|\cdot\|_{\mathcal{J}}$. Now, we investigate derivations on $JB^*$-triples.

**Theorem 5.14.** *Let $r \neq 1$, $\theta$ be a nonnegative real number and $f : \mathcal{J} \to \mathcal{J}$ be a mapping such that*

$$\left\| f(\mu z - \mu x) + \mu f(z - y) + \frac{\mu}{2} f(x + y) \right\|_{\mathcal{J}}$$

$$\leq \left\| 2f\left(z - \frac{x+y}{4}\right) \right\|_{\mathcal{J}} \tag{5.17}$$

*and*

$$\|f(\{xyz\}) - \{f(x)yz\} - \{xf(y)z\} - \{xyf(z)\}\|_{\mathcal{J}}$$

$$\leq \theta(\|x\|_{\mathcal{J}}^{3r} + \|y\|_{\mathcal{J}}^{3r} + \|z\|_{\mathcal{J}}^{3r}) \tag{5.18}$$

*for all $x, y, z \in \mathcal{J}$. Then the mapping $f : \mathcal{J} \to \mathcal{J}$ is a $JB^*$-triple derivation.*

*Proof.* Assume $r > 1$. By the same reasoning as in the proof of Theorem 5.11, the mapping $f : \mathcal{J} \to \mathcal{J}$ is $\mathbb{C}$-linear. It follows from (5.18) that

$$\|f(\{xyz\}) - \{f(x)yz\} - \{xf(y)z\} - \{xyf(z)\}\|_{\mathcal{J}}$$

$$= \lim_{n\to\infty} 8^n \left\| f\left(\frac{\{xyz\}}{8^n}\right) - \left\{ f\left(\frac{x}{2^n}\right)\frac{y}{2^n}\frac{z}{2^n}\right\} \right.$$

$$\left. - \left\{\frac{x}{2^n}f\left(\frac{y}{2^n}\right)\frac{z}{2^n}\right\} - \left\{\frac{x}{2^n}\frac{y}{2^n}f\left(\frac{z}{2^n}\right)\right\} \right\|_{\mathcal{J}}$$

$$\leq \lim_{n\to\infty} \frac{8^n\theta}{8^{nr}}(\|x\|_{\mathcal{J}}^{3r} + \|y\|_{\mathcal{J}}^{3r} + \|z\|_{\mathcal{J}}^{3r})$$

$$= 0$$

for all $x, y, z \in \mathcal{J}$ and so

$$f(\{xyz\}) = \{f(x)yz\} + \{xf(y)z\} + \{xyf(z)\}$$

for all $x, y, z \in \mathcal{J}$. Thus the mapping $f : \mathcal{J} \to \mathcal{J}$ is a $JB^*$-triple derivation.
Similarly, one obtains the result for the case $r < 1$. This completes the proof. $\square$

## 5.2.5   *C\*-Ternary Homomorphisms: Fixed Point Method*

Now, we prove the superstability of $C^*$-ternary homomorphisms by the using fixed
point method.

**Theorem 5.15.** *Let $\varphi : A^3 \to [0, \infty)$ be a function such that there exists $\alpha < 1$
with*

$$\varphi(x, y, z) \leq 8\alpha\varphi\left(\frac{x}{2}, \frac{y}{2}, \frac{z}{2}\right) \tag{5.19}$$

*for all $x, y, z \in A$. Let $f : A \to B$ be a mapping satisfying (5.12) and*

$$\|f([x, y, z]) - [f(x), f(y), f(z)]\|_B \leq \varphi(x, y, z) \tag{5.20}$$

*for all $x, y, z \in A$. Then the mapping $f : A \to B$ is a $C^*$-ternary homomorphism.*

*Proof.* By the same reasoning as in the proof of Theorem 5.11, one can show that
the mapping $f : A \to B$ is $\mathbb{C}$-linear. It follows from (5.19) that

$$\lim_{n\to\infty} \frac{1}{8^n}\varphi(2^n x, 2^n y, 2^n z) = 0 \tag{5.21}$$

for all $x, y, z \in A$. Since $f : A \to B$ is additive, it follows from (5.20) and (5.21) that

$$f([x, y, z]) = [f(x), f(y), f(z)]$$

for all $x, y, z \in A$. Thus the mapping $f : A \to B$ is a $C^*$-ternary homomorphism.
This completes the proof. $\square$

**Theorem 5.16.** *Let $\varphi : A^3 \to [0, \infty)$ be a function such that there exists $\alpha < 1$ with*

$$\varphi(x, y, z) \leq \frac{\alpha}{2}\varphi(2x, 2y, 2z) \tag{5.22}$$

*for all $x, y, z \in A$. Let $f : A \to B$ be a mapping satisfying (5.12) and (5.20). Then the mapping $f : A \to B$ is a $C^*$-ternary homomorphism.*

*Proof.* By the same reasoning as in the proof of Theorem 5.11, one can show that the mapping $f : A \to B$ is $\mathbb{C}$-linear. It follows from (5.22) that

$$\lim_{n \to \infty} 2^n \varphi\left(\frac{x}{2^n}, \frac{y}{2^n}, \frac{z}{2^n}\right) = 0 \tag{5.23}$$

for all $x, y, z \in A$. Since $f : A \to B$ is additive, it follows from (5.20) and (5.23) that

$$f([x, y, z]) = [f(x), f(y), f(z)]$$

for all $x, y, z \in A$. Thus the mapping $f : A \to B$ is a $C^*$-ternary homomorphism. This completes the proof. $\qquad\square$

*Remark 5.17.* Theorem 5.11 follows from Theorems 5.15 and 5.16 by taking

$$\varphi(x, y, z) = \theta(\|x\|^{3r} + \|y\|^{3r} + \|z\|^{3r})$$

for all $x, y, z \in A$.

## 5.2.6  $C^*$-Ternary Derivations: The Fixed Point Method

Now, we prove the superstability of $C^*$-ternary derivations by using the fixed point method.

**Theorem 5.18.** *Let $\varphi : A^3 \to [0, \infty)$ be a function satisfying (5.19). Let $f : A \to A$ be a mapping satisfying (5.14) and*

$$\|f([x, y, z]) - [f(x), y, z] - [x, f(y), z] - [x, y, f(z)]\|_A$$
$$\leq \varphi(x, y, z) \tag{5.24}$$

*for all $x, y, z \in A$. Then the mapping $f : A \to A$ is a $C^*$-ternary derivation.*

*Proof.* The proof is similar to the proof of Theorem 5.15. $\qquad\square$

**Theorem 5.19.** *Let $\varphi : A^3 \to [0, \infty)$ be a function satisfying (5.22). Let $f : A \to A$ be a mapping satisfying (5.14) and (5.24). Then the mapping $f : A \to A$ is a $C^*$-ternary derivation.*

*Proof.* The proof is similar to the proof of Theorem 5.16. □

*Remark 5.20.* Theorem 5.12 follows from Theorems 5.18 and 5.19 by taking

$$\varphi(x, y, z) = \theta(\|x\|^{3r} + \|y\|^{3r} + \|z\|^{3r})$$

for all $x, y, z \in A$.

### 5.2.7 JB*-Triple Homomorphisms: The Fixed Point Method

Now, we prove the superstability of $JB^*$-triple homomorphisms by using the fixed point method.

**Theorem 5.21.** *Let $\varphi : \mathcal{J}^3 \to [0, \infty)$ be a function such that there exists $\alpha < 1$ with*

$$\varphi(x, y, z) \leq 8\alpha\varphi\left(\frac{x}{2}, \frac{y}{2}, \frac{z}{2}\right) \tag{5.25}$$

*for all $x, y, z \in \mathcal{J}$. Let $f : \mathcal{J} \to \mathcal{L}$ be a mapping satisfying (5.15) and*

$$\|f(\{xyz\}) - \{f(x)f(y)f(z)\}\|_{\mathcal{L}} \leq \varphi(x, y, z) \tag{5.26}$$

*for all $x, y, z \in \mathcal{J}$. Then the mapping $f : \mathcal{J} \to \mathcal{L}$ is a JB*-triple homomorphism.*

*Proof.* By the same reasoning as in the proof of Theorem 5.13, one can show that the mapping $f : \mathcal{J} \to \mathcal{L}$ is $\mathbb{C}$-linear. It follows from (5.25) that

$$\lim_{n \to \infty} \frac{1}{8^n} \varphi(2^n x, 2^n y, 2^n z) = 0 \tag{5.27}$$

for all $x, y, z \in \mathcal{J}$. Since $f : \mathcal{J} \to \mathcal{L}$ is additive, it follows from (5.26) and (5.27) that

$$f([x, y, z]) = [f(x), f(y), f(z)]$$

for all $x, y, z \in \mathcal{J}$. Thus the mapping $f : \mathcal{J} \to \mathcal{L}$ is a $JB^*$-triple homomorphism. This completes the proof. □

**Theorem 5.22.** *Let $\varphi : \mathcal{J}^3 \to [0, \infty)$ be a function such that there exists an $\alpha < 1$ with*

$$\varphi(x, y, z) \leq \frac{\alpha}{2} \varphi(2x, 2y, 2z) \tag{5.28}$$

*for all $x, y, z \in \mathcal{J}$. Let $f : \mathcal{J} \to \mathcal{L}$ be a mapping satisfying (5.15) and (5.26). Then the mapping $f : \mathcal{J} \to \mathcal{L}$ is a JB*-triple homomorphism.*

*Proof.* By the same reasoning as in the proof of Theorem 5.13, one can show that the mapping $f : \mathcal{J} \to \mathcal{L}$ is $\mathbb{C}$-linear. It follows from (5.28) that

$$\lim_{n \to \infty} 2^n \varphi \left( \frac{x}{2^n}, \frac{y}{2^n}, \frac{z}{2^n} \right) = 0 \tag{5.29}$$

for all $x, y, z \in \mathcal{J}$. Since $f : \mathcal{J} \to \mathcal{L}$ is additive, it follows from (5.26) and (5.29) that

$$f([x, y, z]) = [f(x), f(y), f(z)]$$

for all $x, y, z \in \mathcal{J}$. Thus the mapping $f : \mathcal{J} \to \mathcal{L}$ is a $C^*$-ternary homomorphism. This completes the proof.                                                                                    □

*Remark 5.23.* Theorem 5.13 follows from Theorems 5.21 and 5.22 by taking

$$\varphi(x, y, z) = \theta(\|x\|^{3r} + \|y\|^{3r} + \|z\|^{3r})$$

for all $x, y, z \in \mathcal{J}$.

### 5.2.8   JB*-Triple Derivations: Fixed Point Method

Now, we prove the superstability of $JB^*$-triple derivations by using the fixed point method.

**Theorem 5.24.** *Let* $\varphi : \mathcal{J}^3 \to [0, \infty)$ *be a function satisfying* (5.25). *Let* $f : \mathcal{J} \to \mathcal{J}$ *be a mapping satisfying* (5.17) *and*

$$\|f(\{xyz\}) - \{f(x)yz\} - \{xf(y)z\} - \{xyf(z)\}\|_{\mathcal{J}}$$
$$\leq \varphi(x, y, z) \tag{5.30}$$

*for all* $x, y, z \in \mathcal{J}$. *Then the mapping* $f : \mathcal{J} \to \mathcal{J}$ *is a JB*-triple derivation.*

*Proof.* The proof is similar to the proof of Theorem 5.21.                                      □

**Theorem 5.25.** *Let* $\varphi : \mathcal{J}^3 \to [0, \infty)$ *be a function satisfying* (5.28). *Let* $f : \mathcal{J} \to \mathcal{J}$ *be a mapping satisfying* (5.17) *and* (5.30). *Then the mapping* $f : \mathcal{J} \to \mathcal{J}$ *is a JB*-triple derivation.*

*Proof.* The proof is similar to the proof of Theorem 5.22.                                      □

*Remark 5.26.* Theorem 5.14 follows from Theorems 5.24 and 5.25 by taking

$$\varphi(x, y, z) = \theta(\|x\|^{3r} + \|y\|^{3r} + \|z\|^{3r})$$

for all $x, y, z \in \mathcal{J}$.

## 5.3  Bi-$\theta$-Derivations in *JB*\*-Triples

In this section, we prove the Hyers-Ulam stability of bi-$\theta$-derivations in *JB*\*-triples (see [237]).

**Definition 5.27 ([97]).** Let $J$ be a complex *JB*\*-triple and $\theta : J \to J$ be a $\mathbb{C}$-linear mapping. A $\mathbb{C}$-bilinear mapping $D : J \times J \to J$ is called a *bi-$\theta$-derivation* on $J$ if

$$D(\{x, y, z\}, w) = \{D(x, w), \theta(y), \theta(z)\} + \{\theta(x), D(y, w), \theta(z)\}$$
$$+\{\theta(x), \theta(y), D(z, w)\}$$

and

$$D(x, \{y, z, w\}) = \{D(x, y), \theta(z), \theta(w)\} + \{\theta(y), D(x, z), \theta(w)\}$$
$$+\{\theta(y), \theta(z), D(x, w)\}$$

for all $x, y, z, w \in J$.

The $w$-variable of the left side in the first equality is $\mathbb{C}$-linear and the $x$-variable of the left side in the second equality is $\mathbb{C}$-linear. But the $w$-variable of the right side in the first equality is not $\mathbb{C}$-linear and the $x$-variable of the right side in the second equality is not $\mathbb{C}$-linear. Thus we correct the definition of bi-$\theta$-derivation as follows:

**Definition 5.28.** Let $J$ be a complex *JB*\*-triple and $\theta : J \to J$ be a $\mathbb{C}$-linear mapping. A $\mathbb{C}$-bilinear mapping $D : J \times J \to J$ is called a *bi-$\theta$-derivation* on $J$ if

$$D(\{x, y, z\}, w) = \{D(x, w), \theta(y), \theta(z)\} + \{\theta(x), D(y, w^*), \theta(z)\}$$
$$+\{\theta(x), \theta(y), D(z, w)\}$$

and

$$D(x, \{y, z, w\}) = \{D(x, y), \theta(z), \theta(w)\} + \{\theta(y), D(x^*, z), \theta(w)\}$$
$$+\{\theta(y), \theta(z), D(x, w)\}$$

for all $x, y, z, w \in J$.

Under the conditions of [97, Theorem 2.5], we can easily show that the mapping $D : J \times J \to J$ must be zero. In particular, if $f$ is bi-additive, then $D$ must be zero. In this section, we correct the statements of the results, and prove the corrected theorems and corollaries.

Throughout this section, assume that $J$ is a *JB*\*-triple with the norm $\|\cdot\|$. For any mapping $f : J \times J \to J$, we define

$$E_{\lambda,\mu}f(x,y,z,w)$$

$$= f(\lambda x + \lambda y, \mu z - \mu w)$$

$$+ f(\lambda x - \lambda y, \mu z + \mu w) - 2\lambda\mu f(x,z) + 2\lambda\mu f(y,w)$$

for all $x, y, z, w \in J$ and $\lambda, \mu \in \mathbb{T}^1 := \{\lambda \in \mathbb{C} : |\lambda| = 1\}$.

**Lemma 5.29 ([21]).** *Let $f : J \times J \to J$ be a mapping satisfying*

$$E_{\lambda,\mu}f(x,y,z,w) = 0$$

*for all $x, y, z, w \in J$ and $\lambda, \mu \in \mathbb{T}^1$. Then the mapping $f : J \times J \to J$ is $\mathbb{C}$-bilinear.*

Now, we prove the Hyers-Ulam stability of bi-$\theta$-derivations on $JB^*$-triples.

**Theorem 5.30.** *Let $p$, $\varepsilon$ be positive real numbers with $p < 1$ and $f : J \times J \to J$ with $f(0,0) = 0$, $h : J \to J$ with $h(0) = 0$ be the mappings such that*

$$\|E_{\lambda,\mu}f(x,y,z,w) + h(\mu a + \mu b) - \mu h(a) - \mu h(b)\|$$

$$\leq \varepsilon(\|x\|^p + \|y\|^p + \|z\|^p + \|w\|^p + \|a\|^p + \|b\|^p) \qquad (5.31)$$

*and*

$$\|f(\{x,y,z\},w) - \{f(x,w),h(y),h(z)\}$$

$$-\{h(x),f(y,w^*),h(z)\} - \{h(x),h(y),f(z,w)\}\|$$

$$+\|f(x,\{y,z,w\}) - \{f(x,y),h(z),h(w)\}$$

$$-\{h(y),f(x^*,z),h(w)\} - \{h(y),h(z),f(x,w)\}\|$$

$$\leq \varepsilon(\|x\|^p + \|y\|^p + \|z\|^p + \|w\|^p) \qquad (5.32)$$

*for all $\lambda, \mu \in \mathbb{T}^1$ and $x, y, z, w \in J$. If the mapping $f : J \times J \to J$ satisfies the following:*

$$\lim_{n\to\infty} \frac{1}{4^n}f(2^n x, 2^n y) = \lim_{n\to\infty}\frac{1}{16^n}f(2^n x, 8^n y)$$

$$= \lim_{n\to\infty}\frac{1}{16^n}f(8^n x, 2^n y), \qquad (5.33)$$

*then there exist a unique $\mathbb{C}$-linear mapping $\theta : J \to J$ and a unique bi-$\theta$-derivation $D : J \times J \to J$ such that*

$$\|h(a) - \theta(a)\| \leq \frac{2\varepsilon}{2 - 2^p}\|a\|^p \qquad (5.34)$$

*and*

$$\|f(x, z) - D(x, z)\| \leq \frac{5\varepsilon}{4 - 2^p}(\|x\|^p + \|z\|^p) \tag{5.35}$$

*for all $a, x, z \in J$.*

*Proof.* Letting $x = y = z = w = 0$ in (5.31), we have

$$\|h(\mu a + \mu b) - \mu h(a) - \mu h(b)\| \leq \varepsilon(\|a\|^p + \|b\|^p)$$

for all $a, b \in J$. By the same reasoning as in the proof of Park [227, Theorem 2.1], one can show that there exists a unique $\mathbb{C}$-linear mapping $\theta : J \to J$ satisfying (5.34) and the mapping $\theta : J \to J$ is given by

$$\theta(a) := \lim_{n \to \infty} \frac{1}{2^n} h(2^n a)$$

for all $a \in J$. Letting $\lambda = \mu = 1, a = b = 0, y = x$ and $w = -z$ in (5.31), we have

$$\|f(2x, 2z) - 2f(x, z) + 2f(x, -z)\| \leq 2\varepsilon(\|x\|^p + \|z\|^p) \tag{5.36}$$

for all $x, z \in J$. Letting $\lambda = \mu = 1, a = b = 0, y = -x$ and $w = z$ in (5.31), we have

$$\|f(2x, 2z) - 2f(x, z) + 2f(-x, z)\| \leq 2\varepsilon(\|x\|^p + \|z\|^p) \tag{5.37}$$

for all $x, z \in J$. Letting $\lambda = \mu = 1, a = b = 0, x = z = 0$ in (5.31), we have

$$\|f(y, -w) + f(-y, w) + 2f(y, w)\| \leq \varepsilon(\|y\|^p + \|w\|^p) \tag{5.38}$$

for all $y, w \in J$. Replacing $y, w$ by $x, z$ in (5.38), respectively, we have

$$\|f(x, -z) + f(-x, z) + 2f(x, z)\| \leq \varepsilon(\|x\|^p + \|z\|^p) \tag{5.39}$$

for all $x, z \in J$. By (5.36) and (5.39), we obtain

$$\|f(2x, 2z) - 4f(x, z) + f(x, -z) - f(-x, z)\|$$
$$\leq 3\varepsilon(\|x\|^p + \|z\|^p) \tag{5.40}$$

for all $x, z \in J$. By (5.36) and (5.37), we obtain

$$\|f(x, -z) - f(-x, z)\| \leq 2\varepsilon(\|x\|^p + \|z\|^p) \tag{5.41}$$

for all $x, z \in J$. By (5.40) and (5.41), we obtain

$$\|f(2x, 2z) - 4f(x, z)\| \leq 5\varepsilon(\|x\|^p + \|z\|^p) \tag{5.42}$$

for all $x, z \in J$. It follows from (5.42) that

$$\left\|\frac{1}{4^l}f(2^l x, 2^l z) - \frac{1}{4^m}f(2^m, 2^m z)\right\| \leq \sum_{j=l}^{m-1} \frac{5\varepsilon}{4}\frac{2^{pj}}{4^j}(\|x\|^p + \|z\|^p) \tag{5.43}$$

for all $x, z \in J$ and $m, l \geq 1$ with $m > l$. This implies that the sequence $\{\frac{1}{4^n}f(2^n x, 2^n z)\}$ is a Cauchy sequence for all $x, z \in J$. Since $J$ is complete, the sequence $\{\frac{1}{4^n}f(2^n x, 2^n z)\}$ converges and so one can define the mapping $D : J \times J \to J$ by

$$D(x, z) := \lim_{n \to \infty} \frac{1}{4^n}f(2^n x, 2^n z)$$

for all $x, z \in J$. Moreover, letting $l = 0$ and passing the limit $m \to \infty$ in (5.43), we have (5.35). Let $a = b = 0$ in (5.31). Then, by the definition of the mapping $D$, we have

$$\|E_{\lambda, \mu}D(x, y, z, w)\| = \lim_{n \to \infty} \frac{1}{4^n}\|E_{\lambda, \mu}f(2^n x, 2^n y, 2^n z, 2^n w)\|$$

$$\leq \lim_{n \to \infty} \frac{2^{pn}}{4^n}\varepsilon(\|x\|^p + \|y\|^p + \|z\|^p + \|w\|^p)$$

$$= 0$$

for all $\lambda, \mu \in \mathbb{T}^1$ and all $x, y, z, w \in J$. By Lemma 5.29, the mapping $D : J \times J \to J$ is $\mathbb{C}$-bilinear. It follows from (5.32) and (5.33) that

$$\|D(\{x, y, z\}, w) - \{D(x, w), \theta(y), \theta(z)\} - \{\theta(x), D(y, w^*), \theta(z)\}$$

$$- \{\theta(x), \theta(y), D(z, w)\}\| + \|D(x, \{y, z, w\}) - \{D(x, y), \theta(z), \theta(w)\}$$

$$- \{\theta(y), D(x^*, z), \theta(w)\} - \{\theta(y), \theta(z), D(x, w)\}\|$$

$$= \lim_{n \to \infty} \left(\left\|\frac{1}{2^{4n}}f(8^n\{x, y, z\}, 2^n w) - \left\{\frac{1}{4^n}f(2^n x, 2^n w), \frac{1}{2^n}h(2^n y), \frac{1}{2^n}h(2^n z)\right\}\right.\right.$$

$$- \left\{\frac{1}{2^n}h(2^n x), \frac{1}{4^n}f(2^n y, 2^n w^*), \frac{1}{2^n}h(2^n z)\right\}$$

$$- \left\{\frac{1}{2^n}h(2^n x), \frac{1}{2^n}h(2^n y), \frac{1}{4^n}f(2^n z, 2^n w)\right\}$$

$$+ \left\|\frac{1}{2^{4n}}f(2^n x, 8^n\{y, z, w\}) - \left\{\frac{1}{4^n}f(2^n x, 2^n y), \frac{1}{2^n}h(2^n z), \frac{1}{2^n}h(2^n w)\right\}\right.$$

$$-\left\{\frac{1}{2^n}h(2^ny), \frac{1}{4^n}f(2^nx^*, 2^nz), \frac{1}{2^n}h(2^nw)\right\}$$

$$-\left\{\frac{1}{2^n}h(2^ny), \frac{1}{2^n}h(2^nz), \frac{1}{4^n}f(2^nx, 2^nw)\right\}\Big\|\Big)$$

$$\leq \lim_{n\to\infty} \frac{2^{pn}\varepsilon}{2^{4n}}(\|x\|^p + \|y\|^p + \|z\|^p + \|w\|^p)$$

$$= 0$$

for all $x, y, z, w \in J$ and so

$$D(\{x, y, z\}, w) = \{D(x, w), \theta(y), \theta(z)\}$$
$$+\{\theta(x), D(y, w^*), \theta(z)\} + \{\theta(x), \theta(y), D(z, w)\}$$

and

$$D(x, \{y, z, w\}) = \{D(x, y), \theta(z), \theta(w)\}$$
$$+\{\theta(y), D(x^*, z), \theta(w)\} + \{\theta(y), \theta(z), D(x, w)\}$$

for all $x, y, z, w \in J$.

Let $T : J \times J \to J$ be another $\mathbb{C}$-bilinear mapping satisfying (5.35). Then we have

$$\|D(x, z) - T(x, z)\|$$

$$= \frac{1}{4^n}\|D(2^nx, 2^nz) - T(2^nx, 2^nz)\|$$

$$\leq \frac{1}{4^n}\|D(2^nx, 2^nz) - f(2^nx, 2^nz)\| + \frac{1}{4^n}\|f(2^nx, 2^nz) - T(2^nx, 2^nz)\|$$

$$\leq 2\frac{2^{pn}}{4^n}\frac{5\varepsilon}{4 - 2^p}(\|x\|^p + \|z\|^p),$$

which tends to zero as $n \to \infty$ for all $x, z \in J$. This proves the uniqueness of $D$. Therefore, the mapping $D : J \times J \to J$ is a unique bi-$\theta$-derivation satisfying (5.35). This completes the proof. $\qquad\square$

Similarly, one can obtain the following theorem.

*Remark 5.31.* Let $p$, $\varepsilon$ be positive real numbers with $p > 4$ and $f : J \times J \to J$ with $f(0, 0) = 0$, $h : J \to J$ with $h(0) = 0$ be the mappings satisfying (5.31), (5.32) and (5.33). Then there exist a unique $\mathbb{C}$-linear mapping $\theta : J \to J$ and a unique bi-$\theta$-derivation $D : J \times J \to J$ such that

$$\|h(a) - \theta(a)\| \leq \frac{2\varepsilon}{2^p - 2}\|a\|^p$$

and

$$\|f(x,z) - D(x,z)\| \leq \frac{5\varepsilon}{2^p - 4}(\|x\|^p + \|z\|^p)$$

for all $a, x, z \in J$.

**Theorem 5.32.** *Let $p$, $\varepsilon$, $\delta$ be nonnegative real numbers with $0 < p < 1$ and $f : J \times J \to J$ with $f(0,0) = 0$, $h : J \to J$ with $h(0) = 0$ be the mappings satisfying (5.33) and*

$$\|E_{\lambda,\mu}f(x,y,z,w) + h(\mu a + \mu b) - \mu h(a) - \mu h(b)\|$$
$$\leq \varepsilon\|x\|^p\|y\|^p\|z\|^p\|w\|^p\|a\|^p\|b\|^p + \delta \qquad (5.44)$$

*and*

$$\|f(\{x,y,z\},w) - \{f(x,w), h(y), h(z)\}$$
$$-\{h(x), f(y,w^*), h(z)\} - \{h(x), h(y), f(z,w)\}\|$$
$$+\|f(x,\{y,z,w\}) - \{f(x,y), h(z), h(w)\}$$
$$-\{h(y), f(x^*,z), h(w)\} - \{h(y), h(z), f(x,w)\}\|$$
$$\leq \varepsilon\|x\|^p\|y\|^p\|z\|^p\|w\|^p + \delta \qquad (5.45)$$

*for all $\lambda, \mu \in \mathbb{T}^1$ and $x, y, z, w \in J$. Then there exist a unique $\mathbb{C}$-linear mapping $\theta : J \to J$ and a unique bi-$\theta$-derivation $D : J \times J \to J$ such that*

$$\|h(a) - \theta(a)\| \leq \delta \qquad (5.46)$$

*and*

$$\|f(x,z) - D(x,z)\| \leq \delta \qquad (5.47)$$

*for all $a, x, z \in J$.*

*Proof.* Letting $x = y = z = w = 0$ in (5.44), we have

$$\|h(\mu a + \mu b) - \mu h(a) - \mu h(b)\| \leq \delta$$

for all $a, b \in J$. By the same reasoning as in the proof of Park [227, Theorem 2.1], one can show that there exists a unique $\mathbb{C}$-linear mapping $\theta : J \to J$ satisfying (5.46) and the mapping $\theta : J \to J$ is given by

$$\theta(a) := \lim_{n \to \infty} \frac{1}{2^n} h(2^n a)$$

for all $a \in J$. Letting $\lambda = \mu = 1$, $a = b = 0$, $y = x$ and $w = -z$ in (5.44), we have

$$\|f(2x, 2z) - 2f(x, z) + 2f(x, -z)\| \le \delta \tag{5.48}$$

for all $x, z \in J$. Letting $\lambda = \mu = 1$, $a = b = 0$, $y = -x$ and $w = z$ in (5.44), we have

$$\|f(2x, 2z) - 2f(x, z) + 2f(-x, z)\| \le \delta \tag{5.49}$$

for all $x, z \in J$. Letting $\lambda = \mu = 1$, $a = b = 0$, $x = z = 0$ in (5.44), we have

$$\|f(y, -w) + f(-y, w) + 2f(y, w)\| \le \delta \tag{5.50}$$

for all $y, w \in J$. Replacing $y, w$ by $x, z$ in (5.50), respectively, we have

$$\|f(x, -z) + f(-x, z) + 2f(x, z)\| \le \delta \tag{5.51}$$

for all $x, z \in J$. By (5.48) and (5.51), we obtain

$$\|f(2x, 2z) - 4f(x, z) + f(x, -z) - f(-x, z)\| \le 2\delta \tag{5.52}$$

for all $x, z \in J$. By (5.48) and (5.49), we obtain

$$\|f(x, -z) - f(-x, z)\| \le \delta \tag{5.53}$$

for all $x, z \in J$. By (5.52) and (5.53), we obtain

$$\|f(2x, 2z) - 4f(x, z)\| \le 3\delta \tag{5.54}$$

for all $x, z \in J$. It follows from (5.54) that

$$\left\| \frac{1}{4^l} f(2^l x, 2^l z) - \frac{1}{4^m} f(2^m, 2^m z) \right\| \le \sum_{j=l}^{m-1} \frac{3\delta}{4} \frac{1}{4^j} \tag{5.55}$$

for all $x, z \in J$ and $m, l \ge 1$ with $m > l$. This implies that the sequence $\left\{ \frac{1}{4^n} f(2^n x, 2^n z) \right\}$ is a Cauchy sequence for all $x, z \in J$. Since $J$ is complete, the sequence $\left\{ \frac{1}{4^n} f(2^n x, 2^n z) \right\}$ converges and so one can define the mapping $D : J \times J \to J$ by

$$D(x, z) := \lim_{n \to \infty} \frac{1}{4^n} f(2^n x, 2^n z)$$

for all $x, z \in J$. Moreover, letting $l = 0$ and passing the limit $m \to \infty$ in (5.55), we have (5.47).

Let $a = b = 0$ in (5.44). Then by the definition of the mapping $D$, we have

$$\|E_{\lambda,\mu}D(x, y, z, w)\| = \lim_{n\to\infty} \frac{1}{4^n} \|E_{\lambda,\mu}f(2^n x, 2^n y, 2^n z, 2^n w)\|$$

$$\leq \lim_{n\to\infty} \frac{\delta}{4^n}$$

$$= 0$$

for all $\lambda, \mu \in \mathbb{T}^1$ and $x, y, z, w \in J$. By Lemma 5.29, the mapping $D : J \times J \to J$ is $\mathbb{C}$-bilinear. It follows from (5.33) and (5.45) that

$$\|D(\{x, y, z\}, w) - \{D(x, w), \theta(y), \theta(z)\} - \{\theta(x), D(y, w^*), \theta(z)\}$$
$$- \{\theta(x), \theta(y), D(z, w)\}\| + \|D(x, \{y, z, w\}) - \{D(x, y), \theta(z), \theta(w)\}$$
$$- \{\theta(y), D(x^*, z), \theta(w)\} - \{\theta(y), \theta(z), D(x, w)\}\|$$

$$= \lim_{n\to\infty} \left( \left\| \frac{1}{2^{4n}} f(8^n\{x, y, z\}, 2^n w) - \left\{ \frac{1}{4^n} f(2^n x, 2^n w), \frac{1}{2^n} h(2^n y), \frac{1}{2^n} h(2^n z) \right\} \right. \right.$$

$$- \left\{ \frac{1}{2^n} h(2^n x), \frac{1}{4^n} f(2^n y, 2^n w^*), \frac{1}{2^n} h(2^n z) \right\}$$

$$- \left\{ \frac{1}{2^n} h(2^n x), \frac{1}{2^n} h(2^n y), \frac{1}{4^n} f(2^n z, 2^n w) \right\}$$

$$+ \left\| \frac{1}{2^{4n}} f(2^n x, 8^n\{y, z, w\}) - \left\{ \frac{1}{4^n} f(2^n x, 2^n y), \frac{1}{2^n} h(2^n z), \frac{1}{2^n} h(2^n w) \right\} \right.$$

$$- \left\{ \frac{1}{2^n} h(2^n y), \frac{1}{4^n} f(2^n x^*, 2^n z), \frac{1}{2^n} h(2^n w) \right\}$$

$$\left. \left. - \left\{ \frac{1}{2^n} h(2^n y), \frac{1}{2^n} h(2^n z), \frac{1}{4^n} f(2^n x, 2^n w) \right\} \right\| \right)$$

$$\leq \lim_{n\to\infty} \left( \frac{2^{4pn}\varepsilon}{2^{4n}} \|x\|^p \|y\|^p \|z\|^p \|w\|^p + \frac{\delta}{2^{4n}} \right)$$

$$= 0$$

for all $x, y, z, w \in J$ and so

$$D(\{x, y, z\}, w) = \{D(x, w), \theta(y), \theta(z)\}$$
$$+ \{\theta(x), D(y, w^*), \theta(z)\} + \{\theta(x), \theta(y), D(z, w)\}$$

and

$$D(x, \{y, z, w\}) = \{D(x, y), \theta(z), \theta(w)\}$$
$$+ \{\theta(y), D(x^*, z), \theta(w)\} + \{\theta(y), \theta(z), D(x, w)\}$$

for all $x, y, z, w \in J$.

Let $T : J \times J \to J$ be another $\mathbb{C}$-bilinear mapping satisfying (5.47). Then we have

$$\|D(x,z) - T(x,z)\|$$
$$= \frac{1}{4^n}\|D(2^n x, 2^n z) - T(2^n x, 2^n z)\|$$
$$\leq \frac{1}{4^n}\|D(2^n x, 2^n z) - f(2^n x, 2^n z)\| + \frac{1}{4^n}\|f(2^n x, 2^n z) - T(2^n x, 2^n z)\|$$
$$\leq \frac{2\delta}{4^n},$$

which tends to zero as $n \to \infty$ for all $x, z \in J$. This proves the uniqueness of $D$. Therefore, the mapping $D : J \times J \to J$ is a unique bi-$\theta$-derivation satisfying (5.47). This completes the proof.                                                                            $\square$

**Theorem 5.33.** *Let $p$, $\varepsilon$ be positive real numbers with $p \neq 1$ and $f : J \times J \to J$ with $f(0,0) = 0$, $h : J \to J$ with $h(0) = 0$ be the mappings such that*

$$\|E_{\lambda,\mu}f(x,y,z,w) + h(\mu a + \mu b) - \mu h(a) - \mu h(b)\|$$
$$\leq \varepsilon\|x\|^p\|y\|^p\|z\|^p\|w\|^p\|a\|^p\|b\|^p \tag{5.56}$$

*and*

$$\|f(\{x,y,z\},w) - \{f(x,w),h(y),h(z)\}$$
$$-\{h(x),f(y,w^*),h(z)\} - \{h(x),h(y),f(z,w)\}\|$$
$$+\|f(x,\{y,z,w\}) - \{f(x,y),h(z),h(w)\}$$
$$-\{h(y),f(x^*,z),h(w)\} - \{h(y),h(z),f(x,w)\}\|$$
$$\leq \varepsilon\|x\|^p\|y\|^p\|z\|^p\|w\|^p \tag{5.57}$$

*for all $\lambda, \mu \in \mathbb{T}^1$ and $x, y, z, w \in J$. Then the mapping $h : J \to J$ is a $\mathbb{C}$-linear mapping and the mapping $f : J \times J \to J$ is a bi-h-derivation.*

*Proof.* Letting $x = y = z = w = 0$ in (5.56), we have

$$\|h(\mu a + \mu b) - \mu h(a) - \mu h(b)\| \leq 0$$

for all $a, b \in J$. By the same reasoning as in the proof of [227, Theorem 2.1], one can show that the mapping $h : J \to J$ is a $\mathbb{C}$-linear mapping. Letting $a = b = 0$ in (5.56), we have

$$\|E_{\lambda,\mu}f(x,y,z,w)\| = 0$$

for all $\lambda, \mu \in \mathbb{T}^1$ and $x, y, z, w \in J$. By Lemma 5.29, the mapping $f : J \times J \to J$ is $\mathbb{C}$-bilinear.

For the case $p < 1$, it follows from (5.57) that

$$\|f(\{x, y, z\}, w) - \{f(x, w), h(y), h(z)\} - \{h(x), f(y, w^*), h(z)\}$$
$$- \{h(x), h(y), f(z, w)\}\| + \|f(x, \{y, z, w\}) - \{f(x, y), h(z), h(w)\}$$
$$- \{h(y), f(x^*, z), h(w)\} - \{h(y), h(z), f(x, w)\}\|$$

$$= \lim_{n \to \infty} \left( \left\| \frac{1}{2^{4n}} f(8^n\{x, y, z\}, 2^n w) - \left\{ \frac{1}{4^n} f(2^n x, 2^n w), \frac{1}{2^n} h(2^n y), \frac{1}{2^n} h(2^n z) \right\} \right. \right.$$

$$- \left\{ \frac{1}{2^n} h(2^n x), \frac{1}{4^n} f(2^n y, 2^n w^*), \frac{1}{2^n} h(2^n z) \right\}$$

$$- \left\{ \frac{1}{2^n} h(2^n x), \frac{1}{2^n} h(2^n y), \frac{1}{4^n} f(2^n z, 2^n w) \right\}$$

$$+ \left\| \frac{1}{2^{4n}} f(2^n x, 8^n\{y, z, w\}) - \left\{ \frac{1}{4^n} f(2^n x, 2^n y), \frac{1}{2^n} h(2^n z), \frac{1}{2^n} h(2^n w) \right\} \right.$$

$$- \left\{ \frac{1}{2^n} h(2^n y), \frac{1}{4^n} f(2^n x^*, 2^n z), \frac{1}{2^n} h(2^n w) \right\}$$

$$\left. \left. - \left\{ \frac{1}{2^n} h(2^n y), \frac{1}{2^n} h(2^n z), \frac{1}{4^n} f(2^n x, 2^n w) \right\} \right\| \right)$$

$$\leq \lim_{n \to \infty} \frac{2^{4pn} \varepsilon}{2^{4n}} \|x\|^p \|y\|^p \|z\|^p \|w\|^p$$

$$= 0$$

for all $x, y, z, w \in J$ and so the mapping $f : J \times J \to J$ is a bi-$h$-derivation.

Similarly, for the case $p > 1$, one can show that the mapping $f : J \times J \to J$ is a bi-$h$-derivation. Therefore, the mapping $h : J \to J$ is a $\mathbb{C}$-linear mapping and the mapping $f : J \times J \to J$ is a bi-$h$-derivation. This completes the proof.    $\square$

# Chapter 6
# Stability of Functional Equations in Multi-Banach Algebras

In this chapter, we extend some results from last chapters in multi-Banach algebras (see [7, 91, 218, 252]).

In Sect. 6.1, we consider the stability of the $m$-variable additive functional equation:

$$\sum_{i=1}^{m} f\left(mx_i + \sum_{j=1, j\neq i}^{m} x_j\right) + f\left(\sum_{i=1}^{m} x_i\right) = 2f\left(\sum_{i=1}^{m} mx_i\right)$$

for each $m \geq 2$, which was presented at Sect. 2.2 of Chap. 2 and, by the fixed point method, we approximate homomorphisms and derivations in multi-Banach algebras.

In Sect. 6.2, by using the fixed point method, we prove the Hyers-Ulam stability of homomorphisms in multi-$C^*$-ternary algebras and derivations on multi-$C^*$-ternary algebras for the additive functional equation:

$$\sum_{i=1}^{m} f\left(mx_i + \sum_{j=1, j\neq i}^{m} x_j\right) + f\left(\sum_{i=1}^{m} x_i\right) = 2f\left(\sum_{i=1}^{m} mx_i\right)$$

for each $m \geq 2$.

In Sect. 6.3, we consider the functional equation (3.97) presented at Sect. 3.5 of Chap. 3 and we use a fixed point method to prove the Hyers-Ulam stability of the functional equation (3.97) in multi-Banach modules over a unital multi-$C^*$-algebra.

© Springer International Publishing Switzerland 2015
Y.J. Cho et al., *Stability of Functional Equations in Banach Algebras*, DOI 10.1007/978-3-319-18708-2_6

As an application, we show that every almost linear bijection $h : A \to B$ of a unital multi-$C^*$-algebra $A$ onto a unital multi-$C^*$-algebra $B$ is a $C^*$-algebra isomorphism when

$$h\left(\frac{2^n}{r^n}uy\right) = h\left(\frac{2^n}{r^n}u\right)h(y)$$

for all unitaries $u \in U(A)$, $y \in A$ and $n \geq 0$.

In Sect. 6.4, we approximate the following additive functional inequality:

$$\left\|\left(\sum_{i=1}^{d+1} f(x_{1i}), \cdots, \sum_{i=1}^{d+1} f(x_{ki})\right)\right\|_k$$

$$\leq \left\|\left(mf\left(\frac{\sum_{i=1}^{d+1} x_{1i}}{m}\right), \cdots, mf\left(\frac{\sum_{i=1}^{d+1} x_{ki}}{m}\right)\right)\right\|_k$$

for all $x_{11}, \cdots, x_{kd+1} \in X$ where $d \geq 2$ is a fixed integer. Also, we investigate homomorphisms in proper multi-$CQ^*$-algebras and derivations on proper multi-$CQ^*$-algebras associated with the above additive functional inequality.

In Sect. 6.5, by using the fixed point method, we prove the Hyers-Ulam stability of homomorphisms and derivations on multi-$C^*$–ternary algebras for the additive functional equation:

$$2f\left(\frac{\sum_{j=1}^{p} x_j}{2} + \sum_{j=1}^{d} y_j\right) = \sum_{j=1}^{p} f(x_j) + 2\sum_{j=1}^{d} f(y_j).$$

## 6.1  Stability of $m$-Variable Additive Mappings

For any mapping $f : A \to B$, we define

$$D_\mu f(x_1, \cdots, x_m)$$

$$:= \sum_{i=1}^{m} \mu f\left(mx_i + \sum_{j=1, j\neq i}^{m} x_j\right) + \mu f\left(\sum_{i=1}^{m} x_i\right) - 2f\left(\mu \sum_{i=1}^{m} mx_i\right)$$

for all $\mu \in \mathbb{T}^1 := \{v \in \mathbb{C} : |v| = 1\}$ and $x_1, \cdots, x_m \in A$.

### 6.1.1  Stability of Homomorphisms in Multi-Banach Algebras

Now, we prove the Hyers-Ulam stability of homomorphisms in multi-Banach algebras for the functional equation $D_\mu f(x_1, \cdots, x_m) = 0$.

**Theorem 6.1.** *Let $((B^k, \| \cdot \|_k) : k \geq 1)$ be a multi-Banach algebra. Let $f : A \to B$ be a mapping for which there exists the functions $\varphi : A^{mk} \to [0, \infty)$ and $\psi : A^{2k} \to [0, \infty)$ such that*

$$\lim_{j \to \infty} m^{-j} \varphi(m^j x_{11}, \cdots, m^j x_{1m}, \cdots, m^j x_{k1}, \cdots, m^j x_{km}) = 0, \tag{6.1}$$

$$\|(D_\mu f(x_{11}, \cdots, x_{1m}), \cdots, D_\mu f(x_{k1}, \cdots, x_{km}))\|_k$$
$$\leq \varphi(x_{11}, \cdots, x_{1m}, \cdots, x_{k1}, \cdots, x_{km}), \tag{6.2}$$

$$\|(f(x_1 y_1) - f(x_1)f(y_1), \cdots, f(x_k y_k) - f(x_k)f(y_k))\|_k$$
$$\leq \psi(x_1, y_1, \cdots, x_k, y_k) \tag{6.3}$$

*and*

$$\lim_{j \to \infty} m^{-2j} \psi(m^j x_1, m^j y_1, \cdots, m^j x_k, m^j y_k) = 0 \tag{6.4}$$

*for all $\mu \in \mathbb{T}^1$ and $x_{11}, \cdots, x_{1m}, \cdots, x_{k1}, \cdots, x_{km}, x_1, \cdots, x_m, x, y_1, \cdots, y_k \in A$. If there exists $L < 1$ such that*

$$\varphi\left( \overbrace{mx_{11}, 0, \cdots, 0}^{m}, \overbrace{mx_{21}, 0, \cdots, 0}^{m}, \cdots, \overbrace{mx_{k1}, 0, \cdots, 0}^{m} \right)$$
$$\leq mL\varphi\left( \overbrace{x_{11}, 0, \cdots, 0}^{m}, \overbrace{x_{21}, 0, \cdots, 0}^{m}, \cdots, \overbrace{x_{k1}, 0, \cdots, 0}^{m} \right)$$

*for all $x_{11}, x_{21}, \cdots, x_{k1} \in A$, then there exists a unique homomorphism $H : A \to B$ such that*

$$\|(f(x_1) - H(x_1), \cdots, f(x_k) - H(x_k))\|_k$$
$$\leq \frac{1}{m - mL} \varphi\left( \overbrace{x_1, 0, \cdots, 0}^{m}, \overbrace{x_2, 0, \cdots, 0}^{m}, \cdots, \overbrace{x_k, 0, \cdots, 0}^{m} \right) \tag{6.5}$$

*for all $x_1, \cdots, x_k \in A$.*

*Proof.* Consider the set $X := \{g : A \to B\}$ and introduce the generalized metric on $X$ as follows:

$$d(g, h) = \inf\{C \in \mathbb{R}_+ : \|(g(x_1) - h(x_1), \cdots, g(x_k) - h(x_k))\|_k$$
$$\leq C\varphi(\overbrace{x_1, 0, \cdots, 0}^{m}, \overbrace{x_2, 0, \cdots, 0}^{m}, \cdots, \overbrace{x_k, 0, \cdots, 0}^{m}), \forall x_1, \cdots, x_k \in A\},$$

which $(X, d)$ is complete.

Now, we consider the linear mapping $J : X \to X$ such that

$$Jg(x) := \frac{1}{m} g(mx)$$

for all $x \in A$. Now, we have

$$d(Jg, Jh) \le Ld(g, h)$$

for all $g, h \in X$. Letting $\mu = 1$, $x_{i1} = x_1$ and $x_{i2} = \cdots = x_{im} = 0$, $1 \le i \le k$, in (6.2), we have

$$\|(f(mx_1) - mf(x_1), \cdots, f(mx_k) - mf(x_k))\|_k$$

$$\le \varphi\left( \overbrace{x_1, 0, \cdots, 0}^{m}, \overbrace{x_2, 0, \cdots, 0}^{m}, \cdots, \overbrace{x_k, 0, \cdots, 0}^{m} \right) \qquad (6.6)$$

for all $x_1, \cdots, x_k \in A$ and so

$$\|(f(x_1) - \frac{1}{m} f(mx_1), \cdots, f(x_k) - \frac{1}{m} f(mx_k))\|_k$$

$$\le \frac{1}{m} \varphi\left( \overbrace{x_1, 0, \cdots, 0}^{m}, \overbrace{x_2, 0, \cdots, 0}^{m}, \cdots, \overbrace{x_k, 0, \cdots, 0}^{m} \right)$$

for all $x_1, \cdots, x_k \in A$. Hence $d(f, Jf) \le \frac{1}{m}$. By Theorem 1.3, there exists a mapping $H : A \to B$ such that

(1) $H$ is a fixed point of $J$, i.e.,

$$H(mx) = mH(x) \qquad (6.7)$$

for all $x \in A$. The mapping $H$ is a unique fixed point of $J$ in the set

$$Y = \{g \in X : d(f, g) < \infty\}.$$

This implies that $H$ is a unique mapping satisfying (6.7) such that there exists $C \in (0, \infty)$ satisfying

$$\|(H(x_1) - f(x_1), \cdots, H(x_k) - f(x_k))\|_k$$

$$\le C\varphi\left( \overbrace{x_1, 0, \cdots, 0}^{m}, \overbrace{x_2, 0, \cdots, 0}^{m}, \cdots, \overbrace{x_k, 0, \cdots, 0}^{m} \right)$$

for all $x_1, \cdots, x_k \in A$;

(2) $d(J^n f, H) \to 0$ as $n \to \infty$. This implies the equality

$$\lim_{n \to \infty} \frac{f(m^n x)}{m^n} = H(x) \tag{6.8}$$

for all $x \in A$;

(3) $d(f, H) \le \frac{1}{1-L} d(f, Jf)$, which implies the inequality

$$d(f, H) \le \frac{1}{m - mL}.$$

This implies that the inequality (6.4) holds.

It follows from (6.1), (6.2) and (6.8) that

$$\left\| \left( \sum_{i=1}^{m} H\left( mx_{1i} + \sum_{j=1, j \ne i}^{m} x_{1j} \right) + H\left( \sum_{i=1}^{m} x_{1i} \right) - 2H\left( \sum_{i=1}^{m} mx_{1i} \right), \right.$$

$$\left. \cdots, \sum_{i=1}^{m} H\left( mx_{ki} + \sum_{j=1, j \ne i}^{m} x_{kj} \right) + H\left( \sum_{i=1}^{m} x_{ki} \right) - 2H\left( \sum_{i=1}^{m} mx_{ki} \right) \right) \right\|_k$$

$$= \lim_{n \to \infty} \frac{1}{m^n} \left\| \left( \sum_{i=1}^{m} f\left( m^{n+1} x_{1i} + \sum_{j=1, j \ne i}^{m} m^n x_{1j} \right) \right. \right.$$

$$+ f\left( \sum_{i=1}^{m} m^n x_{1i} \right) - 2f\left( \sum_{i=1}^{m} m^{n+1} x_{1i} \right),$$

$$\cdots, \sum_{i=1}^{m} f\left( m^{n+1} x_{ki} + \sum_{j=1, j \ne i}^{m} m^n x_{kj} \right)$$

$$\left. \left. + f\left( \sum_{i=1}^{m} m^n x_{ki} \right) - 2f\left( \sum_{i=1}^{m} m^{n+1} x_{ki} \right) \right) \right\|_k$$

$$\le \lim_{n \to \infty} \frac{1}{m^n} \varphi(m^n x_{11}, \cdots, m^n x_{1m}, \cdots, m^n x_{k1}, \cdots, m^n x_{km})$$

$$= 0$$

for all $x_{11}, \cdots, x_{1m}, \cdots, x_{k1}, \cdots, x_{km} \in A$ and so

$$\sum_{i=1}^{m} H\left( mx_i + \sum_{j=1, j \ne i}^{m} x_j \right) + H\left( \sum_{i=1}^{m} x_i \right) = 2H\left( \sum_{i=1}^{m} mx_i \right) \tag{6.9}$$

for all $x_1, \cdots, x_m \in A$. So $H$ is additive. By a similar method to above, we get

$$\mu H(mx) = H(m\mu x)$$

for all $\mu \in \mathbb{T}^1$ and $x \in A$. Thus one can show that the mapping $H : A \to B$ is $\mathbb{C}$-linear. It follows from (6.3), (6.4) and (6.8) that

$$\|(H(x_1y_1) - H(x_1)H(y_1), \cdots, H(x_ky_k) - H(x_k)H(y_k))\|_k$$

$$= \lim_{n\to\infty} \frac{1}{m^{2n}} \|(f(m^{2n}x_1y_1) - f(m^n x_1)f(m^n y_1),$$

$$\cdots, f(m^{2n}x_ky_k) - f(m^n x_k)f(m^n y_k))\|_k$$

$$\leq \lim_{n\to\infty} \frac{1}{m^{2n}} \psi(m^n x_1, m^n y_1, \cdots, m^n x_k, m^n y_k)$$

$$= 0$$

for all $x_1, y_1, \cdots, x_k, y_k \in A$ and so

$$H(xy) = H(x)H(y)$$

for all $x, y \in A$. Thus $H : A \to B$ is a homomorphism satisfying (6.5). This completes the proof. $\qquad\square$

**Corollary 6.2.** *Let $((B^k, \|\cdot\|_k) : k \geq 1)$ be a multi-Banach algebra. Let $r < 1$ and $\theta$ be nonnegative real numbers and $f : A \to B$ be a mapping such that*

$$\|(D_\mu f(x_{11}, \cdots, x_{1m}), \cdots, D_\mu f(x_{k1}, \cdots, x_{km}))\|_k$$

$$\leq \theta \Big( \sum_{j=1}^{m} \|x_{1j}\|_A^r + \cdots + \sum_{j=1}^{m} \|x_{km}\|_A^r \Big) \tag{6.10}$$

*and*

$$\|(f(x_1y_1) - f(x_1)f(y_1), \cdots, f(x_ky_k) - f(x_k)f(y_k))\|_k$$

$$\leq \theta \Big( \|x_1\|_A^r \cdot \|y_1\|_A^r + \cdots + \|x_k\|_A^r \cdot \|y_k\|_A^r \Big) \tag{6.11}$$

*for all $\mu \in \mathbb{T}^1$ and $x_{11}, \cdots, x_{1m}, \cdots, x_{k1}, \cdots, x_{km}, x_1, \cdots, x_m, x, y_1, \cdots, y_k \in A$. Then there exists a unique homomorphism $H : A \to B$ such that*

$$\|(f(x_1) - H(x_1), \cdots, f(x_k) - H(x_k))\|_k$$

$$\leq \frac{\theta}{m - m^r} \Big( \|x_1\|_A^r + \|y_1\|_A^r + \cdots + \|x_k\|_A^r + \|y_k\|_A^r \Big)$$

*for all $x_1, \cdots, x_k \in A$.*

*Proof.* The proof follows from Theorem 6.1 by taking

$$\varphi(x_{11}, \cdots, x_{1m}, \cdots, x_{k1}, \cdots, x_{km}) := \theta\left(\sum_{j=1}^{m} \|x_{1j}\|_A^r + \cdots + \sum_{j=1}^{m} \|x_{km}\|_A^r\right)$$

and

$$\psi(x_1, y_1, \cdots, x_k, y_k) := \theta\left(\|x_1\|_A^r \cdot \|y_1\|_A^r + \cdots + \|x_k\|_A^r \cdot \|y_k\|_A^r\right)$$

for all $x_{11}, \cdots, x_{1m}, \cdots, x_{k1}, \cdots, x_{km}, x_1, \cdots, x_m, x, y_1, \cdots, y_k \in A$ and $L = m^{r-1}$.

$\square$

**Theorem 6.3.** *Let* $((B^k, \|\cdot\|_k) : k \geq 1)$ *be a multi-Banach algebra. Let* $f : A \to B$ *be a mapping for which there are functions* $\varphi : A^{mk} \to [0, \infty)$ *and* $\psi : A^{2k} \to [0, \infty)$ *such that*

$$\lim_{j \to \infty} m^j \varphi(m^{-j}x_{11}, \cdots, m^{-j}x_{1m}, \cdots, m^{-j}x_{k1}, \cdots, m^{-j}x_{km}) = 0, \quad (6.12)$$

$$\|(D_\mu f(x_{11}, \cdots, x_{1m}), \cdots, D_\mu f(x_{k1}, \cdots, x_{km}))\|_k$$
$$\leq \varphi(x_{11}, \cdots, x_{1m}, \cdots, x_{k1}, \cdots, x_{km}), \quad (6.13)$$

$$\|(f(x_1 y_1) - f(x_1)f(y_1), \cdots, f(x_k y_k) - f(x_k)f(y_k))\|_k$$
$$\leq \psi(x_1, y_1, \cdots, x_k, y_k) \quad (6.14)$$

*and*

$$\lim_{j \to \infty} m^{2j} \psi(m^{-j}x_1, m^{-j}y_1, \cdots, m^{-j}x_k, m^{-j}y_k) = 0 \quad (6.15)$$

*for all* $\mu \in \mathbb{T}^1$ *and* $x_{11}, \cdots, x_{1m}, \cdots, x_{k1}, \cdots, x_{km}, x_1, \cdots, x_m, x, y_1, \cdots, y_k \in A$. *If there exists* $L < 1$ *such that*

$$\varphi\left(\overbrace{x_{11}, 0, \cdots, 0}^{m}, \overbrace{x_{21}, 0, \cdots, 0}^{m}, \cdots, \overbrace{x_{k1}, 0, \cdots, 0}^{m}\right)$$
$$\leq \frac{L}{m} \varphi\left(\overbrace{mx_{11}, 0, \cdots, 0}^{m}, \overbrace{mx_{21}, 0, \cdots, 0}^{m}, \cdots, \overbrace{mx_{k1}, 0, \cdots, 0}^{m}\right)$$

*for all* $x_{11}, x_{21}, \cdots, x_{k1} \in A$, *then there exists a unique homomorphism* $H : A \to B$ *such that*

$$\|(f(x_1) - H(x_1), \cdots, f(x_k) - H(x_k))\|_k$$
$$\leq \frac{L}{m - mL} \varphi\left(\overbrace{x_1, 0, \cdots, 0}^{m}, \overbrace{x_2, 0, \cdots, 0}^{m}, \cdots, \overbrace{x_k, 0, \cdots, 0}^{m}\right) \quad (6.16)$$

*for all* $x_1, \cdots, x_k \in A$.

*Proof.* We consider the linear mapping $J : X \rightarrow X$ such that

$$Jg(x) := mg\left(\frac{x}{m}\right)$$

for all $x \in A$. It follows from (6.6) that

$$\left\|\left(f(x_1) - mf\left(\frac{x_1}{m}\right), \cdots, f(x_k) - mf\left(\frac{x_k}{m}\right)\right)\right\|_k$$

$$\leq \varphi\left(\overbrace{\frac{x_1}{m}, 0, \cdots, 0}^{m}, \overbrace{\frac{x_2}{m}, 0, \cdots, 0}^{m}, \cdots, \overbrace{\frac{x_k}{m}, 0, \cdots, 0}^{m}\right) \qquad (6.17)$$

$$\leq \frac{L}{m}\varphi\left(\overbrace{x_1, 0, \cdots, 0}^{m}, \overbrace{x_2, 0, \cdots, 0}^{m}, \cdots, \overbrace{x_k, 0, \cdots, 0}^{m}\right)$$

for all $x_1, \cdots, x_k \in A$. Hence we have

$$d(f, Jf) \leq \frac{L}{m}.$$

By Theorem 1.3, there exists a mapping $H : A \rightarrow B$ such that

(1) $H$ is a fixed point of $J$, i.e.,

$$H(mx) = mH(x) \qquad (6.18)$$

for all $x \in A$. The mapping $H$ is a unique fixed point of $J$ in the set

$$Y = \{g \in X : d(f, g) < \infty\}.$$

This implies that $H$ is a unique mapping satisfying (6.18) such that there exists $C \in (0, \infty)$ satisfying

$$\|(H(x_1) - f(x_1), \cdots, H(x_k) - f(x_k))\|_k$$

$$\leq C\varphi\left(\overbrace{x_1, 0, \cdots, 0}^{m}, \overbrace{x_2, 0, \cdots, 0}^{m}, \cdots, \overbrace{x_k, 0, \cdots, 0}^{m}\right)$$

for all $x_1, \cdots, x_k \in A$;

(2) $d(J^n f, H) \rightarrow 0$ as $n \rightarrow \infty$. This implies the equality

$$\lim_{n \rightarrow \infty} m^n f\left(\frac{x}{m^n}\right) = H(x)$$

for all $x \in A$;

(3) $d(f, H) \le \frac{1}{1-L}d(f, Jf)$, which implies the inequality

$$d(f, H) \le \frac{L}{m - mL},$$

which implies that the inequality (6.16) holds.

The rest of the proof is similar to the proof of Theorem 6.1. This completes the proof. $\qquad\square$

**Corollary 6.4.** *Let $((B^k, \|\cdot\|_k) : k \ge 1)$ be a multi-Banach algebra. Let $r > 1$, $\theta$ be nonnegative real numbers and $f : A \to B$ be a mapping such that*

$$\|(D_\mu f(x_{11}, \cdots, x_{1m}), \cdots, D_\mu f(x_{k1}, \cdots, x_{km}))\|_k$$

$$\le \theta \Big( \sum_{j=1}^{m} \|x_{1j}\|_A^r + \cdots + \sum_{j=1}^{m} \|x_{km}\|_A^r \Big) \tag{6.19}$$

*and*

$$\|(f(x_1 y_1) - f(x_1)f(y_1), \cdots, f(x_k y_k) - f(x_k)f(y_k))\|_k$$

$$\le \theta \left( \|x_1\|_A^r \cdot \|y_1\|_A^r + \cdots + \|x_k\|_A^r \cdot \|y_k\|_A^r \right) \tag{6.20}$$

*for all $\mu \in \mathbb{T}^1$ and $x_{11}, \cdots, x_{1m}, \cdots, x_{k1}, \cdots, x_{km}, x_1, \cdots, x_m, x, y_1, \cdots, y_k \in A$. Then there exists a unique homomorphism $H : A \to B$ such that*

$$\|(f(x_1) - H(x_1), \cdots, f(x_k) - H(x_k))\|_k$$

$$\le \frac{\theta}{m^r - m} \left( \|x_1\|_A^r + \|y_1\|_A^r + \cdots + \|x_k\|_A^r + \|y_k\|_A^r \right)$$

*for all $x_1, \cdots, x_k \in A$.*

*Proof.* The proof follows from Theorem 6.3 by taking

$$\varphi(x_{11}, \cdots, x_{1m}, \cdots, x_{k1}, \cdots, x_{km}) := \theta \Big( \sum_{j=1}^{m} \|x_{1j}\|_A^r + \cdots + \sum_{j=1}^{m} \|x_{km}\|_A^r \Big),$$

$$\psi(x_1, y_1, \cdots, x_k, y_k) := \theta \left( \|x_1\|_A^r \cdot \|y_1\|_A^r + \cdots + \|x_k\|_A^r \cdot \|y_k\|_A^r \right)$$

for all $x_{11}, \cdots, x_{1m}, \cdots, x_{k1}, \cdots, x_{km}, x_1, \cdots, x_m, x, y_1, \cdots, y_k \in A$ and $L = m^{1-r}$.

$\qquad\square$

### 6.1.2   Stability of Derivations in Multi-Banach Algebras

Now, we prove the Hyers-Ulam stability of derivations in multi-Banach algebras for the following functional equation:

$$D_\mu f(x_1, \cdots, x_m) = 0$$

for all $\mu \in \mathbb{T}^1 := \{v \in \mathbb{C} : |v| = 1\}$ and $x_1, \cdots, x_m \in A$.

**Theorem 6.5.** *Let $((A^k, \| \cdot \|_k) : k \geq 1)$ be a multi-Banach algebra. Let $f : A \to A$ be a mapping for which there exist the functions $\varphi : A^{mk} \to [0, \infty)$ and $\psi : A^{2k} \to [0, \infty)$ such that*

$$\lim_{j \to \infty} m^{-j} \varphi(m^j x_{11}, \cdots, m^j x_{1m}, \cdots, m^j x_{k1}, \cdots, m^j x_{km}) = 0, \tag{6.21}$$

$$\|(D_\mu f(x_{11}, \cdots, x_{1m}), \cdots, D_\mu f(x_{k1}, \cdots, x_{km}))\|_k$$
$$\leq \varphi(x_{11}, \cdots, x_{1m}, \cdots, x_{k1}, \cdots, x_{km}), \tag{6.22}$$

$$\|(f(x_1 y_1) - f(x_1)y_1 - x_1 f(y_1), \cdots, f(x_k y_k) - f(x_k)y_k - x_k f(y_k))\|_k$$
$$\leq \psi(x_1, y_1, \cdots, x_k, y_k) \tag{6.23}$$

*and*

$$\lim_{j \to \infty} m^{-2j} \psi(m^j x_1, m^j y_1, \cdots, m^j x_k, m^j y_k) = 0 \tag{6.24}$$

*for all $\mu \in \mathbb{T}^1$ and $x_{11}, \cdots, x_{1m}, \cdots, x_{k1}, \cdots, x_{km}, x_1, \cdots, x_m, x, y_1, \cdots, y_k \in A$. If there exists $L < 1$ such that*

$$\varphi\Big( \overbrace{mx_{11}, 0, \cdots, 0}^{m}, \overbrace{mx_{21}, 0, \cdots, 0}^{m}, \cdots, \overbrace{mx_{k1}, 0, \cdots, 0}^{m} \Big)$$
$$\leq mL\varphi\Big( \overbrace{x_{11}, 0, \cdots, 0}^{m}, \overbrace{x_{21}, 0, \cdots, 0}^{m}, \cdots, \overbrace{x_{k1}, 0, \cdots, 0}^{m} \Big)$$

*for all $x_{11}, x_{21}, \cdots, x_{k1} \in A$, then there exists a unique derivation $\delta : A \to A$ such that*

$$\|(f(x_1) - \delta(x_1), \cdots, f(x_k) - \delta(x_k))\|_k$$
$$\leq \frac{1}{m - mL} \varphi\Big( \overbrace{x_1, 0, \cdots, 0}^{m}, \overbrace{x_2, 0, \cdots, 0}^{m}, \cdots, \overbrace{x_k, 0, \cdots, 0}^{m} \Big) \tag{6.25}$$

*for all $x_1, \cdots, x_k \in A$.*

*Proof.* By the same reasoning as in the proof of Theorem 6.1, there exists a unique $\mathbb{C}$-linear mapping $\delta : A \to A$ satisfying (6.24). The mapping $\delta : A \to A$ is given by

$$\delta(x) = \lim_{n\to\infty} \frac{f(m^n x)}{m^n} \tag{6.26}$$

for all $x \in A$. It follows from (6.21), (6.24) and (6.26) that

$$\|(\delta(x_1 y_1) - \delta(x_1)y_1 - x_1\delta(y_1), \cdots, \delta(x_k y_k) - \delta(x_k)y_k - x_k\delta(y_k))\|_k$$

$$= \lim_{n\to\infty} \frac{1}{m^{2n}} \|(f(m^{2n}x_1 y_1) - f(m^n x_1) \cdot m^n y_1 - m^n x_1 f(m^n y_1),$$

$$\cdots, f(m^{2n} x_k y_k) - f(m^n x_k) \cdot m^n y_k - m^n x_k f(m^n y_k))\|_k$$

$$\leq \lim_{n\to\infty} m^{-2n} \psi(m^n x_1, m^n y_1, \cdots, m^n x_k, m^n y_k)$$

$$= 0$$

for all $x, y \in A$ and so

$$\delta(xy) = \delta(x)y + x\delta(y)$$

for all $x, y \in A$. Thus $\delta : A \to A$ is a derivation satisfying (6.23). This completes the proof. $\qquad\square$

**Corollary 6.6.** *Let $((A^k, \|\cdot\|_k) : k \geq 1)$ be a multi-Banach algebra. Let $r < 1$, $\theta$ be nonnegative real numbers and $f : A \to A$ be a mapping such that*

$$\|(D_\mu f(x_{11}, \cdots, x_{1m}), \cdots, D_\mu f(x_{k1}, \cdots, x_{km}))\|_k$$

$$\leq \theta \left( \sum_{j=1}^m \|x_{1j}\|_A^r + \cdots + \sum_{j=1}^m \|x_{kj}\|_A^r \right) \tag{6.27}$$

*and*

$$\|(f(x_1 y_1) - f(x_1)y_1 - x_1 f(y_1), \cdots, f(x_k y_k) - f(x_k)y_k - x_k f(y_k))\|_k$$

$$\leq \theta \left( \|x_1\|_A^r \cdot \|y_1\|_A^r + \cdots + \|x_k\|_A^r \cdot \|y_k\|_A^r \right) \tag{6.28}$$

*for all $\mu \in \mathbb{T}^1$ and $x_{11}, \cdots, x_{1m}, \cdots, x_{k1}, \cdots, x_{km}, x_1, \cdots, x_m, x, y_1, \cdots, y_k \in A$. Then there exists a unique derivation $\delta : A \to A$ such that*

$$\|(f(x_1) - \delta(x_1), \cdots, f(x_k) - \delta(x_k))\|_k \leq \frac{\theta}{m - m^r} \|x\|_A^{2r}$$

*for all $x \in A$.*

*Proof.* The proof follows from Theorem 6.5 by taking

$$\varphi(x_{11},\cdots,x_{1m},\cdots,x_{k1},\cdots,x_{km}) := \theta\Big(\sum_{j=1}^{m}\|x_{1j}\|_A^r + \cdots + \sum_{j=1}^{m}\|x_{kj}\|_A^r\Big),$$

$$\psi(x_1,y_1,\cdots,x_k,y_k) := \theta\left(\|x_1\|_A^r \cdot \|y_1\|_A^r + \cdots + \|x_k\|_A^r \cdot \|y_k\|_A^r\right)$$

for all $x,y \in A$ and $L = m^r$.                                              □

*Remark 6.7.* Let $((A^k,\|\cdot\|_k) : k \geq 1)$ be a multi-Banach algebra. Let $f : A \to A$ be a mapping for which there exist the functions $\varphi : A^{mk} \to [0,\infty)$ and $\psi : A^{2k} \to [0,\infty)$ such that

$$\lim_{j\to\infty} m^j\varphi(m^{-j}x_{11},\cdots,m^{-j}x_{1m},\cdots,m^{-j}x_{k1},\cdots,m^{-j}x_{km}) = 0, \qquad (6.29)$$

$$\|(D_\mu f(x_{11},\cdots,x_{1m}),\cdots,D_\mu f(x_{k1},\cdots,x_{km}))\|_k$$

$$\leq \varphi(x_{11},\cdots,x_{1m},\cdots,x_{k1},\cdots,x_{km}), \qquad (6.30)$$

$$\|(f(x_1y_1)-f(x_1)y_1-x_1f(y_1),\cdots,f(x_ky_k)-f(x_k)y_k-x_kf(y_k))\|_k$$

$$\leq \psi(x_1,y_1,\cdots,x_k,y_k) \qquad (6.31)$$

and

$$\lim_{j\to\infty} m^{2j}\psi(m^{-j}x_1,m^{-j}y_1,\cdots,m^{-j}x_k,m^{-j}y_k) = 0 \qquad (6.32)$$

for all $\mu \in \mathbb{T}^1$ and $x_{11},\cdots,x_{1m},\cdots,x_{k1},\cdots,x_{km},x_1,\cdots,x_m,x,y_1,\cdots,y_k \in A$. If there exists $L < 1$ such that

$$\varphi\Big(\overbrace{mx_{11},0,\cdots,0}^{m},\overbrace{mx_{21},0,\cdots,0}^{m},\cdots,\overbrace{mx_{k1},0,\cdots,0}^{m}\Big)$$

$$\leq \frac{L}{m}\varphi\Big(\overbrace{x_{11},0,\cdots,0}^{m},\overbrace{x_{21},0,\cdots,0}^{m},\cdots,\overbrace{x_{k1},0,\cdots,0}^{m}\Big)$$

for all $x_{11},x_{21},\cdots,x_{k1} \in A$, then there exists a unique derivation $\delta : A \to A$ such that

$$\|(f(x_1)-\delta(x_1),\cdots,f(x_k)-\delta(x_k))\|_k$$

$$\leq \frac{L}{m-mL}\varphi\Big(\overbrace{x_1,0,\cdots,0}^{m},\overbrace{x_2,0,\cdots,0}^{m},\cdots,\overbrace{x_k,0,\cdots,0}^{m}\Big) \qquad (6.33)$$

for all $x_1,\cdots,x_k \in A$.

**Corollary 6.8.** *Let $((A^k, \|\cdot\|_k) : k \geq 1)$ be a multi-Banach algebra. Let $r > 1$, $\theta$ be nonnegative real numbers and $f : A \to A$ be a mapping such that*

$$\|(D_\mu f(x_{11}, \cdots, x_{1m}), \cdots, D_\mu f(x_{k1}, \cdots, x_{km}))\|_k$$

$$\leq \theta \left( \sum_{j=1}^{m} \|x_{1j}\|_A^r + \cdots + \sum_{j=1}^{m} \|x_{kj}\|_A^r \right)$$

*and*

$$\|(f(x_1 y_1) - f(x_1)y_1 - x_1 f(y_1), \cdots, f(x_k y_k) - f(x_k)y_k - x_k f(y_k))\|_k$$

$$\leq \theta \left( \|x_1\|_A^r \cdot \|y_1\|_A^r + \cdots + \|x_k\|_A^r \cdot \|y_k\|_A^r \right)$$

*for all $\mu \in \mathbb{T}^1$ and $x_{11}, \cdots, x_{1m}, \cdots, x_{k1}, \cdots, x_{km}, x_1, \cdots, x_m, x, y_1, \cdots, y_k \in A$.*
*Then there exists a unique derivation $\delta : A \to A$ such that*

$$\|(f(x_1) - \delta(x_1), \cdots, f(x_k) - \delta(x_k))\|_k \leq \frac{\theta}{m^r - m} \|x\|_A^{2r}$$

*for all $x \in A$.*

*Proof.* The proof follows from Remark 6.7 by taking

$$\varphi(x_{11}, \cdots, x_{1m}, \cdots, x_{k1}, \cdots, x_{km}) := \theta \left( \sum_{j=1}^{m} \|x_{1j}\|_A^r + \cdots + \sum_{j=1}^{m} \|x_{kj}\|_A^r \right),$$

$$\psi(x_1, y_1, \cdots, x_k, y_k) := \theta \left( \|x_1\|_A^r \cdot \|y_1\|_A^r + \cdots + \|x_k\|_A^r \cdot \|y_k\|_A^r \right)$$

for all $x, y \in A$ and $L = m^{1-r}$. $\qquad\qquad\qquad\qquad\qquad\qquad\qquad\qquad\qquad\square$

## 6.2  Ternary Jordan Homomorphisms and Derivations in Multi-$C^*$-Ternary Algebras

In this section, using the fixed point method, we prove the Hyers-Ulam stability of homomorphisms in multi-$C^*$-ternary algebras and derivations on multi-$C^*$-ternary algebras for the following additive functional equation:

$$\sum_{i=1}^{m} f \left( mx_i + \sum_{j=1, j\neq i}^{m} x_j \right) + f \left( \sum_{i=1}^{m} x_i \right) = 2f \left( \sum_{i=1}^{m} mx_i \right)$$

for each $m \geq 2$.

Throughout this section, assume that $A, B$ are $C^*$-ternary algebras.

## 6.2.1   Stability of Homomorphisms in Multi-C*-Ternary Algebras

For any mapping $f : A \to B$, we define

$$D_\mu f(x_1, \cdots, x_m)$$

$$:= \sum_{i=1}^m \mu f\left(mx_i + \sum_{j=1, j \neq i}^m x_j\right) + \mu f\left(\sum_{i=1}^m x_i\right) - 2f\left(\mu \sum_{i=1}^m mx_i\right)$$

for all $\mu \in \mathbb{T}^1 := \{v \in \mathbb{C} : |v| = 1\}$ and $x_1, \cdots, x_m \in A$.

Using Theorem 1.3, we prove the Hyers-Ulam stability of homomorphisms in multi-$C^*$ ternary algebras for the functional equation

$$D_\mu f(x_1, \cdots, x_m) = 0.$$

**Theorem 6.9.** *Let* $((B^k, \| \cdot \|_k) : k \geq 1)$ *be a multi-C\*-ternary algebra. Let* $f : A \to B$ *be a mapping for which there exist the functions* $\varphi : A^{mk} \to [0, \infty)$ *and* $\psi : A^{2k} \to [0, \infty)$ *such that*

$$\lim_{j \to \infty} m^{-j} \varphi(m^j x_{11}, \cdots, m^j x_{1m}, \cdots, m^j x_{k1}, \cdots, m^j x_{km}) = 0, \qquad (6.34)$$

$$\|(D_\mu f(x_{11}, \cdots, x_{1m}), \cdots, D_\mu f(x_{k1}, \cdots, x_{km}))\|_k$$

$$\leq \varphi(x_{11}, \cdots, x_{1m}, \cdots, x_{k1}, \cdots, x_{km}), \qquad (6.35)$$

$$\|(f([x_1, y_1, z_1]) - [f(x_1), f(y_1), f(z_1)],$$

$$\cdots, f([x_k, y_k, z_k]) - [f(x_k), f(y_k), f(z_k)])\|_k$$

$$\leq \psi(x_1, y_1, z_1, \cdots, x_k, y_k, z_k) \qquad (6.36)$$

*and*

$$\lim_{j \to \infty} m^{-3j} \psi(m^j x_1, m^j y_1, m^j z_1, \cdots, m^j x_k, m^j y_k, m^j z_k) = 0 \qquad (6.37)$$

*for all* $\mu \in \mathbb{T}^1$ *and* $x_{11}, \cdots, x_{1m}, \cdots, x_{k1}, \cdots, x_{km}, x_1, \cdots, x_k, y_1, \cdots, y_k,$ $z_1, \cdots, z_k \in A$. *If there exists* $L < 1$ *such that*

$$\varphi\Big(\overbrace{mx_{11}, 0, \cdots, 0}^{m}, \overbrace{mx_{21}, 0, \cdots, 0}^{m}, \cdots, \overbrace{mx_{k1}, 0, \cdots, 0}^{m}\Big)$$

$$\leq mL\varphi\Big(\overbrace{x_{11}, 0, \cdots, 0}^{m}, \overbrace{x_{21}, 0, \cdots, 0}^{m}, \cdots, \overbrace{x_{k1}, 0, \cdots, 0}^{m}\Big)$$

*for all $x_{11}, x_{21}, \cdots, x_{k1} \in A$, then there exists a unique homomorphism $H : A \to B$ such that*

$$\|(f(x_1) - H(x_1), \cdots, f(x_k) - H(x_k))\|_k$$

$$\leq \frac{1}{m - mL} \varphi\left( \overbrace{x_1, 0, \cdots, 0}^{m}, \overbrace{x_2, 0, \cdots, 0}^{m}, \cdots, \overbrace{x_k, 0, \cdots, 0}^{m} \right) \qquad (6.38)$$

*for all $x_1, \cdots, x_k \in A$.*

*Proof.* Consider the set $X := \{g : A \to B\}$ and introduce the generalized metric on $X$ as follows:

$$d(g, h) = \inf\{C \in \mathbb{R}_+ : \|(g(x_1) - h(x_1), \cdots, g(x_k) - h(x_k))\|_k$$

$$\leq C\varphi(\overbrace{x_1, 0, \cdots, 0}^{m}, \overbrace{x_2, 0, \cdots, 0}^{m}, \cdots, \overbrace{x_k, 0, \cdots, 0}^{m}), \ \forall x_1, \cdots, x_k \in A\},$$

which $(X, d)$ is complete.

Now, we consider the linear mapping $J : X \to X$ such that

$$Jg(x) := \frac{1}{m} g(mx)$$

for all $x \in A$. Now, we have

$$d(Jg, Jh) \leq Ld(g, h)$$

for all $g, h \in X$. Letting $\mu = 1$, $x_{i1} = x_i$ and $x_{i2} = \cdots = x_{im} = 0$ ($1 \leq i \leq k$) in (6.35), we have

$$\|(f(mx_1) - mf(x_1), \cdots, f(mx_k) - mf(x_k))\|_k$$

$$\leq \varphi\left( \overbrace{x_1, 0, \cdots, 0}^{m}, \overbrace{x_2, 0, \cdots, 0}^{m}, \cdots, \overbrace{x_k, 0, \cdots, 0}^{m} \right) \qquad (6.39)$$

for all $x_1, \cdots, x_k \in A$. Thus we have

$$\|(f(x_1) - \frac{1}{m} f(mx_1), \cdots, f(x_k) - \frac{1}{m} f(mx_k))\|_k$$

$$\leq \frac{1}{m} \varphi\left( \overbrace{x_1, 0, \cdots, 0}^{m}, \overbrace{x_2, 0, \cdots, 0}^{m}, \cdots, \overbrace{x_k, 0, \cdots, 0}^{m} \right)$$

for all $x_1, \cdots, x_k \in A$. Hence $d(f, Jf) \leq \frac{1}{m}$. By Theorem 1.3, there exists a mapping $H : A \to B$ such that

(1) $H$ is a fixed point of $J$, i.e.,

$$H(mx) = mH(x) \tag{6.40}$$

for all $x \in A$. The mapping $H$ is a unique fixed point of $J$ in the set

$$Y = \{g \in X : d(f, g) < \infty\}.$$

This implies that $H$ is a unique mapping satisfying (6.40) such that there exists $C \in (0, \infty)$ satisfying the following:

$$\|(H(x_1) - f(x_1), \cdots, H(x_k) - f(x_k))\|_k$$

$$\leq C\varphi\left(\overbrace{x_1, 0, \cdots, 0}^{m}, \overbrace{x_2, 0, \cdots, 0}^{m}, \cdots, \overbrace{x_k, 0, \cdots, 0}^{m}\right)$$

for all $x_1, \cdots, x_k \in A$;

(2) $d(J^n f, H) \to 0$ as $n \to \infty$. This implies the equality

$$\lim_{n \to \infty} \frac{f(m^n x)}{m^n} = H(x) \tag{6.41}$$

for all $x \in A$;

(3) $d(f, H) \leq \frac{1}{1-L} d(f, Jf)$, which implies the inequality

$$d(f, H) \leq \frac{1}{m - mL}.$$

This implies that the inequality (6.38) holds.

It follows from (6.34), (6.35) and (6.41) that

$$\left\| \left( \sum_{i=1}^{m} H\left(mx_{1i} + \sum_{j=1, j \neq i}^{m} x_{1j}\right) + H\left(\sum_{i=1}^{m} x_{1i}\right) - 2H\left(\sum_{i=1}^{m} mx_{1i}\right), \right.$$

$$\left. \cdots, \sum_{i=1}^{m} H\left(mx_{ki} + \sum_{j=1, j \neq i}^{m} x_{kj}\right) + H\left(\sum_{i=1}^{m} x_{ki}\right) - 2H\left(\sum_{i=1}^{m} mx_{ki}\right) \right) \right\|_k$$

$$= \lim_{n \to \infty} \frac{1}{m^n} \left\| \left( \sum_{i=1}^{m} f\left(m^{n+1} x_{1i} + \sum_{j=1, j \neq i}^{m} m^n x_{1j}\right) \right.\right.$$

$$\left.\left. + f\left(\sum_{i=1}^{m} m^n x_{1i}\right) - 2f\left(\sum_{i=1}^{m} m^{n+1} x_{1i}\right), \right.\right.$$

$$\cdots, \sum_{i=1}^{m} f\left(m^{n+1}x_{ki} + \sum_{j=1,j\neq i}^{m} m^n x_{kj}\right)$$

$$+ f\left(\sum_{i=1}^{m} m^n x_{ki}\right) - 2f\left(\sum_{i=1}^{m} m^{n+1}x_{ki}\right)\bigg)\bigg\|_k$$

$$\leq \lim_{n\to\infty} \frac{1}{m^n} \varphi(m^n x_{11}, \cdots, m^n x_{1m}, \cdots, m^n x_{k1}, \cdots, m^n x_{km}) = 0$$

for all $x_{11}, \cdots, x_{1m}, \cdots, x_{k1}, \cdots, x_{km} \in A$ and so

$$\sum_{i=1}^{m} H\left(mx_i + \sum_{j=1,j\neq i}^{m} x_j\right) + H\left(\sum_{i=1}^{m} x_i\right) = 2H\left(\sum_{i=1}^{m} mx_i\right) \qquad (6.42)$$

for all $x_1, \cdots, x_m \in A$. Thus $H$ is additive. By the similar method, we have

$$\mu H(mx) = H(m\mu x)$$

for all $\mu \in \mathbb{T}^1$ and $x \in A$. Thus one can show that the mapping $H : A \to B$ is $\mathbb{C}$-linear. It follows from (6.36), (6.37) and (6.41) that

$$\|(H([x_1, y_1, z_1]) - [H(x_1), H(y_1), H(z_1)],$$

$$\cdots, H([x_k, y_k, z_k]) - [H(x_k), H(y_k), H(z_k)])\|_k$$

$$= \lim_{n\to\infty} \frac{1}{m^{3n}} \|(f([m^n x_1, m^n y_1, m^n z_1]) - [f(m^n x_1), f(m^n y_1), f(m^n z_1)],$$

$$\cdots, f([m^n x_k, m^n y_k, m^n z_k]) - [f(m^n x_k), f(m^n y_k), f(m^n z_k)])\|_k$$

$$\leq \lim_{n\to\infty} \frac{1}{m^{3n}} \psi(m^n x_1, m^n y_1, \cdots, m^n x_k, m^n y_k)$$

$$= 0$$

for all $x_1, y_1, \cdots, x_k, y_k \in A$ and so

$$H([x, y, z]) = [H(x), H(y), H(z)]$$

for all $x, y \in A$. Thus $H : A \to B$ is a homomorphism satisfying (6.38). This completes the proof.  □

**Corollary 6.10.** *Let $((B^k, \| \cdot \|_k) : k \geq 1)$ be a multi-$C^*$-ternary algebra. Let $r < 1$, $\theta$ be nonnegative real numbers and $f : A \to B$ be a mapping such that*

$$\|(D_\mu f(x_{11}, \cdots, x_{1m}), \cdots, D_\mu f(x_{k1}, \cdots, x_{km}))\|_k$$

$$\leq \theta \left( \sum_{j=1}^{m} \|x_{1j}\|_A^r + \cdots + \sum_{j=1}^{m} \|x_{kj}\|_A^r \right) \qquad (6.43)$$

*and*

$$\|(f([x_1, y_1, z_1]) - [f(x_1), f(y_1), f(z_1)],$$

$$\cdots, f([x_k, y_k, z_k]) - [f(x_k), f(y_k), f(z_k)])\|_k$$

$$\leq \theta \left( \|x_1\|_A^r \cdot \|y_1\|_A^r \cdot \|z_1\|_A^r + \cdots + \|x_k\|_A^r \cdot \|y_k\|_A^r \cdot \|z_k\|_A^r \right) \qquad (6.44)$$

*for all $\mu \in \mathbb{T}^1$ and $x_{11}, \cdots, x_{1m}, \cdots, x_{k1}, \cdots, x_{km}, x_1, \cdots, x_k, y_1, \cdots, y_k,$*
*$z_1, \cdots, z_k \in A$. Then there exists a unique homomorphism $H : A \to B$ such that*

$$\|(f(x_1) - H(x_1), \cdots, f(x_k) - H(x_k))\|_k$$

$$\leq \frac{\theta}{m - m^r} \left( \|x_1\|_A^r + \cdots + \|x_k\|_A^r \right)$$

*for all $x_1, \cdots, x_k \in A$.*

*Proof.* The proof follows from Theorem 6.9 by taking

$$\varphi(x_{11}, \cdots, x_{1m}, \cdots, x_{k1}, \cdots, x_{km}) := \theta \left( \sum_{j=1}^{m} \|x_{1j}\|_A^r + \cdots + \sum_{j=1}^{m} \|x_{kj}\|_A^r \right),$$

$$\psi(x_1, y_1, z_1, \cdots, x_k, y_k, z_k)$$

$$:= \theta \left( \|x_1\|_A^r \cdot \|y_1\|_A^r \cdot \|z_1\|_A^r + \cdots + \|x_k\|_A^r \cdot \|y_k\|_A^r \cdot \|z_k\|_A^r \right)$$

for all $x_{11}, \cdots, x_{1m}, \cdots, x_{k1}, \cdots, x_{km}, x_1, \cdots, x_k, y_1, \cdots, y_k, z_1, \cdots, z_k \in A$ and
$L = m^{r-1}$. $\qquad \Box$

**Theorem 6.11.** *Let $((B^k, \| \cdot \|_k) : k \geq 1)$ be a multi-$C^*$-ternary algebra. Let*
*$f : A \to B$ be a mapping for which there exist the functions $\varphi : A^{mk} \to [0, \infty)$*
*and $\psi : A^{2k} \to [0, \infty)$ satisfying the inequalities (6.35) and (6.36) such that*

$$\lim_{j \to \infty} m^j \varphi(m^j x_{11}, \cdots, m^{-j} x_{1m}, \cdots, m^{-j} x_{k1}, \cdots, m^{-j} x_{km}) = 0 \qquad (6.45)$$

*and*

$$\lim_{j \to \infty} m^{3j} \psi(m^{-j} x_1, m^{-j} y_1, m^{-j} z_1, \cdots, m^{-j} x_k, m^{-j} y_k, m^{-j} z_k) = 0 \qquad (6.46)$$

*for all $\mu \in \mathbb{T}^1$ and $x_{11}, \cdots, x_{1m}, \cdots, x_{k1}, \cdots, x_{km}, x_1, \cdots, x_m, y_1, \cdots, y_k,$*
*$z_1, \cdots, z_k \in A$. If there exists $L < 1$ such that*

$$\varphi\Big(\overbrace{x_{11}, 0, \cdots, 0}^{m}, \overbrace{x_{21}, 0, \cdots, 0}^{m}, \cdots, \overbrace{x_{k1}, 0, \cdots, 0}^{m}\Big)$$

$$\leq \frac{L}{m}\varphi\Big(\overbrace{mx_{11}, 0, \cdots, 0}^{m}, \overbrace{mx_{21}, 0, \cdots, 0}^{m}, \cdots, \overbrace{mx_{k1}, 0, \cdots, 0}^{m}\Big)$$

*for all $x_{11}, x_{21}, \cdots, x_{k1} \in A$, then there exists a unique homomorphism $H : A \to B$*
*such that*

$$\|(f(x_1) - H(x_1), \cdots, f(x_k) - H(x_k))\|_k$$

$$\leq \frac{L}{m - mL}\varphi\Big(\overbrace{x_1, 0, \cdots, 0}^{m}, \overbrace{x_2, 0, \cdots, 0}^{m}, \cdots, \overbrace{x_k, 0, \cdots, 0}^{m}\Big) \qquad (6.47)$$

*for all $x_1, \cdots, x_k \in A$.*

*Proof.* We consider the linear mapping $J : X \to X$ defined by

$$Jg(x) := mg\left(\frac{x}{m}\right)$$

for all $x \in A$. It follows from (6.39) that

$$\left\|\Big(f(x_1) - mf\left(\frac{x_1}{m}\right), \cdots, f(x_k) - mf\left(\frac{x_k}{m}\right)\Big)\right\|_k$$

$$\leq \varphi\Big(\overbrace{\frac{x_1}{m}, 0, \cdots, 0}^{m}, \overbrace{\frac{x_2}{m}, 0, \cdots, 0}^{m}, \cdots, \overbrace{\frac{x_k}{m}, 0, \cdots, 0}^{m}\Big)$$

$$\leq \frac{L}{m}\varphi\Big(\overbrace{x_1, 0, \cdots, 0}^{m}, \overbrace{x_2, 0, \cdots, 0}^{m}, \cdots, \overbrace{x_k, 0, \cdots, 0}^{m}\Big) \qquad (6.48)$$

for all $x_1, \cdots, x_k \in A$. Hence we have

$$d(f, Jf) \leq \frac{L}{m}.$$

By Theorem 1.3, there exists a mapping $H : A \to B$ such that

(1)  $H$ is a fixed point of $J$, i.e.,

$$H(mx) = mH(x) \qquad (6.49)$$

for all $x \in A$. The mapping $H$ is a unique fixed point of $J$ in the set

$$Y = \{g \in X : d(f, g) < \infty\}.$$

This implies that $H$ is a unique mapping satisfying (6.49) such that there exists $C \in (0, \infty)$ satisfying the following:

$$\|(H(x_1) - f(x_1), \cdots, H(x_k) - f(x_k))\|_k$$

$$\leq C\varphi\Big(\overbrace{x_1, 0, \cdots, 0}^{m}, \overbrace{x_2, 0, \cdots, 0}^{m}, \cdots, \overbrace{x_k, 0, \cdots, 0}^{m}\Big)$$

for all $x_1, \cdots, x_k \in A$;

(2) $d(J^n f, H) \to 0$ as $n \to \infty$. This implies the equality

$$\lim_{n \to \infty} m^n f\left(\frac{x}{m^n}\right) = H(x)$$

for all $x \in A$;

(3) $d(f, H) \leq \frac{1}{1-L} d(f, Jf)$, which implies the inequality

$$d(f, H) \leq \frac{L}{m - mL},$$

which implies that the inequality (6.47) holds.

The rest of the proof is similar to the proof of Theorem 6.9. This completes the proof. □

**Corollary 6.12.** *Let $((B^k, \| \cdot \|_k) : k \geq 1)$ be a multi-$C^*$-ternary algebra. Let $r > 1$, $\theta$ be nonnegative real numbers and $f : A \to B$ be a mapping such that satisfying (6.43) and (6.44). Then there exists a unique homomorphism $H : A \to B$ such that*

$$\|(f(x_1) - H(x_1), \cdots, f(x_k) - H(x_k))\|_k$$

$$\leq \frac{\theta}{m^r - m}\left(\|x_1\|_A^r + \cdots + \|x_k\|_A^r\right)$$

*for all $x_1, \cdots, x_k \in A$.*

*Proof.* The proof follows from Theorem 6.11 by taking

$$\varphi(x_{11}, \cdots, x_{1m}, \cdots, x_{k1}, \cdots, x_{km}) := \theta\Big(\sum_{j=1}^{m} \|x_{1j}\|_A^r + \cdots + \sum_{j=1}^{m} \|x_{kj}\|_A^r\Big),$$

$$\psi(x_1, y_1, z_1, \cdots, x_k, y_k, z_k)$$

$$:= \theta\left(\|x_1\|_A^r \cdot \|y_1\|_A^r \cdot \|z_1\|_A^r + \cdots + \|x_k\|_A^r \cdot \|y_k\|_A^r \cdot \|z_k\|_A^r\right)$$

for all $x_{11}, \cdots, x_{1m}, \cdots, x_{k1}, \cdots, x_{km}, x_1, \cdots, x_k, y_1, \cdots, y_k, z_1, \cdots, z_k \in A$ and $L = m^{1-r}$. □

## 6.2.2 Stability of Derivations in Multi-$C^*$-Ternary Algebras

Now, we prove the Hyers-Ulam stability of derivations on multi-$C^*$-ternary algebras for the following functional equation:

$$D_\mu f(x_1, \cdots, x_m) = 0$$

for all $\mu \in \mathbb{T}^1 := \{v \in \mathbb{C} : |v| = 1\}$ and $x_1, \cdots, x_m \in A$.

**Theorem 6.13.** *Let $((A^k, \| \cdot \|_k) : k \geq 1)$ be a multi-$C^*$-ternary algebra. Let $f : A \to A$ be a mapping for which there exist the functions $\varphi : A^{mk} \to [0, \infty)$ and $\psi : A^{3k} \to [0, \infty)$ satisfying (6.34), (6.35) and (6.37) such that*

$$\begin{aligned}
&\|(f([x_1, y_1, z_1]) - [f(x_1), y_1, z_1] - [x_1, f(y_1), z_1] - [x_1, y_1, f(z_1)], \\
&\quad \cdots, f([x_k, y_k, z_k]) - [f(x_k), y_k, z_k] \\
&\quad - [x_k, f(y_k), z_k] - [x_k, y_k, f(z_k)])\|_k \\
&\leq \psi(x_1, y_1, z_1, \cdots, x_k, y_k, z_k)
\end{aligned} \tag{6.50}$$

*for all $\mu \in \mathbb{T}^1$ and $x_{11}, \cdots, x_{1m}, \cdots, x_{k1}, \cdots, x_{km}, x_1, \cdots, x_k, y_1, \cdots, y_k,$
$z_1, \cdots, z_k \in A$. If there exists $L < 1$ such that*

$$\varphi\left( \overbrace{mx_{11}, 0, \cdots, 0}^{m}, \overbrace{mx_{21}, 0, \cdots, 0}^{m}, \cdots, \overbrace{mx_{k1}, 0, \cdots, 0}^{m} \right)$$

$$\leq mL\varphi\left( \overbrace{x_{11}, 0, \cdots, 0}^{m}, \overbrace{x_{21}, 0, \cdots, 0}^{m}, \cdots, \overbrace{x_{k1}, 0, \cdots, 0}^{m} \right)$$

*for all $x_{11}, x_{21}, \cdots, x_{k1} \in A$, then there exists a unique derivation $\delta : A \to A$ such that*

$$\begin{aligned}
&\|(f(x_1) - \delta(x_1), \cdots, f(x_k) - \delta(x_k))\|_k \\
&\leq \frac{1}{m - mL}\varphi\left( \overbrace{x_1, 0, \cdots, 0}^{m}, \overbrace{x_2, 0, \cdots, 0}^{m}, \cdots, \overbrace{x_k, 0, \cdots, 0}^{m} \right)
\end{aligned} \tag{6.51}$$

*for all $x_1, \cdots, x_k \in A$.*

*Proof.* By the same reasoning as in the proof of Theorem 6.9, there exists a unique $\mathbb{C}$-linear mapping $\delta : A \to A$ satisfying (6.50) and the mapping $\delta : A \to A$ is given by

$$\delta(x) = \lim_{n \to \infty} \frac{f(m^n x)}{m^n} \tag{6.52}$$

for all $x \in A$. It follows from (6.34), (6.37) and (6.52) that

$$\|(\delta([x_1, y_1, z_1]) - [\delta(x_1), y_1, z_1] - [x_1, \delta(y_1), z_1] - [x_1, y_1, f(z_1)],$$
$$\cdots, \delta([x_k, y_k, z_k]) - [\delta(x_k), y_k, z_k]$$
$$-[x_k, \delta(y_k), z_k] - [x_k, y_k, \delta(z_k)])\|_k$$
$$= \lim_{n \to \infty} \frac{1}{m^{3n}} \|(f([m^n x_1, m^n y_1, m^n z_1]) - [f(m^n x_1), m^n y_1, m^n z_1]$$
$$-[m^n x_1, f(m^n y_1), m^n z_1] - [m^n x_1, m^n y_1, f(m^n z_1)],$$
$$\cdots, f([m^n x_k, m^n y_k, m^n z_k]) - [f(m^n x_k), m^n y_k, m^n z_k]$$
$$-[m^n x_k, f(m^n y_k), m^n z_k] - [m^n x_k, m^n y_k, f(m^n z_k)])\|_k$$
$$\leq \lim_{n \to \infty} m^{-3n} \psi(m^n x_1, m^n y_1, m^n z_1, \cdots, m^n x_k, m^n y_k, m^n z_k)$$
$$= 0$$

for all $x_1, \cdots, x_m, y_1, \cdots, y_k, z_1, \cdots, z_k \in A$ and so

$$\delta([x, y, z]) = [\delta(x), y, z] + [x, \delta(y), z] + [x, y, \delta(z)]$$

for all $x, y, z \in A$. Thus $\delta : A \to A$ is a derivation satisfying (6.50). This completes the proof.  $\square$

**Corollary 6.14.** *Let $((A^k, \|\cdot\|_k) : k \geq 1)$ be a multi-$C^*$-ternary algebra. Let $r < 1$, $\theta$ be nonnegative real numbers and $f : A \to A$ be a mapping such that*

$$\|(D_\mu f(x_{11}, \cdots, x_{1m}), \cdots, D_\mu f(x_{k1}, \cdots, x_{km}))\|_k$$
$$\leq \theta \left( \sum_{j=1}^{m} \|x_{1j}\|_A^r + \cdots + \sum_{j=1}^{m} \|x_{kj}\|_A^r \right) \tag{6.53}$$

*and*

$$\|(f([x_1, y_1, z_1]) - [f(x_1), y_1, z_1] - [x_1, f(y_1), z_1] - [x_1, y_1, f(z_1)],$$
$$\cdots, f([x_k, y_k, z_k]) - [f(x_k), y_k, z_k]$$
$$-[x_k, f(y_k), z_k] - [x_k, y_k, f(z_k)])\|_k \tag{6.54}$$
$$\leq \theta \left( \|x_1\|_A^r \cdot \|y_1\|_A^r \cdot \|z_1\|_A^r + \cdots + \|x_k\|_A^r \cdot \|y_k\|_A^r \cdot \|y_k\|_A^r \right)$$

*for all $\mu \in \mathbb{T}^1$ and $x_{11}, \cdots, x_{1m}, \cdots, x_{k1}, \cdots, x_{km}, x_1, \cdots, x_k, y_1, \cdots, y_k,$ $z_1, \cdots, z_k \in A$. Then there exists a unique derivation $\delta : A \to A$ such that*

$$\|(f(x_1) - \delta(x_1), \cdots, f(x_k) - \delta(x_k))\|_k \leq \frac{\theta}{m^r - m} \left( \|x_1\|_A^r + \cdots + \|x_k\|_A^r \right)$$

*for all $x_1, \cdots, x_k \in A$.*

*Proof.* The proof follows from Theorem 6.50 by taking

$$\varphi(x_{11},\cdots,x_{1m},\cdots,x_{k1},\cdots,x_{km}) := \theta\Big(\sum_{j=1}^{m}\|x_{1j}\|_A^r + \cdots + \sum_{j=1}^{m}\|x_{kj}\|_A^r\Big),$$

$$\psi(x_1,y_1,z_1,\cdots,x_k,y_k,z_k)$$
$$:= \theta\left(\|x_1\|_A^r \cdot \|y_1\|_A^r \cdot \|z_1\|_A^r + \cdots + \|x_k\|_A^r \cdot \|y_k\|_A^r \cdot \|y_k\|_A^r\right)$$

for all $x_1,\cdots,x_k,y_1,\cdots,y_k,z_1,\cdots,z_k \in A$ and $L = m^r$. This completes the proof.  □

*Remark 6.15.* Let $((A^k,\|\cdot\|_k) : k \geq 1)$ be a multi-$C^*$-ternary algebra. Let $f : A \to A$ be a mapping for which there exist the functions $\varphi : A^{mk} \to [0,\infty)$ and $\psi : A^{3k} \to [0,\infty)$ satisfying (6.35), (6.45), (6.46) and (6.50) for all $\mu \in \mathbb{T}^1$ and $x_{11},\cdots,x_{1m},\cdots,x_{k1},\cdots,x_{km},x_1,\cdots,x_k,y_1,\cdots,y_k,z_1,\cdots,z_k \in A$. If there exists $L < 1$ such that

$$\varphi\Big(\overbrace{mx_{11},0,\cdots,0}^{m},\overbrace{mx_{21},0,\cdots,0}^{m},\cdots,\overbrace{mx_{k1},0,\cdots,0}^{m}\Big)$$
$$\leq \frac{L}{m}\varphi\Big(\overbrace{x_{11},0,\cdots,0}^{m},\overbrace{x_{21},0,\cdots,0}^{m},\cdots,\overbrace{x_{k1},0,\cdots,0}^{m}\Big)$$

for all $x_{11},x_{21},\cdots,x_{k1} \in A$, then there exists a unique derivation $\delta : A \to A$ such that

$$\|(f(x_1) - \delta(x_1),\cdots,f(x_k) - \delta(x_k))\|_k$$
$$\leq \frac{L}{m - mL}\varphi\Big(\overbrace{x_1,0,\cdots,0}^{m},\overbrace{x_2,0,\cdots,0}^{m},\cdots,\overbrace{x_k,0,\cdots,0}^{m}\Big)$$

for all $x_1,\cdots,x_k \in A$.

**Corollary 6.16.** *Let $((A^k,\|\cdot\|_k) : k \geq 1)$ be a multi-$C^*$-ternary algebra. Let $r > 1$, $\theta$ be nonnegative real numbers and $f : A \to A$ be a mapping satisfying (6.53) and (6.54). Then there exists a unique derivation $\delta : A \to A$ such that*

$$\|(f(x_1) - \delta(x_1),\cdots,f(x_k) - \delta(x_k))\|_k \leq \frac{\theta}{m^r - m}\left(\|x_1\|_A^r + \cdots + \|x_k\|_A^r\right)$$

*for all $x_1,\cdots,x_k \in A$.*

*Proof.* The proof follows from Theorem 6.15 by taking

$$\varphi(x_{11}, \cdots, x_{1m}, \cdots, x_{k1}, \cdots, x_{km}) := \theta \Big( \sum_{j=1}^{m} \|x_{1j}\|_A^r + \cdots + \sum_{j=1}^{m} \|x_{kj}\|_A^r \Big),$$

$$\psi(x_1, y_1, z_1, \cdots, x_k, y_k, z_k)$$
$$:= \theta \left( \|x_1\|_A^r \cdot \|y_1\|_A^r \cdot \|z_1\|_A^r + \cdots + \|x_k\|_A^r \cdot \|y_k\|_A^r \cdot \|y_k\|_A^r \right)$$

for all $x_1, \cdots, x_k, y_1, \cdots, y_k, z_1, \cdots, z_k \in A$ and $L = m^{1-r}$. This completes the proof. $\qquad\qquad\square$

## 6.3    Generalized Additive Mappings and Isomorphisms in Multi-$C^*$-Algebras

Let $X$ and $Y$ be vector spaces. It is shown that, if an odd mapping $f : X \to Y$ satisfies the functional equation (3.97), then the odd mapping $f$ is additive. Also, we use the fixed point method to prove the Hyers-Ulam stability of the functional equation (3.97) in multi-Banach modules over a unital multi-$C^*$-algebra. As an application, we show that every almost linear bijection $h : A \to B$ of a unital multi-$C^*$-algebra $A$ onto a unital multi-$C^*$-algebra $B$ is a $C^*$-algebra isomorphism when

$$h\Big(\frac{2^n}{r^n} uy\Big) = h\Big(\frac{2^n}{r^n} u\Big) h(y)$$

for all unitaries $u \in U(A)$, $y \in A$ and $n \geq 0$.

### 6.3.1    *Stability of Odd Functional Equations in Multi-Banach Modules over a Multi-$C^*$-Algebra*

We assume that $((A^k, \| \cdot \|_k) : k \geq 1)$ is a unital multi-$C^*$-algebra and $((X^k, \| \cdot \|_k) : k \geq 1)$, $((Y^k, \| \cdot \|_k) : k \geq 1)$ are multi-Banach left modules over $((A^k, \| \cdot \|_k) : k \geq 1)$. Moreover, by $U(A)$, we denote the unitary group of $A$. For any mapping $f : X \to Y$, we set

$$D_u f(x_1, \cdots, x_d)$$

$$:= rf\left(\frac{\sum_{j=1}^d u x_j}{r}\right) + \sum_{\substack{\iota(j) \ = 0, 1 \\ \sum_{j=1}^d \iota(j) \ = l}} rf\left(\frac{\sum_{j=1}^d (-1)^{\iota(j)} u x_j}{r}\right)$$

$$- (_{d-1}C_l -_{d-1}C_{l-1} + 1) \sum_{j=1}^d u f(x_j)$$

for all $u \in U(A)$ and $x_1, \cdots, x_d \in X$.

**Theorem 6.17.** *Let $r \neq 2$ and $f : X \to Y$ be an odd mapping such that, for each $k \geq 1$, there exists a function $\varphi_k : X^{kd} \to [0, \infty)$ such that*

$$\lim_{j \to \infty} \frac{r^j}{2^j} \varphi_k\left(\frac{2^j}{r^j} x_{11}, \cdots, \frac{2^j}{r^j} x_{1d}, \cdots, \frac{2^j}{r^j} x_{k1}, \cdots, \frac{2^j}{r^j} x_{kd}\right) = 0 \tag{6.55}$$

*and*

$$\|(D_u f(x_{11}, \cdots, x_{1d}), \cdots, D_u f(x_{k1}, \cdots, x_{kd}))\|_k$$
$$\leq \varphi_k(x_{11}, \cdots, x_{1d}, \cdots, x_{k1}, \cdots, x_{kd}) \tag{6.56}$$

*for all $u \in U(A)$ and $x_{11}, \cdots, x_{1d}, \cdots, x_{k1}, \cdots, x_{kd} \in X$. If there exists $L < 1$ such that*

$$\varphi_k\left(\overbrace{\frac{2}{r} x_{11}, \frac{2}{r} x_{11}}^{d}, \cdots, 0, \cdots, \overbrace{\frac{2}{r} x_{k1}, \frac{2}{r} x_{k1}}^{d}, \cdots, 0\right)$$
$$\leq \frac{2}{r} L \varphi_k\left(\overbrace{x_{11}, x_{11}, \cdots, 0}^{d}, \cdots, \overbrace{x_{k1}, x_{k1}, \cdots, 0}^{d}\right)$$

*for all $k \geq 1$ and $x_{11}, \cdots, x_{k1} \in X$, then there exist a unique A-linear generalized additive mapping $\Lambda : X \to Y$ such that*

$$\|(\Lambda(x_1) - f(x_1), \cdots, \Lambda(x_k) - f(x_k))\|_k \tag{6.57}$$
$$\leq \frac{1}{2(_{d-2}C_l -_{d-2}C_{l-2} + 1)(1 - L)} \varphi_k\left(x_1, x_1, \underbrace{0, \ldots, 0}_{d-2 \text{ times}}, \ldots, x_k, x_k, \underbrace{0, \ldots, 0}_{d-2 \text{ times}}\right)$$

*for all $k \geq 1$ and $x_1, \cdots, x_k \in X$.*

*Proof.* Put $S := \{\Lambda : X \to Y\}$ and

$$d(\Lambda, h) = \inf\{C \in \mathbb{R}_+ : \|(\Lambda(x_1) - h(x_1), \cdots, \Lambda(x_k) - h(x_k))\|_k$$

$$\leq C\varphi_k(\overbrace{x_1, x_1, 0, \cdots, 0}^{d}, \cdots, \overbrace{x_k, x_k, 0, \cdots, 0}^{d}), \ k \geq 1, \ x_1, \cdots, x_k \in X\}$$

for all $\Lambda, h \in S$ which $(S, d)$ is a complete generalized metric space. Define a mapping $J : S \to S$ by

$$J\Lambda(x) := \frac{r}{2}\Lambda\left(\frac{2}{r}x\right)$$

for all $\Lambda \in S$ and $x \in X$. Now, we have

$$d(J\Lambda, Jh) \leq Ld(\Lambda, h)$$

for all $\Lambda, h \in S$. For a fixed $k \geq 1$, putting $u = 1 \in U(A)$, $x_{i1} = x_{i2} = x_1$ and $x_{i3} = \cdots = x_{id} = 0$ for each $i \in \{1, \cdots, k\}$ in (6.56), we have

$$\left\|\left(rf\left(\frac{2}{r}x_1\right) - 2f(x_1), \cdots, rf\left(\frac{2}{r}x_k\right) - 2f(x_k)\right)\right\|_k$$

$$\leq \frac{1}{{}_{d-2}C_l - {}_{d-2}C_{l-2} + 1}\varphi_k\left(x_1, x_1, \underbrace{0, \cdots, 0}_{d-2 \text{ times}}, \cdots, x_k, x_k, \underbrace{0, \cdots, 0}_{d-2 \text{ times}}\right)$$

since $f$ is odd and $t :=_{d-2}C_l - _{d-2}C_{l-2} + 1 =_{d-1}C_l - _{d-1}C_{l-1} + 1$. Thus we have

$$\left\|\left(f(x_1) - \frac{r}{2}f\left(\frac{2}{r}x_1\right), \cdots, f(x_k) - \frac{r}{2}f\left(\frac{2}{r}x_k\right)\right)\right\|_k$$

$$\leq \frac{1}{2t}\varphi_k\left(x_1, x_1, \underbrace{0, \cdots, 0}_{d-2 \text{ times}}, \cdots, x_k, x_k, \underbrace{0, \cdots, 0}_{d-2 \text{ times}}\right)$$

for all $x_1, \cdots, x_k \in X$ and so

$$d(f, Jf) \leq \frac{1}{2t}. \tag{6.58}$$

Consequently, by Theorem 1.3, there exists a mapping $\Lambda : X \to Y$ such that

(1)　$\Lambda$ is a fixed point of $J$, i.e.,

$$\Lambda\left(\frac{2}{r}x\right) = \frac{2}{r}\Lambda(x) \tag{6.59}$$

for all $x \in X$ and $\Lambda$ is unique in the set

$$Y = \{\Lambda \in X : d(f, \Lambda) < \infty\}.$$

This means that $\Lambda$ is a unique mapping satisfying (6.59) such that there exists $C \in (0, \infty)$ with

$$\|(\Lambda(x_1) - f(x_1), \cdots, \Lambda(x_k) - f(x_k))\|_k$$

$$\leq C\varphi_k\Big(x_1, x_1, \underbrace{0, \cdots, 0}_{d-2 \text{ times}}, \cdots, x_k, x_k, \underbrace{0, \cdots, 0}_{d-2 \text{ times}}\Big)$$

for all $k \geq 1$ and $x_1, \cdots, x_k \in X$;

(2) $d(J^n f, \Lambda) \to 0$ as $n \to \infty$. This implies the equality

$$\lim_{n \to \infty} \frac{r^n}{2^n} f\left(\frac{2^n}{r^n} x\right) = \Lambda(x) \tag{6.60}$$

for all $x \in X$;

(3) $d(f, \Lambda) \leq \frac{1}{1-L} d(f, Jf)$ which together with (6.58) gives

$$d(f, \Lambda) \leq \frac{1}{2t - 2tL}$$

and so the inequality (6.57) holds for all $x_1, \cdots, x_k \in X$.

Next, note that the fact that the mapping $f$ is odd and (6.60) imply that $\Lambda$ is odd. Moreover, by (6.55) and (6.56), we have

$$\|(D_1 \Lambda(x_{11}, \cdots, x_{1d}), \cdots, D_1 \Lambda(x_{k1}, \cdots, x_{kd}))\|_k$$

$$= \lim_{n \to \infty} \frac{r^n}{2^n} \left\|\left(D_1 f\left(\frac{2^n}{r^n} x_{11}, \cdots, \frac{2^n}{r^n} x_{1d}\right), \cdots, D_1 f\left(\frac{2^n}{r^n} x_{k1}, \cdots, \frac{2^n}{r^n} x_{kd}\right)\right)\right\|_k$$

$$\leq \lim_{n \to \infty} \frac{r^n}{2^n} \varphi_k\left(\frac{2^n}{r^n} x_{11}, \cdots, \frac{2^n}{r^n} x_{1d}, \cdots, \frac{2^n}{r^n} x_{k1}, \cdots, \frac{2^n}{r^n} x_{kd}\right)$$

$$= 0$$

for all $k \geq 1$ and $x_{11}, \cdots, x_{1d}, \cdots, x_{k1}, \cdots, x_{kd} \in X$ and so $\Lambda$ is a generalized additive mapping.

For any fixed $u \in U(A)$ and $x \in X$, using (6.55) and (6.56), we have

$$\left\|(D_u \Lambda(x, \underbrace{0, \cdots, 0}_{d-1 \text{ times}}), \cdots, D_u \Lambda(x, \underbrace{0, \cdots, 0}_{d-1 \text{ times}}))\right\|_k$$

$$= \lim_{n \to \infty} \frac{r^n}{2^n} \left\|\left(D_u f\left(\frac{2^n}{r^n} x, \underbrace{0, \cdots, 0}_{d-1 \text{ times}}\right), \cdots, D_u f\left(\frac{2^n}{r^n} x, \underbrace{0, \cdots, 0}_{d-1 \text{ times}}\right)\right)\right\|_k$$

$$\leq \lim_{n \to \infty} \frac{r^n}{2^n} \varphi_k\left(\frac{2^n}{r^n} x, \underbrace{0, \cdots, 0}_{d-1 \text{ times}}, \cdots, \frac{2^n}{r^n} x, \underbrace{0, \cdots, 0}_{d-1 \text{ times}}\right)$$

$$= 0$$

and so

$$(_{d-1}C_l - _{d-1}C_{l-1} + 1)r\Lambda\left(\frac{ux}{r}\right) = (_{d-1}C_l - _{d-1}C_{l-1} + 1)u\Lambda(x).$$

Since $\Lambda$ is a generalized additive mapping, from Lemma 3.64 it follows that $\Lambda$ is additive and so

$$\Lambda(ux) = r\Lambda\left(\frac{ux}{r}\right) = u\Lambda(x)$$

for all $u \in U(A)$ and $x \in X$. It is straight forward to show that $\Lambda$ is an $A$-linear mapping (see also Theorem 3.1 in [20]). This completes the proof. □

**Corollary 6.18.** *Let $r \neq 2$ and $\theta, p \in (0, \infty)$. Assume also that $p > 1$ for $r > 2$ and $p < 1$ for $r < 2$. If $f : X \to Y$ is an odd mapping such that*

$$\|(D_u f(x_{11}, \cdots, x_{1d}), \cdots, D_u f(x_{k1}, \cdots, x_{kd}))\|_k$$

$$\leq \theta\left(\sum_{j=1}^{d} \|x_{1j}\|^p + \ldots + \sum_{j=1}^{d} \|x_{kj}\|^p\right)$$

*for all $u \in U(A)$, $k \geq 1$ and $x_{11}, \cdots, x_{1d}, \cdots, x_{k1}, \cdots, x_{kd} \in X$, then there exists a unique $A$-linear generalized additive mapping $\Lambda : X \to Y$ such that*

$$\|(\Lambda(x_1) - f(x_1), \cdots, \Lambda(x_k) - f(x_k))\|_k$$

$$\leq \frac{r^{p-1}\theta}{(r^{p-1} - 2^{p-1})(_{d-2}C_l - _{d-2}C_{l-2} + 1)}(\|x_1\|^p + \cdots + \|x_k\|^p)$$

*for all $k \geq 1$ and $x_1, \cdots, x_k \in X$.*

*Proof.* Taking $L = \frac{2^{p-1}}{r^{p-1}}$ and

$$\varphi_k(x_{11}, \cdots, x_{1d}, \cdots, x_{k1}, \cdots, x_{kd}) = \theta\left(\sum_{j=1}^{d} \|x_{1j}\|^p + \ldots + \sum_{j=1}^{d} \|x_{kj}\|^p\right)$$

for all $k \geq 1$ and $x_{11}, \cdots, x_{1d}, \cdots, x_{k1}, \cdots, x_{kd} \in X$ in Theorem 6.17, we get the desired assertion. □

**Theorem 6.19.** *Let $r \neq 2$. Let $f : X \to Y$ be an odd mapping for which there exists a function $\varphi : X^{kd} \to [0, \infty)$ such that*

$$\lim_{j \to \infty} \frac{2^j}{r^j}\varphi\left(\frac{r^j}{2^j}x_{11}, \cdots, \frac{r^j}{2^j}x_{1d}, \cdots, \frac{r^j}{2^j}x_{k1}, \cdots, \frac{r^j}{2^j}x_{kd}\right) = 0$$

*and*

$$\|(D_u f(x_{11}, \cdots, x_{1d}), \cdots, D_u f(x_{k1}, \cdots, x_{kd}))\|_k$$

$$\leq \varphi(x_{11}, \cdots, x_{1d}, \cdots, x_{k1}, \cdots, x_{kd})$$

*for all* $u \in U(A)$ *and* $x_{11}, \cdots, x_{1d}, \cdots, x_{k1}, \cdots, x_{kd} \in X$. *If there exists* $L < 1$ *such that*

$$\varphi\left( \overbrace{\frac{r}{2}x_{11}, \frac{r}{2}x_{11}, \cdots, 0}^{d}, \overbrace{\frac{r}{2}x_{21}, \frac{r}{2}x_{21}, \cdots, 0}^{d}, \cdots, \overbrace{\frac{r}{2}x_{k1}, \frac{r}{2}x_{k1}, \cdots, 0}^{d} \right)$$

$$\leq \frac{r}{2} L \varphi\left( \overbrace{x_{11}, x_{11}, \cdots, 0}^{d}, \overbrace{x_{21}, x_{21}, \cdots, 0}^{d}, \cdots, \overbrace{x_{k1}, x_{k1}, \cdots, 0}^{d} \right)$$

*for all* $x_{11}, x_{21}, \cdots, x_{k1} \in X$, *then there exists a unique A-linear generalized additive mapping* $\Lambda : X \to Y$ *such that*

$$\sup_{k \geq 1} \|(\Lambda(x_1) - f(x_1), \cdots, \Lambda(x_k) - f(x_k))\|_k$$

$$\leq \sup_{k \geq 1} \frac{L}{2(_{d-2}C_l - _{d-2}C_{l-2} + 1)(1 - L)} \varphi\left( x_1, x_1, \underbrace{0, \cdots, 0}_{d - 2 \, times}, \right.$$

$$\left. \cdots, x_k, x_k, \underbrace{0, \cdots, 0}_{d - 2 \, times} \right)$$

*for all* $x_1, \cdots, x_k \in X$.

*Proof.* Note that $f(0) = 0$ and $f(-x) = -f(x)$ for all $x \in X$ since $f$ is an odd mapping. Let $u = 1 \in U(A)$. Putting $x_{i1} = x_{i2} = x_1$ and $x_{i3} = \cdots = x_{im} = 0$ $(1 \leq i \leq k)$ in (6.56), we have

$$\left\| \left( rf\left(\frac{2}{r}x_1\right) - 2f(x_1), \cdots, rf\left(\frac{2}{r}x_k\right) - 2f(x_k) \right) \right\|_k$$

$$\leq \frac{1}{_{d-2}C_l - _{d-2}C_{l-2} + 1} \varphi\left( x_1, x_1, \underbrace{0, \cdots, 0}_{d - 2 \, times}, \cdots, x_k, x_k, \underbrace{0, \cdots, 0}_{d - 2 \, times} \right).$$

Letting $t := _{d-2}C_l - _{d-2}C_{l-2} + 1$, we have

$$\left\| \left( f(x_1) - \frac{2}{r}f\left(\frac{r}{2}x_1\right), \cdots, f(x_k) - \frac{2}{r}f\left(\frac{r}{2}x_k\right) \right) \right\|_k$$

$$\leq \frac{1}{rt}\varphi\Big(\frac{r}{2}x_1,\frac{r}{2}x_1,\underbrace{0,\cdots,0}_{d-2\ \text{times}},\cdots,\frac{r}{2}x_k,\frac{r}{2}x_k,\underbrace{0,\cdots,0}_{d-2\ \text{times}}\Big)$$

$$\leq \frac{L}{2t}\varphi\Big(x_1,x_1,\underbrace{0,\cdots,0}_{d-2\ \text{times}},\cdots,x_k,x_k,\underbrace{0,\cdots,0}_{d-2\ \text{times}}\Big)$$

for all $x_1,\cdots,x_k \in X$.

The rest of the proof is similar to the proof of Theorem 6.17. This completes the proof.                                                                                                      □

**Corollary 6.20.** *Let $r < 2$ and $\theta$, $p > 1$ be positive real numbers or let $r > 2$ and $\theta$, $p < 1$ be positive real numbers. Let $f : X \to Y$ be an odd mapping such that*

$$\|(D_u f(x_{11},\cdots,x_{1d}),\cdots,D_u f(x_{k1},\cdots,x_{kd}))\|_k$$

$$\leq \theta\Big(\sum_{j=1}^{d}\|x_{1j}\|^p + \cdots + \sum_{j=1}^{d}\|x_{kj}\|^p\Big)$$

*for all $u \in U(A)$ and $x_{11},\cdots,x_{1d},\cdots,x_{k1},\cdots,x_{kd} \in X$. Then there exists a unique A-linear generalized additive mapping $\Lambda : X \to Y$ such that*

$$\sup_{k\geq 1}\|(\Lambda(x_1)-f(x_1),\cdots,\Lambda(x_k)-f(x_k))\|_k$$

$$\leq \sup_{k\geq 1}\frac{r^{p-1}\theta}{(2^{p-1}-r^{p-1})(_{d-2}C_l -_{d-2}C_{l-2}+1)}(\|x_1\|^p+\cdots+\|x_k\|^p)$$

*for all $x \in X$.*

*Proof.* Define

$$\varphi(x_{11},\cdots,x_{1d},\cdots,x_{k1},\cdots,x_{kd}) = \theta\Big(\sum_{j=1}^{d}\|x_{1j}\|^p + \cdots + \sum_{j=1}^{d}\|x_{kj}\|^p\Big)$$

and put $L = \frac{r^{p-1}}{2^{p-1}}$ in Theorem 6.19. Then we get the desired result.                              □

Now, we investigate the Hyers–Ulam stability of linear mappings for the case $d = 2$.

**Theorem 6.21.** *Let $r \neq 2$. Let $f : X \to Y$ be an odd mapping for which there exists a function $\varphi : X^{2k} \to [0, \infty)$ such that*

$$\lim_{j\to\infty}\frac{r^j}{2j}\varphi\Big(\frac{2^j}{r^j}x_1,\frac{2^j}{r^j}y_1,\cdots,\frac{2^j}{r^j}x_k,\frac{2^j}{r^j}y_k\Big) = 0$$

*and*

$$\left\|\left(rf\left(\frac{ux_1 + uy_1}{r}\right) - uf(x_1) - uf(y_1),\right.\right.$$

$$\left.\left.\cdots, rf\left(\frac{ux_k + uy_k}{r}\right) - uf(x_k) - uf(y_k)\right)\right\|_k$$

$$\leq \varphi(x_1, y_1, \cdots, x_k, y_k) \tag{6.61}$$

*for all $u \in U(A)$ and $x_1, \cdots x_k, y_1 \cdots, y_k \in X$. If there exists $L < 1$ such that*

$$\varphi\left(\frac{2}{r}x_1, \frac{2}{r}x_1, \frac{2}{r}x_2, \frac{2}{r}x_2, \cdots, \frac{2}{r}x_k, \frac{2}{r}x_k\right) \leq \frac{2}{r}L\varphi(x_1, x_1, x_2, x_2, \cdots, x_k, x_k)$$

*for all $x_1, \cdots x_k \in X$. Then there exists a unique A-linear generalized additive mapping $\Lambda : X \to Y$ such that*

$$\sup_{k \geq 1} \|(\Lambda(x_1) - f(x_1), \cdots, \Lambda(x_k) - f(x_k))\|_k$$

$$\leq \sup_{k \geq 1} \frac{L}{2(1 - L)}\varphi(x_1, x_1, \cdots, x_k, x_k)$$

*for all $x_1, \cdots, x_k \in X$.*

*Proof.* Let $u = 1 \in U(A)$. Putting $x_i = y_i$ ($1 \leq i \leq k$) in (6.61), we have

$$\left\|\left(rf\left(\frac{2}{r}x_1\right) - 2f(x_1), \cdots, rf\left(\frac{2}{r}x_k\right) - 2f(x_k)\right)\right\|_k$$

$$\leq \varphi(x_1, x_1, \cdots, x_k, x_k)$$

for all $x \in X$ and so

$$\left\|\left(f(x_1) - \frac{r}{2}f\left(\frac{2}{r}x_1\right), \cdots, f(x_k) - \frac{r}{2}f\left(\frac{2}{r}x_k\right)\right)\right\|_k$$

$$\leq \frac{1}{2}\varphi(x_1, x_1, \cdots, x_k, x_k)$$

for all $x \in X$.

The rest of the proof is the same as in the proof of Theorem 6.17. This completes the proof. □

**Corollary 6.22.** *Let $r > 2$ and $\theta$, $p > 1$ be positive real numbers or let $r < 2$ and $\theta$, $p < 1$ be positive real numbers. Let $f : X \to Y$ be an odd mapping such that*

$$\left\|\left(rf\left(\frac{ux_1 + uy_1}{r}\right) - uf(x_1) - uf(y_1),\right.\right.$$

$$\left.\left.\cdots, rf\left(\frac{ux_k + uy_k}{r}\right) - uf(x_k) - uf(y_k)\right)\right\|_k$$

$$\leq \theta \sum_{j=1}^{k} (\|x_j\|^p + \|y_j\|^p)$$

*for all* $u \in U(A)$ *and* $x_1, \cdots x_k \in X$. *Then there exists a unique A-linear generalized additive mapping* $\Lambda : X \to Y$ *such that*

$$\sup_{k \geq 1} \|(\Lambda(x_1) - f(x_1), \cdots, \Lambda(x_k) - f(x_k))\|_k \leq \sup_{k \geq 1} \frac{r^{p-1}\theta}{r^{p-1} - 2^{p-1}} \sum_{j=1}^{k} \|x_j\|^p$$

*for all* $x_1, \cdots x_k \in X$.

*Proof.* Define

$$\varphi(x_1, y_1, \cdots, x_k, y_k) = \theta \sum_{j=1}^{k} (\|x_j\|^p + \|y_j\|^p)$$

and apply Theorem 6.21. Then we get the desired result.  □

**Theorem 6.23.** *Let* $r \neq 2$. *Let* $f : X \to Y$ *be an odd mapping for which there exists a function* $\varphi : X^{2k} \to [0, \infty)$ *such that*

$$\lim_{j \to \infty} \frac{2^j}{r^j} \varphi\left(\frac{r^j}{2^j}x_1, \frac{r^j}{2^j}y_1, \cdots, \frac{r^j}{2^j}x_k, \frac{r^j}{2^j}y_k\right) = 0$$

*and*

$$\left\|\left(rf\left(\frac{ux_1 + uy_1}{r}\right) - uf(x_1) - uf(y_1),\right.\right.$$

$$\left.\left.\cdots, rf\left(\frac{ux_k + uy_k}{r}\right) - uf(x_k) - uf(y_k)\right)\right\|_k$$

$$\leq \varphi(x_1, y_1, \cdots, x_k, y_k) \tag{6.62}$$

*for all* $u \in U(A)$ *and* $x_1, \cdots x_k, y_1 \cdots, y_k \in X$. *If there exists* $L < 1$ *such that*

$$\varphi\left(\frac{r}{2}x_1, \frac{r}{2}x_1, \frac{r}{2}x_2, \frac{r}{2}x_2, \cdots, \frac{r}{2}x_k, \frac{r}{2}x_k\right) \leq \frac{r}{2}L\varphi(x_1, x_1, x_2, x_2, \cdots, x_k, x_k)$$

*for all* $x_1, \cdots x_k \in X$. *Then there exists a unique A-linear generalized additive mapping* $\Lambda : X \to Y$ *such that*

$$\sup_{k\geq 1} \|(\Lambda(x_1) - f(x_1), \cdots, \Lambda(x_k) - f(x_k))\|_k$$

$$\leq \sup_{k\geq 1} \frac{1}{2(1-L)} \varphi(x_1, x_1, \cdots, x_k, x_k)$$

for all $x_1, \cdots, x_k \in X$.

*Proof.* Let $u = 1 \in U(A)$. Putting $x_i = y_i$ $(1 \leq i \leq k)$ in (6.62), we have

$$\left\| \left( rf\left(\frac{2}{r}x_1\right) - 2f(x_1), \cdots, rf\left(\frac{2}{r}x_k\right) - 2f(x_k)\right)\right\|_k$$

$$\leq \varphi(x_1, x_1, \cdots, x_k, x_k)$$

for all $x_1, \cdots, x_k \in X$ and so

$$\left\| \left( f(x_1) - \frac{2}{r}f\left(\frac{r}{2}x_1\right), \cdots, f(x_k) - \frac{2}{r}f\left(\frac{r}{2}x_k\right)\right)\right\|_k$$

$$\leq \frac{1}{r}\varphi\left(\frac{r}{2}x_1, \frac{r}{2}x_1, \cdots, \frac{r}{2}x_k, \frac{r}{2}x_k\right)$$

$$\leq \frac{1}{2}L\varphi(x_1, x_1, \cdots, x_k, x_k)$$

for all $x_1, \cdots, x_k \in X$.

The rest of the proof is similar to the proof of Theorem 6.17. This completes the proof.                                                                               □

**Corollary 6.24.** *Let $r > 2$ and $\theta$, $p > 1$ be positive real numbers or let $r < 2$ and $\theta$, $p < 1$ be positive real numbers. Let $f : X \to Y$ be an odd mapping such that*

$$\left\| \left( rf\left(\frac{ux_1 + uy_1}{r}\right) - uf(x_1) - uf(y_1), \right.\right.$$

$$\left.\left. \cdots, rf\left(\frac{ux_k + uy_k}{r}\right) - uf(x_k) - uf(y_k)\right)\right\|_k$$

$$\leq \theta \sum_{j=1}^{k} (\|x_j\|^p + \|y_j\|^p)$$

*for all $u \in U(A)$ and $x_1, \cdots x_k \in X$. Then there exists a unique A-linear generalized additive mapping $\Lambda : X \to Y$ such that*

$$\sup_{k\geq 1} \|(\Lambda(x_1) - f(x_1), \cdots, \Lambda(x_k) - f(x_k))\|_k \leq \sup_{k\geq 1} \frac{r^{p-1}\theta}{2^{p-1} - r^{p-1}} \sum_{j=1}^{k} \|x_j\|^p$$

*for all $x_1, \cdots x_k \in X$.*

*Proof.* Define

$$\varphi(x_1, y_1, \cdots, x_k, y_k) = \theta \sum_{j=1}^{k} (\|x_j\|^p + \|y_j\|^p)$$

and apply Theorem 6.23. Then we get the desired result.                                □

## 6.3.2   Isomorphisms in Unital Multi-C*-Algebras

Assume that $A$ and $B$ are unital multi-$C^*$-algebras with the unit $e$. Let $U(A)$ be the set of unitary elements in $A$.

Now, we investigate $C^*$-algebra isomorphisms in unital multi-$C^*$-algebras.

**Theorem 6.25.** *Let $r \neq 2$. Let $h : A \to B$ be an odd bijective mapping satisfying*

$$h\left(\frac{2^n}{r^n} uy\right) = h\left(\frac{2^n}{r^n} u\right) h(y)$$

*for all $u \in U(A)$, $y \in A$ and $n \geq 0$ for which there exists a function $\varphi : A^{kd} \to [0, \infty)$ such that*

$$\lim_{j \to \infty} \frac{r^j}{2^j} \varphi\left(\frac{2^j}{r^j} x_{11}, \cdots, \frac{2^j}{r^j} x_{1d}, \cdots, \frac{2^j}{r^j} x_{k1}, \cdots, \frac{2^j}{r^j} x_{kd}\right) = 0,$$

$$\|(D_\mu h(x_{11}, \cdots, x_{1d}), \cdots, D_\mu h(x_{k1}, \cdots, x_{kd}))\|_k$$

$$\leq \varphi(x_{11}, \cdots, x_{1d}, \cdots, x_{k1}, \cdots, x_{kd})$$

*and*

$$\left\|\left(h\left(\frac{2^n}{r^n} u_1^*\right) - h\left(\frac{2^n}{r^n} u_1\right)^*, \cdots, h\left(\frac{2^n}{r^n} u_k^*\right) - h\left(\frac{2^n}{r^n} u_k\right)^*\right)\right\|_k$$

$$\leq \varphi\left(\underbrace{\frac{2^n}{r^n} u_1, \cdots, \frac{2^n}{r^n} u_1}_{d \ times}, \cdots, \underbrace{\frac{2^n}{r^n} u_k, \cdots, \frac{2^n}{r^n} u_k}_{d \ times}\right)$$

*for all $\mu \in S^1 := \{\lambda \in \mathbb{C} : |\lambda| = 1\}$, $u_1, \cdots, u_k \in U(A)$, $n \geq 0$ and $x_{11}, \cdots, x_{kd} \in A$. Assume that $\lim_{n \to \infty} \frac{r^n}{2^n} h(\frac{2^n}{r^n} e)$ is invertible. Then the odd bijective mapping $h : A \to B$ is a $C^*$-algebra isomorphism.*

*Proof.* Consider the multi-$C^*$-algebras $A$ and $B$ as left Banach modules over the unital multi-$C^*$-algebra $\mathbb{C}$. By Theorem 6.17, there exists a unique $\mathbb{C}$-linear generalized additive mapping $H : A \to B$ such that

$$\sup_{k\geq 1} \|(h(x_1) - H(x_1), \cdots, h(x_k) - H(x_k))\|_k$$

$$\leq \sup_{k\geq 1} \frac{1}{2(_{d-2}C_l -_{d-2} C_{l-2} + 1)} \varphi \Big(x_1, x_1, \underbrace{0, \cdots, 0}_{d-2\ \text{times}}, \cdots, x_k, x_k, \underbrace{0, \cdots, 0}_{d-2\ \text{times}}\Big)$$

for all $x_1, \cdots, x_k \in A$ in which $H : A \to B$ is given by

$$H(x) = \lim_{n\to\infty} \frac{r^n}{2^n} h\Big(\frac{2^n}{r^n} x\Big)$$

for all $x \in A$.

The rest of the proof is easy. This completes the proof.          □

**Corollary 6.26.** *Let $r > 2$ and $\theta$, $p > 1$ be positive real numbers or let $r < 2$ and $\theta$, $p < 1$ be positive real numbers. Let $h : A \to B$ be an odd bijective mapping satisfying*

$$h\Big(\frac{2^n}{r^n} uy\Big) = h\Big(\frac{2^n}{r^n} u\Big) h(y)$$

*for all $u \in U(A)$, $y \in A$ and $n \geq 0$ such that*

$$\|(D_\mu h(x_{11}, \cdots, x_{1d}), \cdots, D_\mu h(x_{k1}, \cdots, x_{kd}))\|_k$$

$$\leq \theta \sum_{j=1}^{d} (\|x_{1j}\|^p + \cdots + \|x_{kj}\|^p)$$

*and*

$$\left\| \left( h\Big(\frac{2^n}{r^n} u_1^*\Big) - h\Big(\frac{2^n}{r^n} u_1\Big)^*, \cdots, h\Big(\frac{2^n}{r^n} u_k^*\Big) - h\Big(\frac{2^n}{r^n} u_k\Big)^* \right) \right\|_k$$

$$\leq kd \frac{2^{pn}}{r^{pn}} \theta$$

*for all $\mu \in S^1$, $u \in U(A)$, $n \geq 0$ and $x_{11}, \cdots, x_{kd} \in A$. Assume that $\lim_{n\to\infty} \frac{r^n}{2^n} h(\frac{2^n}{r^n} e)$ is invertible. Then the odd bijective mapping $h : A \to B$ is a $C^*$-algebra isomorphism.*

*Proof.* Define

$$\varphi(x_{11}, \cdots, x_{1d}, \cdots, x_{k1}, \cdots, x_{kd}) = \theta \sum_{j=1}^{d} (\|x_{1j}\|^p + \cdots + \|x_{kj}\|^p)$$

and apply Theorem 6.25. Then we get the desired result.          □

## 6.4 Additive Functional Inequalities in Proper Multi-$CQ^*$-Algebras

In this section, we approximate the following additive functional inequality:

$$\left\| \left( \sum_{i=1}^{d+1} f(x_{1i}), \cdots, \sum_{i=1}^{d+1} f(x_{ki}) \right) \right\|_k$$

$$\leq \left\| \left( mf\left( \frac{\sum_{i=1}^{d+1} x_{1i}}{m} \right), \cdots, mf\left( \frac{\sum_{i=1}^{d+1} x_{ki}}{m} \right) \right) \right\|_k \qquad (6.63)$$

for all $x_{11}, \cdots, x_{kd+1} \in X$, where $d \geq 2$ is a fixed integer. Also, we investigate homomorphisms in proper multi-$CQ^*$-algebras and derivations on proper multi-$CQ^*$-algebras associated with the above additive functional inequality.

### 6.4.1 Stability of $\mathbb{C}$-Linear Mappings in Multi-Banach Spaces

Now, we investigate the Hyers-Ulam stability of $\mathbb{C}$-linear mappings in multi-Banach spaces associated with the multi-additive functional inequality (6.63).

In this section, we assume that $(X, \| \cdot \|)$ and $(Y, \| \cdot \|)$ are Banach spaces such that $(X^k, \| \cdot \|_k)$ and $(Y^k, \| \cdot \|_k)$ are multi-Banach spaces.

**Lemma 6.27.** *Let* $f : X \to Y$ *be a mapping satisfying* (6.63) *in which* $f(0) = 0$. *Then* $f$ *is additive.*

*Proof.* Letting $x_3 = \cdots = x_{d+1} = 0$ and replacing $x_1$ by $x$ and $x_2$ by $-x$ in (6.63), we have

$$\|f(x) + f(-x)\| \leq \|mf(0)\| = 0$$

for all $x \in X$. Hence $f(-x) = -f(x)$ for all $x \in X$. Replacing $x_1$ by $x$, $x_2$ by $y$ and $x_3$ by $-x - y$ and putting $x_4 = \cdots = x_{d+1} = 0$ in (6.63), we have

$$\|f(x) + f(y) - f(x + y)\| = \|f(x) + f(y) + f(-x - y)\|$$
$$\leq \|mf(0)\|_Y$$
$$= 0$$

for all $x, y \in X$. Thus we have

$$f(x + y) = f(x) + f(y)$$

for all $x, y \in X$. This completes the proof.    □

**Theorem 6.28.** *Let* $f : X \to Y$ *be a mapping. If there exists a function* $\varphi : X^{kd+k} \to [0, \infty)$ *satisfying the following:*

$$\left\| \left( \sum_{i=1}^{d+1} f(x_{1i}), \cdots, \sum_{i=1}^{d+1} f(x_{ki}) \right) \right\|_k$$
$$\leq \left\| \left( mf \left( \frac{\sum_{i=1}^{d+1} x_{1i}}{m} \right), \cdots, mf \left( \frac{\sum_{i=1}^{d+1} x_{ki}}{m} \right) \right) \right\|_k$$
$$+ \varphi(x_{11}, \cdots, x_{1\,d+1}, \cdots, x_{k1}, \cdots, x_{k\,d+1}) \tag{6.64}$$

*and*

$$\tilde{\varphi}(x_{11}, \cdots, x_{1\,d+1}, \cdots, x_{k1}, \cdots, x_{k\,d+1})$$
$$:= \sum_{j=0}^{\infty} \sup_{k \geq 1} d^j \varphi \left( d^{-j-1} x_{11}, \cdots, d^{-j-1} x_{1\,d+1}, \cdots, d^{-j-1} x_{k1}, \cdots, d^{-j-1} x_{k\,d+1} \right)$$
$$< \infty \tag{6.65}$$

*for all* $x_{11}, \cdots, x_{k\,d+1} \in X$, *then there exists a unique additive mapping* $L : X \to Y$ *such that*

$$\sup_{k \geq 1} \| (f(x_1) - L(x_1), \cdots, f(x_k) - L(x_k)) \|_k$$
$$\leq \sup_{k \geq 1} \tilde{\varphi}(x_1, x_1, \cdots, -dx_1, \cdots, x_k, x_k, \cdots, -dx_k) \tag{6.66}$$

*for all* $x_1, \cdots, x_k \in X$.

*Proof.* Since $\tilde{\varphi}(0, \cdots, 0) < \infty$ in (6.65), we have $\varphi(0, \cdots, 0) = 0$ and so $f(0) = 0$. Replacing $x_{i1}, \cdots, x_{id}$ by $x_i$ and $x_{id+1}$ by $-dx_i$ $(1 \leq i \leq k)$, respectively, in (6.64), since $f(0) = 0$, we have

$$\| (df(x_1) - f(dx_1), \cdots, df(x_k) - f(dx_k)) \|_k$$
$$= \| (df(x_1) + f(-dx_1), \cdots, df(x_k) + f(-dx_k)) \|_k$$
$$\leq \| (mf(0), \cdots, mf(0)) \|_k$$
$$+ \varphi(x_1, x_1, \cdots, -dx_1, \cdots, x_k, x_k, \cdots, -dx_k)$$

for all $x_1, \cdots, x_k \in X$. From the above inequality, we have

$$\left\| \left( f(x_1) - df \left( \frac{x_1}{d} \right), \cdots, f(x_k) - df \left( \frac{x_k}{d} \right) \right) \right\|_k$$
$$\leq \varphi \left( \frac{x_1}{d}, \frac{x_1}{d}, \cdots, -x_1, \cdots, \frac{x_k}{d}, \frac{x_k}{d}, \cdots, -x_k \right)$$

for all $x_1, \cdots, x_k \in X$. Replacing $x_i$ by $d^{-n}x_i$ $(1 \leq i \leq k)$ in the above inequality, we have

$$\left\| \left( d^n f \left( \frac{x_1}{d^n} \right) - d^{n+1} f \left( \frac{x_1}{d^{n+1}} \right), \cdots, d^n f \left( \frac{x_k}{d^n} \right) - d^{n+1} f \left( \frac{x_k}{d^{n+1}} \right) \right) \right\|_k$$
$$\leq d^n \varphi \left( \frac{x_1}{d^{n+1}}, \frac{x_1}{d^{n+1}}, \cdots, -\frac{x_1}{d^n}, \cdots, \frac{x_k}{d^{n+1}}, \frac{x_k}{d^{n+1}}, \cdots, -\frac{x_k}{d^n} \right).$$

From the above inequality, we have

$$\sup_{k \geq 1} \left\| \left( d^n f \left( \frac{x_1}{d^n} \right) - d^q f \left( \frac{x_1}{d^q} \right), \cdots, d^n f \left( \frac{x_k}{d^n} \right) - d^q f \left( \frac{x_k}{d^q} \right) \right) \right\|_k$$
$$\leq \sum_{j=q}^{n-1} \sup_{k \geq 1} \left\| \left( d^{j+1} f \left( \frac{x_1}{d^{j+1}} \right) - d^j f \left( \frac{x_1}{d^j} \right), \cdots, d^{j+1} f \left( \frac{x_k}{d^{j+1}} \right) - d^j f \left( \frac{x_k}{d^j} \right) \right) \right\|_k$$
$$\leq \sum_{j=q}^{n-1} \sup_{k \geq 1} d^j \varphi \left( \frac{x_1}{d^{j+1}}, \frac{x_1}{d^{j+1}}, \cdots, -\frac{dx_1}{d^{j+1}}, \cdots, \frac{x_k}{d^{j+1}}, \frac{x_k}{d^{j+1}}, \cdots, -\frac{dx_k}{d^{j+1}} \right)$$

for all $x_1, \cdots, x_k \in X$ and $q, n \geq 1$ with $q < n$. From (6.64), the sequence $\{ d^n f \left( \frac{x}{d^n} \right) \}$ is a Cauchy sequence for all $x \in X$ and so it is convergent in the complete multi-norm $Y$. Thus we can define a mapping $L : X \to Y$ by

$$L(x) := \lim_{n \to \infty} d^n f \left( \frac{x}{d^n} \right)$$

for all $x \in X$.

In order to prove that $L$ satisfies (6.66), if we put $q = 0$ and let $n \to \infty$ in the above inequality, then we obtain

$$\sup_{k \geq 1} \| (f(x_1) - L(x_1), \cdots, f(x_k) - L(x_k)) \|_k$$
$$\leq \sum_{j=0}^{n-1} \sup_{k \geq 1} d^j \varphi \left( \frac{x_1}{d^{j+1}}, \frac{x_1}{d^{j+1}}, \cdots, -\frac{dx_1}{d^{j+1}}, \cdots, \frac{x_k}{d^{j+1}}, \frac{x_k}{d^{j+1}}, \cdots, -\frac{dx_k}{d^{j+1}} \right)$$
$$= \sup_{k \geq 1} \tilde{\varphi}(x_1, x_1, \cdots, -dx_1, \cdots, x_k, x_k, \cdots, -dx_k)$$

for all $x_1, \cdots, x_k \in X$. Replacing $x_{ij}$ by $\frac{x_{1j}}{d^n}$ $(1 \leq i \leq k$ and $1 \leq j \leq d + 1)$, respectively, and multiplying by $d^{n+1}$ in (6.64), we have

$$\left\| \sum_{i=1}^{d+1} d^n f\left(\frac{x_{1i}}{d^{n+1}}\right) \right\|$$

$$\leq \left\| m d^n f\left(\frac{\sum_{i=1}^{d+1} x_{1i}}{m d^{n+1}}\right) \right\| + d^n \varphi\left(\frac{x_{11}}{d^{n+1}}, \cdots, \frac{x_{1d+1}}{d^{n+1}}, \cdots, \frac{x_{11}}{d^{n+1}}, \cdots, \frac{x_{1d+1}}{d^{n+1}}\right)$$

for all $x_{1j} \in X$ $(1 \leq j \leq d+1)$. Since (6.65) gives

$$\lim_{n\to\infty} d^n \sup_{k\geq 1} \varphi\left(\frac{x_{11}}{d^{n+1}}, \cdots, \frac{x_{1d+1}}{d^{n+1}}, \cdots, \frac{x_{11}}{d^{n+1}}, \cdots, \frac{x_{1d+1}}{d^{n+1}}\right) = 0$$

for all $x_{1j} \in X$ $(1 \leq j \leq d+1)$, letting $n \to \infty$ in the above inequality, we have

$$\left\| \sum_{i=1}^{d+1} L(x_{1i}) \right\| \leq \left\| m L\left(\frac{\sum_{i=1}^{d+1} x_{1i}}{m}\right) \right\| \tag{6.67}$$

and so $L$ is additive by Lemma 6.27.

Now, to prove the uniqueness of $L$, let $L' : X \to Y$ be another additive mapping satisfying (6.66). Since $L$ and $L'$ are additive, we have

$$\|L(x) - L'(x)\|$$

$$= d^n \left\| L\left(\frac{x}{d^n}\right) - L'\left(\frac{x}{d^n}\right) \right\|$$

$$\leq d^n \left( \left\| L\left(\frac{x}{d^n}\right) - f\left(\frac{x}{d^n}\right) \right\| + \left\| L'\left(\frac{x}{d^n}\right) - f\left(\frac{x}{d^n}\right) \right\| \right)$$

$$\leq d^n \cdot 2\tilde\varphi\left(\frac{x}{d^n}, \cdots, \frac{x}{d^n}, \frac{-dx}{d^n}, \cdots, \frac{x}{d^n}, \cdots, \frac{x}{d^n}, \frac{-dx}{d^n}\right)$$

$$= 2 \sum_{j=0}^{\infty} \sup_{k\geq 1} d^{n+j} \varphi\left(A\right),$$

where

$$\varphi(A)$$

$$= \varphi\left( \overbrace{\underbrace{\frac{x}{d^{n+j+1}}, \cdots, \frac{x}{d^{n+j+1}}, \frac{-dx}{d^{n+j+1}}}_{d+1}, \cdots, \underbrace{\frac{x}{d^{n+j+1}}, \cdots, \frac{x}{d^{n+j+1}}, \frac{-dx}{d^{n+j+1}}}_{d+1}}^{k} \right),$$

which goes to zero as $n \to \infty$ for all $x \in X$ by (6.65). Consequently, $L$ is the unique additive mapping satisfying (6.66). This completes the proof. $\qquad\square$

**Corollary 6.29.** *Let* $f : X \to Y$ *be a mapping with* $f(0) = 0$. *If there exists a function* $\varphi : X^{kd+k} \to [0, \infty)$ *satisfying* (6.64) *and*

$$\tilde{\varphi}(x_{11}, \cdots, x_{kd+k}) := \sum_{j=0}^{\infty} \sup_{k \in \mathbb{N}} \frac{1}{d^{j+1}} \varphi \left( d^j x_{11}, \cdots, d^j x_{kd+1} \right) < \infty \qquad (6.68)$$

*for all* $x_{11}, \cdots, x_{kd+1} \in X$, *then there exists a unique additive mapping* $L : X \to Y$ *such that*

$$\sup_{k \geq 1} \| (f(x_1) - L(x_1), \cdots, f(x_k) - L(x_k)) \|_k$$

$$\leq \sup_{k \geq 1} \tilde{\varphi}(x_1, x_1, \cdots, -dx_1, \cdots, x_k, x_k, \cdots, -dx_k) \qquad (6.69)$$

*for all* $x_1, \cdots, x_k \in X$.

*Proof.* The proof is same as in the corresponding part of the proof of Theorem 6.28. □

**Lemma 6.30.** *Let* $f : X \to Y$ *be a mapping satisfying the following:*

$$\left\| \sum_{i=1}^{d} f(x_i) + \mu f(x_{d+1}) \right\| \leq \left\| mf \left( \frac{\sum_{i=1}^{d} x_i + \mu x_{d+1}}{m} \right) \right\| \qquad (6.70)$$

*for all* $\mu \in \mathbb{T}^1$ *and* $x_1, \cdots, x_{d+1} \in X$. *Then* $f$ *is* $\mathbb{C}$-*linear.*

*Proof.* If we put $\mu = 1$ in (6.70), then $f$ is additive by Lemma 6.27.

Putting $x_1 = x$, $x_i = 0$ $(2 \leq i \leq d)$ and $x_{d+1} = -x$, respectively, we get $f(\mu x) + \mu f(-x) = 0$ and so $f(\mu x) = \mu f(x)$ for all $\mu \in \mathbb{T}^1$ and $x \in X$. Thus we have

$$f(\mu x + \bar{\mu} x) = f(\mu x) + f(\bar{\mu} x) = \mu f(x) + \bar{\mu} f(x)$$

for all $\mu \in \mathbb{T}^1$ and $x \in X$ and so $f(tx) = tf(x)$ for any real number $t$ with $|t| \leq 1$ and $x \in X$.

On the other hand, since $f(mx) = mf(x)$, we get $f(m^n x) = m^n f(x)$ for all $n \in \mathbb{N}$. So, for any real number $t$, there exists a positive integer $n$ with $|t| \leq m^n$. Thus we have

$$f(tx) = f\left( m^n \cdot \frac{t}{m^n} x \right) = m^n f\left( \frac{t}{m^n} x \right) = m^n \cdot \frac{t}{m^n} f(x) = tf(x).$$

Now, we consider any $\alpha \in \mathbb{C}$ with $\alpha = t + si$ for some real numbers $t, s$. Since $f(ix) = if(x)$ holds, we have

$$f(\alpha x) = f(tx) + f(six) = tf(x) + sf(ix) = tf(x) + sif(x) = \alpha f(x)$$

and so $f$ is $\mathbb{C}$-linear. This completes the proof. □

**Theorem 6.31.** *Let $f : X \rightarrow Y$ be a mapping. If there exists a function $\varphi : X^{kd+k} \rightarrow [0, \infty)$ satisfying (6.64) and*

$$\left\| \left( \sum_{i=1}^{d} f(x_{1i}) + \mu f(x_{1d+1}), \cdots, \sum_{i=1}^{d} f(x_{ki}) + \mu f(x_{kd+1}) \right) \right\|_k$$

$$\leq \left\| \left( mf\left( \frac{\sum_{i=1}^{d} x_{1i} + \mu x_{1d+1}}{m} \right), \cdots, mf\left( \frac{\sum_{i=1}^{d} x_{ki} + \mu x_{kd+1}}{m} \right) \right) \right\|_k$$

$$+ \varphi(x_{11}, \cdots, x_{kd+1}) \tag{6.71}$$

*for all $\mu \in \mathbb{T}^1$ and $x_{11}, \cdots, x_{kd+1} \in X$, then there exists a unique $\mathbb{C}$-linear mapping $L : X \rightarrow Y$ satisfying (6.66).*

*Proof.* If we put $\mu = 1$ in (6.71), then, by Theorem 6.28, there exists a unique additive mapping $L : X \rightarrow Y$ defined by

$$L(x) := \lim_{n \to \infty} d^n f\left( \frac{x}{d^n} \right)$$

for all $x \in X$ which satisfies (6.66). By the similar method to the corresponding part of the proof of Theorem 6.28, $L$ satisfies the following:

$$\left\| \sum_{i=1}^{d} L(x_i) + \mu L(x_{d+1}) \right\| \leq \left\| mL\left( \frac{\sum_{i=1}^{d} x_i + \mu x_{d+1}}{m} \right) \right\|$$

for all $\mu \in \mathbb{T}^1$ and $x_1, \cdots, x_{d+1} \in X$. Thus Lemma 6.30 gives that $L$ is $\mathbb{C}$-linear. This completes the proof. $\qquad \square$

**Corollary 6.32.** *Let $f : X \rightarrow Y$ be a mapping with $f(0) = 0$. If there exists a function $\varphi : X^{kd+k} \rightarrow [0, \infty)$ satisfying (6.68) and (6.70), then there exists a unique $\mathbb{C}$-linear mapping $L : X \rightarrow Y$ satisfying (6.71).*

*Proof.* The proof is same as in the corresponding part of the proof of Theorem 6.31. $\qquad \square$

### 6.4.2   Stability of Homomorphisms in Proper Multi-$CQ^*$-Algebras

Now, we investigate the Hyers-Ulam stability of isomorphisms in proper multi-$CQ^*$-algebras associated with the additive functional inequality.

We assume that $(A, \| \cdot \|)$ and $(B, \| \cdot \|)$ are Banach algebras such that $(A^k, \| \cdot \|_k)$ and $(B^k, \| \cdot \|_k)$ are multi-Banach algebras.

**Theorem 6.33.** *Let* $f : A \to B$ *be a mapping. Suppose that there exists a function* $\varphi : A^{kd+k} \to [0, \infty)$ *satisfying (6.65) and*

$$\left\| \left( \sum_{i=1}^{d} f(x_{1i}) + \mu f(x_{1d+1}), \cdots, \sum_{i=1}^{d} f(x_{ki}) + \mu f(x_{kd+1}) \right) \right\|_k$$

$$\leq \left\| \left( mf\left( \frac{\sum_{i=1}^{d} x_{1i} + \mu x_{1d+1}}{m} \right), \cdots, mf\left( \frac{\sum_{i=1}^{d} x_{ki} + \mu x_{kd+1}}{m} \right) \right) \right\|_k$$

$$+ \varphi(x_{11}, \cdots, x_{kd+1}) \tag{6.72}$$

*for all* $\mu \in \mathbb{T}^1$ *and* $x_{11}, \cdots, x_{kd+1} \in A$. *If, in addition, there exists a function* $\phi : A^{2k} \to [0, \infty)$ *satisfying the following:*

$$\| (f(x_1 y_1) - f(x_1)f(y_1), \cdots, f(x_k y_k) - f(x_k)f(y_k)) \|_k$$

$$\leq \phi(x_1, y_1, \cdots, x_k, y_k) \tag{6.73}$$

*and*

$$\lim_{n \to \infty} \sup_{k \geq 1} d^{2n} \phi(d^{-n}x_1, d^{-n}y_1, \cdots, d^{-n}x_k, d^{-n}y_k) = 0 \tag{6.74}$$

*for all* $x_1, \cdots, x_k, y_1, \cdots, y_k \in A$ *whenever the multiplication is defined, then there exists a unique proper* $CQ^*$-*algebra homomorphism* $h : A \to B$ *such that*

$$\sup_{k \geq 1} \| (f(x_1) - h(x_1), \cdots, f(x_k) - h(x_k)) \|_k$$

$$\leq \sup_{k \geq 1} \tilde{\varphi}(x_1, x_1, \cdots, -dx_1, \cdots, x_k, x_k, \cdots, -dx_k) \tag{6.75}$$

*for all* $x_1, \cdots, x_k \in A$.

*Proof.* By Theorem 6.31, we have a unique $\mathbb{C}$-linear mapping $h : A \to B$ defined by

$$h(x) := \lim_{n \to \infty} d^n f\left( \frac{x}{d^n} \right)$$

for all $x \in A$ which satisfies (6.75).

Now, we show that $h(xy) = h(x)h(y)$ for all $x, y \in A$ whenever the multiplication is defined. Replacing $x_i, y_i$ by $d^{-n}x_i, d^{-n}y_i$ $(1 \leq i \leq k)$, respectively, and multiplying by $d^{2n}$ in (6.73), we have

$$\| (d^{2n}[f(d^{-n}x_1 d^{-n}y_1) - f(d^{-n}x_1)f(d^{-n}y_1)],$$

$$\cdots, d^{2n}[f(d^{-n}x_k d^{-n}y_k) - f(d^{-n}x_k)f(d^{-n}y_k)]) \|_k$$

$$\leq d^{2n} \phi(d^{-n}x_1, d^{-n}y_1, \cdots, d^{-n}x_k, d^{-n}y_k) \tag{6.76}$$

for all $x_1, \cdots, x_k, y_1, \cdots, y_k \in A$ whenever the multiplication is defined. Also, we have

$$\lim_{n\to\infty} d^{2n}f(d^{-n}xd^{-n}y) = \lim_{n\to\infty} d^{2n}f(d^{-2n}xy) = h(xy)$$

and

$$\lim_{n\to\infty} d^{2n}f(d^{-n}x)f(d^{-n}y) = \lim_{n\to\infty} d^n f(d^{-n}x) \cdot \lim_{n\to\infty} d^n f(d^{-n}y)$$

$$= h(x)h(y)$$

for all $x, y \in A$ whenever the multiplication is defined. If we let $n \to \infty$ in the above inequality, then (6.74) gives $h(xy) = h(x)h(y)$ for all $x, y \in A$ whenever the multiplication is defined. This completes the proof.  □

**Corollary 6.34.** *Let $\theta$, $p$ be nonnegative real numbers with $p > 1$ and $f : A \to B$ be a mapping satisfying the following:*

$$\left\| \left( \sum_{i=1}^{d} f(x_{1i}) + \mu f(x_{1d+1}), \cdots, \sum_{i=1}^{d} f(x_{ki}) + \mu f(x_{kd+1}) \right) \right\|_k$$

$$\leq \left\| \left( mf\left( \frac{\sum_{i=1}^{d} x_{1i} + \mu x_{1d+1}}{m} \right), \cdots, mf\left( \frac{\sum_{i=1}^{d} x_{ki} + \mu x_{kd+1}}{m} \right) \right) \right\|_k$$

$$+ \theta \cdot \sum_{l=1}^{k} \sum_{i=1}^{d+1} \|x_{li}\|^p \tag{6.77}$$

*for all $\mu \in \mathbb{T}^1$ and $x_{11}, \cdots, x_{kd+1} \in A$. If, in addition,*

$$\| (f(x_1 y_1) - f(x_1)f(y_1), \cdots, f(x_k y_k) - f(x_k)f(y_k)) \|_k$$

$$\leq \theta \cdot \sum_{i=1}^{d+1} (\|x_i\|^{2p} + \|y_i\|^{2p}) \tag{6.78}$$

*for all $x_1, \cdots, x_k, y_1, \cdots, y_k \in A$ whenever the multiplication is defined. Then there exists a unique proper $CQ^*$-algebra homomorphism $h : A \to B$ such that*

$$\sup_{k\geq 1} \| (f(x_1) - h(x_1), \cdots, f(x_k) - h(x_k)) \|_k \leq \sup_{k\geq 1} \frac{d^{p-1} + 1}{d^{p-1} - 1} \theta \sum_{l=1}^{k} \|x_l\|^p$$

*for all $x_1, \cdots, x_k \in A$.*

*Proof.* Let $\varphi : A^{kd+k} \to [0, \infty)$ be a mapping defined by

$$\varphi(x_{11}, \cdots, x_{1\,d+1}, \cdots, x_{k1}, \cdots, x_{k\,d+1}) = \theta \cdot \sum_{l=1}^{k} \sum_{i=1}^{d+1} \|x_{li}\|^{p}.$$

When $p > 1$, we have

$$\tilde{\varphi}(x_{11}, \cdots, x_{1\,d+1}, \cdots, x_{k1}, \cdots, x_{k\,d+1})$$

$$:= \sum_{j=0}^{\infty} \sup_{k \geq 1} d^{j} \varphi(d^{-j-1}x_{11}, \cdots, d^{-j-1}x_{1\,d+1},$$

$$\cdots, d^{-j-1}x_{k1}, \cdots, d^{-j-1}x_{k\,d+1})$$

$$= \frac{1}{d} \sum_{j=0}^{\infty} \frac{d^{j+1}}{d^{(j+1)p}} \sup_{k \geq 1} \theta \cdot \sum_{l=1}^{k} \sum_{i=1}^{d+1} \|x_{li}\|^{p}$$

$$= \frac{\theta}{d^{p} - d} \sup_{k \geq 1} \sum_{l=1}^{k} \sum_{i=1}^{d+1} \|x_{li}\|^{p}.$$

In addition, let $\phi : A^{2k} \to [0, \infty)$ be a mapping defined by

$$\phi(x_1, y_1, \cdots, x_k, y_k) = \theta \cdot \sum_{i=1}^{d+1} (\|x_i\|^{2p} + \|y_i\|^{2p}).$$

When $p > 1$, we have

$$\lim_{n \to \infty} d^{2n} \phi(d^{-n}x_1, d^{-n}y_1, \cdots, d^{-n}x_k, d^{-n}y_k)$$

$$= \lim_{n \to \infty} \frac{d^{2n}}{d^{2pn}} \theta \cdot \sum_{i=1}^{d+1} (\|x_i\|^{2p} + \|y_i\|^{2p})$$

$$= 0$$

for all $x_1, \cdots, x_k, y_1, \cdots, y_k \in A$. By applying Theorem 6.33, there exists a unique proper $CQ^*$-algebra homomorphism $h : A \to B$ such that

$$\sup_{k \geq 1} \|(f(x_1) - h(x_1), \cdots, f(x_k) - h(x_k))\|_k \leq \sup_{k \geq 1} \frac{d^p + d}{d^p - d} \theta \sum_{l=1}^{k} \|x_l\|^{p}$$

for all $x_1, \cdots, x_k \in A$. This completes the proof.                    □

**Corollary 6.35.** *Let $\theta$, $p$ be nonnegative real numbers with $p > 1$ and $f : A \to B$ be a mapping satisfying the following:*

$$\left\| \left( \sum_{i=1}^{d} f(x_{1i}) + \mu f(x_{1d+1}), \cdots, \sum_{i=1}^{d} f(x_{ki}) + \mu f(x_{kd+1}) \right) \right\|_k$$

$$\leq \left\| \left( mf\left( \frac{\sum_{i=1}^{d} x_{1i} + \mu x_{1d+1}}{m} \right), \cdots, mf\left( \frac{\sum_{i=1}^{d} x_{ki} + \mu x_{kd+1}}{m} \right) \right) \right\|_k$$

$$+ \theta \cdot \sum_{l=1}^{k} \sum_{i=1}^{d+1} \|x_{li}\|^p \tag{6.79}$$

*for all $\mu \in \mathbb{T}^1$ and $x_{11}, \cdots, x_{kd+1} \in A$. If, in addition,*

$$\|(f(x_1y_1) - f(x_1)f(y_1), \cdots, f(x_ky_k) - f(x_k)f(y_k))\|_k$$

$$\leq \theta \cdot \sum_{i=1}^{d+1} (\|x_i\|^p \cdot \|y_i\|^p) \tag{6.80}$$

*for all $x_1, \cdots, x_k, y_1, \cdots, y_k \in A$ whenever the multiplication is defined. Then there exists a unique proper $CQ^*$-algebra homomorphism $h : A \to B$ such that*

$$\sup_{k \geq 1} \|(f(x_1) - h(x_1), \cdots, f(x_k) - h(x_k))\|_k \leq \sup_{k \geq 1} \frac{d^{p-1} + 1}{d^{p-1} - 1} \theta \sum_{l=1}^{k} \|x_l\|^p$$

*for all $x_1, \cdots, x_k \in A$.*

*Proof.* Let $\varphi : A^{kd+k} \to [0, \infty)$ be a mapping defined by

$$\varphi(x_{11}, \cdots, x_{1d+1}, \cdots, x_{k1}, \cdots, x_{kd+1}) = \theta \cdot \sum_{l=1}^{k} \sum_{i=1}^{d+1} \|x_{li}\|^p.$$

When $p > 1$, we have

$$\tilde{\varphi}(x_{11}, \cdots, x_{1d+1}, \cdots, x_{k1}, \cdots, x_{kd+1})$$

$$:= \sum_{j=0}^{\infty} \sup_{k \geq 1} d^j \varphi(d^{-j-1} x_{11}, \cdots, d^{-j-1} x_{1d+1},$$

$$\cdots, d^{-j-1} x_{k1}, \cdots, d^{-j-1} x_{kd+1})$$

$$= \frac{1}{d} \sum_{j=0}^{\infty} \frac{d^{j+1}}{d^{(j+1)p}} \sup_{k \geq 1} \theta \cdot \sum_{l=1}^{k} \sum_{i=1}^{d+1} \|x_{li}\|^p$$

$$= \frac{\theta}{d^p - d} \sup_{k \geq 1} \sum_{l=1}^{k} \sum_{i=1}^{d+1} \|x_{li}\|^p.$$

In addition, let $\phi : A^{2k} \to [0, \infty)$ be a mapping defined by

$$\phi(x_1, y_1, \cdots, x_k, y_k) = \theta \cdot \sum_{i=1}^{d+1} (\|x_i\|^p \cdot \|y_i\|^p).$$

When $p > 1$, we have

$$\lim_{n \to \infty} d^{2n}\phi(d^{-n}x_1, d^{-n}y_1, \cdots, d^{-n}x_k, d^{-n}y_k)$$

$$= \lim_{n \to \infty} \frac{d^{2n}}{d^{2pn}} \theta \cdot \sum_{i=1}^{d+1} (\|x_i\|^p \cdot \|y_i\|^p)$$

$$= 0$$

for all $x_1, \cdots, x_k, y_1, \cdots, y_k \in A$. By applying Theorem 6.33, there exists a unique proper $CQ^*$-algebra homomorphism $h : A \to B$ such that

$$\sup_{k \geq 1} \|(f(x_1) - h(x_1), \cdots, f(x_k) - h(x_k))\|_k \leq \sup_{k \geq 1} \frac{d^p + d}{d^p - d} \theta \sum_{l=1}^{k} \|x_l\|^p$$

for all $x_1, \cdots, x_k \in A$. This completes the proof.                                   $\square$

*Remark 6.36.* Let $f : A \to B$ be a mapping with $f(0) = 0$. Suppose that there exists a function $\varphi : A^{kd+k} \to [0, \infty)$ satisfying (6.68) and

$$\left\| \left( \sum_{i=1}^{d} f(x_{1i}) + \mu f(x_{1d+1}), \cdots, \sum_{i=1}^{d} f(x_{ki}) + \mu f(x_{kd+1}) \right) \right\|_k$$

$$\leq \left\| \left( mf\left( \frac{\sum_{i=1}^{d} x_{1i} + \mu x_{1d+1}}{m} \right), \cdots, mf\left( \frac{\sum_{i=1}^{d} x_{ki} + \mu x_{kd+1}}{m} \right) \right) \right\|_k$$

$$+ \varphi(x_{11}, \cdots, x_{kd+1}) \tag{6.81}$$

for all $\mu \in \mathbb{T}^1$ and $x_{11}, \cdots, x_{kd+1} \in A$. If, in addition, there exists a function $\phi : A^{2k} \to [0, \infty)$ satisfying the following:

$$\|(f(x_1 y_1) - f(x_1)f(y_1), \cdots, f(x_k y_k) - f(x_k)f(y_k))\|_k$$

$$\leq \phi(x_1, y_1, \cdots, x_k, y_k) \tag{6.82}$$

and

$$\lim_{n \to \infty} \sup_{k \geq 1} d^{-2n} \phi(d^n x_1, d^n y_1, \cdots, d^n x_k, d^n y_k) = 0 \tag{6.83}$$

for all $x_1, \cdots, x_k, y_1, \cdots, y_k \in A$ whenever the multiplication is defined, then there exists a unique proper $CQ^*$-algebra homomorphism $h : A \to B$ satisfying the following:

$$\sup_{k \geq 1} \|(f(x_1) - h(x_1), \cdots, f(x_k) - h(x_k))\|_k$$

$$\leq \sup_{k \geq 1} \tilde{\varphi}(x_1, x_1, \cdots, -dx_1, \cdots, x_k, x_k, \cdots, -dx_k) \tag{6.84}$$

for all $x_1, \cdots, x_k \in A$.

**Corollary 6.37.** *Let $\theta, p$ be nonnegative real numbers with $p < 1$ and $f : A \to B$ be a mapping satisfying the following:*

$$\left\|\left(\sum_{i=1}^{d} f(x_{1i}) + \mu f(x_{1d+1}), \cdots, \sum_{i=1}^{d} f(x_{ki}) + \mu f(x_{kd+1})\right)\right\|_k$$

$$\leq \left\|\left(mf\left(\frac{\sum_{i=1}^{d} x_{1i} + \mu x_{1d+1}}{m}\right), \cdots, mf\left(\frac{\sum_{i=1}^{d} x_{ki} + \mu x_{kd+1}}{m}\right)\right)\right\|_k$$

$$+ \theta \cdot \sum_{l=1}^{k} \sum_{i=1}^{d+1} \|x_{li}\|^p \tag{6.85}$$

*for all $\mu \in \mathbb{T}^1$ and $x_{11}, \cdots, x_{kd+1} \in A$. If, in addition,*

$$\|(f(x_1 y_1) - f(x_1)f(y_1), \cdots, f(x_k y_k) - f(x_k)f(y_k))\|_k$$

$$\leq \theta \cdot \sum_{i=1}^{d+1} (\|x_i\|^{2p} + \|y_i\|^{2p}) \tag{6.86}$$

*for all $x_1, \cdots, x_k, y_1, \cdots, y_k \in A$ whenever the multiplication is defined, then there exists a unique proper $CQ^*$-algebra homomorphism $h : A \to B$ satisfying*

$$\sup_{k \geq 1} \|(f(x_1) - h(x_1), \cdots, f(x_k) - h(x_k))\|_k \leq \sup_{k \geq 1} \frac{d + d^p}{d - d^p} \theta \sum_{l=1}^{k} \|x_l\|^p$$

*for all $x_1, \cdots, x_k \in A$.*

*Proof.* Let $\phi : A^{2k} \to [0, \infty)$ be a mapping defined by

$$\phi(x_1, y_1, \cdots, x_k, y_k) = \theta \cdot \sum_{i=1}^{d+1} (\|x_i\|^{2p} + \|y_i\|^{2p}).$$

When $p < 1$, we have

$$\lim_{n \to \infty} d^{-2n} \phi(d^n x_1, d^n y_1, \cdots, d^n x_k, d^n y_k)$$

$$= \lim_{n \to \infty} \frac{d^{2np}}{d^{2n}} \theta \cdot \sum_{i=1}^{d+1} (\|x_i\|^{2p} + \|y_i\|^{2p})$$

$$= 0$$

for all $x_1, \cdots, x_k, y_1, \cdots, y_k \in A$. By Remark 6.36, there exists a unique proper $CQ^*$-algebra homomorphism $h : A \to B$ such that

$$\sup_{k \geq 1} \|(f(x_1) - h(x_1), \cdots, f(x_k) - h(x_k))\|_k \leq \sup_{k \geq 1} \frac{d + d^p}{d - d^p} \theta \sum_{l=1}^{k} \|x_l\|^p$$

for all $x_1, \cdots, x_k \in A$. This completes the proof.                    □

**Corollary 6.38.** *Let $\theta$, $p$ be nonnegative real numbers with $p < 1$ and $f : A \to B$ be a mapping satisfying the following:*

$$\left\| \left( \sum_{i=1}^{d} f(x_{1i}) + \mu f(x_{1d+1}), \cdots, \sum_{i=1}^{d} f(x_{ki}) + \mu f(x_{kd+1}) \right) \right\|_k$$

$$\leq \left\| \left( mf\left( \frac{\sum_{i=1}^{d} x_{1i} + \mu x_{1d+1}}{m} \right), \cdots, mf\left( \frac{\sum_{i=1}^{d} x_{ki} + \mu x_{kd+1}}{m} \right) \right) \right\|_k$$

$$+ \theta \cdot \sum_{l=1}^{k} \sum_{i=1}^{d+1} \|x_{li}\|^p \tag{6.87}$$

*for all $\mu \in \mathbb{T}^1$ and $x_{11}, \cdots, x_{kd+1} \in A$. If, in addition,*

$$\|(f(x_1 y_1) - f(x_1)f(y_1), \cdots, f(x_k y_k) - f(x_k)f(y_k))\|_k$$

$$\leq \theta \cdot \sum_{i=1}^{d+1} (\|x_i\|^p \cdot \|y_i\|^p) \tag{6.88}$$

*for all $x_1, \cdots, x_k, y_1, \cdots, y_k \in A$ whenever the multiplication is defined, then there exists a unique proper $CQ^*$-algebra homomorphism $h : A \to B$ such that*

$$\sup_{k \geq 1} \|(f(x_1) - h(x_1), \cdots, f(x_k) - h(x_k))\|_k \leq \sup_{k \geq 1} \frac{d + d^p}{d - d^p} \theta \sum_{l=1}^{k} \|x_l\|^p$$

for all $x_1, \cdots, x_k \in A$.

*Proof.* Let $\phi : A^{2k} \to [0, \infty)$ be a mapping defined by

$$\phi(x_1, y_1, \cdots, x_k, y_k) = \theta \cdot \sum_{i=1}^{d+1} (\|x_i\|^p \cdot \|y_i\|^p).$$

When $p < 1$, we have

$$\lim_{n \to \infty} d^{-2n} \phi(d^n x_1, d^n y_1, \cdots, d^n x_k, d^n y_k)$$

$$= \lim_{n \to \infty} \frac{d^{2pn}}{d^{2n}} \theta \cdot \sum_{i=1}^{d+1} (\|x_i\|^p \cdot \|y_i\|^p)$$

$$= 0$$

for all $x_1, \cdots, x_k, y_1, \cdots, y_k \in A$. By Remark 6.36, there exists a unique proper $CQ^*$-algebra homomorphism $h : A \to B$ such that

$$\sup_{k \geq 1} \|(f(x_1) - h(x_1), \cdots, f(x_k) - h(x_k))\|_k \leq \sup_{k \geq 1} \frac{d + d^p}{d - d^p} \theta \sum_{l=1}^{k} \|x_l\|^p$$

for all $x_1, \cdots, x_k \in A$. This completes the proof. $\qquad \square$

### 6.4.3 Stability of Derivations in Proper $CQ^*$-Algebras

Now, we investigate the Hyers-Ulam stability of derivations on proper multi-$CQ^*$-algebras associated with the additive functional inequality.

In this section, we assume that $(A, \| \cdot \|)$ is a Banach algebra such that $(A^k, \| \cdot \|_k)$ is a multi-Banach algebra.

**Theorem 6.39.** *Let $f : A \to A$ be a mapping. Suppose that there exists a function $\varphi : A^{kd+k} \to [0, \infty)$ satisfying (6.65) and*

$$\left\| \left( \sum_{i=1}^{d} f(x_{1i}) + \mu f(x_{1d+1}), \cdots, \sum_{i=1}^{d} f(x_{ki}) + \mu f(x_{kd+1}) \right) \right\|_k$$

$$\leq \left\| \left( mf\left( \frac{\sum_{i=1}^{d} x_{1i} + \mu x_{1d+1}}{m} \right), \cdots, mf\left( \frac{\sum_{i=1}^{d} x_{ki} + \mu x_{kd+1}}{m} \right) \right) \right\|_k$$

$$+ \varphi(x_{11}, \cdots, x_{kd+1}) \tag{6.89}$$

*for all $\mu \in \mathbb{T}^1$ and $x_{11}, \cdots, x_{kd+1} \in A$. If, in addition, there exists a function $\psi : A^{2k} \to [0, \infty)$ satisfying the following:*

$$\|(f(x_1 y_1) - f(x_1)y_1 - x_1 f(y_1)), \cdots, f(x_k y_k) - f(x_k)y_k - x_k f(y_k))\|_k$$
$$\leq \psi(x_1, y_1, \cdots, x_k, y_k) \tag{6.90}$$

*and*

$$\lim_{\substack{n \to \infty \\ k \in \mathbb{N}}} \sup d^{2n} \psi(d^{-n} x_1, d^{-n} y_1, \cdots, d^{-n} x_k, d^{-n} y_k) = 0 \tag{6.91}$$

*for all $x_1, \cdots, x_k, y_1, \cdots, y_k \in A$ whenever the multiplication is defined, then there exists a unique derivation $\delta : A \to A$ satisfying the following:*

$$\sup_{k \geq 1} \|(f(x_1) - \delta(x_1), \cdots, f(x_k) - \delta(x_k))\|_k$$
$$\leq \sup_{k \geq 1} \tilde{\varphi}(x_1, x_1, \cdots, -dx_1, \cdots, x_k, x_k, \cdots, -dx_k) \tag{6.92}$$

*for all $x_1, \cdots, x_k \in A$.*

*Proof.* By Theorem 6.31, we have a unique $\mathbb{C}$-linear mapping $\delta : A \to A$ defined by

$$\delta(x) := \lim_{n \to \infty} d^n f\left(\frac{x}{d^n}\right)$$

for all $x \in A$ which satisfies (6.92).

Now, we show that $\delta(xy) = \delta(x)\delta(y)$ for all $x, y \in A$ whenever the multiplication is defined. Replacing $x_i, y_i$ by $d^{-n} x_i, d^{-n} y_i$ $(1 \leq i \leq k)$, respectively, and multiplying by $d^{2n}$ in (6.90), we have

$$\|(d^{2n}[f(d^{-n} x_1 d^{-n} y_1) - d^{-n} f(d^{-n} x_1)y_1 - d^{-n} x_1 f(d^{-n} y_1)],$$
$$\cdots, d^{2n}[f(d^{-n} x_k d^{-n} y_k) - d^{-n} f(d^{-n} x_k)y_k - d^{-n} x_k f(d^{-n} y_k)])\|_k$$
$$\leq d^{2n} \psi(d^{-n} x_1, d^{-n} y_1, \cdots, d^{-n} x_k, d^{-n} y_k) \tag{6.93}$$

for all $x_1, \cdots, x_k, y_1, \cdots, y_k \in A$ whenever the multiplication is defined. Also, we have

$$\lim_{n \to \infty} d^{2n} f(d^{-n} x d^{-n} y) = \lim_{n \to \infty} d^{2n} f(d^{-2n} xy) = \delta(xy),$$

$$\lim_{n \to \infty} d^{2n} f(d^{-n} x) d^{-n} y = \lim_{n \to \infty} d^n f(d^{-n} x) \cdot y = \delta(x)y$$

and

$$\lim_{n \to \infty} d^{2n} d^{-n} x f(d^{-n} y) = \lim_{n \to \infty} x \cdot d^n f(d^{-n} y) = x\delta(y)$$

for all $x, y \in A$ whenever the multiplication is defined. If we let $n \to \infty$ in the above inequality, then (6.93) gives

$$\delta(xy) = \delta(x)y - x\delta(y)$$

for all $x, y \in A$ whenever the multiplication is defined. This completes the proof.   $\square$

**Corollary 6.40.** *Let* $\theta$, $p$ *be nonnegative real numbers with* $p > 1$ *and* $f : A \to A$ *be a mapping such that*

$$\left\| \left( \sum_{i=1}^{d} f(x_{1i}) + \mu f(x_{1d+1}), \cdots, \sum_{i=1}^{d} f(x_{ki}) + \mu f(x_{kd+1}) \right) \right\|_k$$

$$\leq \left\| \left( mf\left( \frac{\sum_{i=1}^{d} x_{1i} + \mu x_{1d+1}}{m} \right), \cdots, mf\left( \frac{\sum_{i=1}^{d} x_{ki} + \mu x_{kd+1}}{m} \right) \right) \right\|_k$$

$$+ \theta \cdot \sum_{l=1}^{k} \sum_{i=1}^{d+1} \| x_{li} \|^p \tag{6.94}$$

*for all* $\mu \in \mathbb{T}^1$ *and* $x_{11}, \cdots, x_{kd+1} \in A$. *If, in addition,*

$$\| (f(x_1 y_1) - f(x_1)y_1 - x_1 f(y_1), \cdots, f(x_k y_k) - f(x_k)y_k - x_k f(y_k)) \|_k$$

$$\leq \theta \cdot \sum_{i=1}^{d+1} (\| x_i \|^{2p} + \| y_i \|^{2p}) \tag{6.95}$$

*for all* $x_1, \cdots, x_k, y_1, \cdots, y_k \in A$ *whenever the multiplication is defined, then there exists a unique derivation* $\delta : A \to A$ *satisfying the following:*

$$\sup_{k \geq 1} \| (f(x_1) - \delta(x_1), \cdots, f(x_k) - \delta(x_k)) \|_k \leq \sup_{k \geq 1} \frac{d^p + d}{d^p - d} \theta \sum_{l=1}^{k} \| x_l \|^p$$

*for all* $x_1, \cdots, x_k \in A$.

*Proof.* The proof is same to the proof given in Corollary 6.34.   $\square$

**Corollary 6.41.** *Let* $\theta$, $p$ *be nonnegative real numbers with* $p > 1$ *and* $f : A \to A$ *be a mapping such that*

$$\left\| \left( \sum_{i=1}^{d} f(x_{1i}) + \mu f(x_{1d+1}), \cdots, \sum_{i=1}^{d} f(x_{ki}) + \mu f(x_{kd+1}) \right) \right\|_k$$

$$\leq \left\| \left( mf\left( \frac{\sum_{i=1}^{d} x_{1i} + \mu x_{1d+1}}{m} \right), \cdots, mf\left( \frac{\sum_{i=1}^{d} x_{ki} + \mu x_{kd+1}}{m} \right) \right) \right\|_k$$

$$+ \theta \cdot \sum_{l=1}^{k} \sum_{i=1}^{d+1} \| x_{li} \|^p \tag{6.96}$$

*for all $\mu \in \mathbb{T}^1$ and $x_{11}, \cdots, x_{kd+1} \in A$. If, in addition,*

$$\|(f(x_1 y_1) - f(x_1)y_1 - x_1 f(y_1), \cdots, f(x_k y_k) - f(x_k)y_k - x_k f(y_k))\|_k$$

$$\leq \theta \cdot \sum_{i=1}^{d+1} (\|x_i\|^p \cdot \|y_i\|^p) \tag{6.97}$$

*for all $x_1, \cdots, x_k, y_1, \cdots, y_k \in A$ whenever the multiplication is defined, then there exists a unique derivation $\delta : A \to A$ satisfying the following:*

$$\sup_{k \geq 1} \|(f(x_1) - \delta(x_1), \cdots, f(x_k) - \delta(x_k))\|_k \leq \sup_{k \geq 1} \frac{d^p + d}{d^p - d} \theta \sum_{l=1}^{k} \|x_l\|^p$$

*for all $x_1, \cdots, x_k \in A$.*

*Remark 6.42.* Let $f : A \to A$ be a mapping with $f(0) = 0$. Suppose that there exists a function $\varphi : A^{kd+k} \to [0, \infty)$ satisfying (6.68) and (6.89). If, in addition, there exists a function $\psi : A^{2k} \to [0, \infty)$ such that

$$\|(f(x_1 y_1) - f(x_1)f(y_1), \cdots, f(x_k y_k) - f(x_k)f(y_k))\|_k$$

$$\leq \psi(x_1, y_1, \cdots, x_k, y_k) \tag{6.98}$$

and

$$\lim_{n \to \infty} \sup_{k \geq 1} d^{-2n} \psi(d^n x_1, d^n y_1, \cdots, d^n x_k, d^n y_k) = 0 \tag{6.99}$$

for all $x_1, \cdots, x_k, y_1, \cdots, y_k \in A$ whenever the multiplication is defined, then there exists a unique derivation $\delta : A \to A$ such that

$$\sup_{k \geq 1} \|(f(x_1) - \delta(x_1), \cdots, f(x_k) - \delta(x_k))\|_k$$

$$\leq \sup_{k \geq 1} \tilde{\varphi}(x_1, x_1, \cdots, -dx_1, \cdots, x_k, x_k, \cdots, -dx_k) \tag{6.100}$$

for all $x_1, \cdots, x_k \in A$.

**Corollary 6.43.** *Let $\theta, p$ be nonnegative real numbers with $p < 1$ and $f : A \to A$ be a mapping satisfying the following:*

$$\left\|\left(\sum_{i=1}^{d}f(x_{1i})+\mu f(x_{1d+1}),\cdots,\sum_{i=1}^{d}f(x_{ki})+\mu f(x_{kd+1})\right)\right\|_k$$

$$\leq\left\|\left(mf\left(\frac{\sum_{i=1}^{d}x_{1i}+\mu x_{1d+1}}{m}\right),\cdots,mf\left(\frac{\sum_{i=1}^{d}x_{ki}+\mu x_{kd+1}}{m}\right)\right)\right\|_k$$

$$+\,\theta\cdot\sum_{l=1}^{k}\sum_{i=1}^{d+1}\|x_{li}\|^p \tag{6.101}$$

for all $\mu\in\mathbb{T}^1$ and $x_{11},\cdots,x_{kd+1}\in A$. If, in addition,

$$\|(f(x_1y_1)-f(x_1)y_1-x_1f(y_1),\cdots,f(x_ky_k)-f(x_k)y_k-x_kf(y_k))\|_k$$

$$\leq\theta\cdot\sum_{i=1}^{d+1}(\|x_i\|^{2p}+\|y_i\|^{2p}) \tag{6.102}$$

for all $x_1,\cdots,x_k,y_1,\cdots,y_k\in A$ whenever the multiplication is defined. Then there exists a unique derivation $\delta:A\rightarrow A$ satisfying

$$\sup_{k\geq1}\|(f(x_1)-\delta(x_1),\cdots,f(x_k)-\delta(x_k))\|_k\leq\sup_{k\geq1}\frac{d+d^p}{d-d^p}\theta\sum_{l=1}^{k}\|x_l\|^p$$

for all $x_1,\cdots,x_k\in A$.

**Corollary 6.44.** *Let $\theta,p$ be nonnegative real numbers with $p<1$ and $f:A\rightarrow A$ be a mapping satisfying the following:*

$$\left\|\left(\sum_{i=1}^{d}f(x_{1i})+\mu f(x_{1d+1}),\cdots,\sum_{i=1}^{d}f(x_{ki})+\mu f(x_{kd+1})\right)\right\|_k$$

$$\leq\left\|\left(mf\left(\frac{\sum_{i=1}^{d}x_{1i}+\mu x_{1d+1}}{m}\right),\cdots,mf\left(\frac{\sum_{i=1}^{d}x_{ki}+\mu x_{kd+1}}{m}\right)\right)\right\|_k$$

$$+\,\theta\cdot\sum_{l=1}^{k}\sum_{i=1}^{d+1}\|x_{li}\|^p \tag{6.103}$$

for all $\mu\in\mathbb{T}^1$ and $x_{11},\cdots,x_{kd+1}\in A$. If, in addition,

$$\|(f(x_1y_1)-f(x_1)y_1-x_1f(y_1),\cdots,f(x_ky_k)-f(x_k)y_k-x_kf(y_k))\|_k$$

$$\leq\theta\cdot\sum_{i=1}^{d+1}(\|x_i\|^p\cdot\|y_i\|^p) \tag{6.104}$$

*for all* $x_1, \cdots, x_k, y_1, \cdots, y_k \in A$ *whenever the multiplication is defined, then there exists a unique derivation* $\delta : A \to A$ *such that*

$$\sup_{k \geq 1} \| (f(x_1) - \delta(x_1), \cdots, f(x_k) - \delta(x_k)) \|_k \leq \sup_{k \geq 1} \frac{d + d^p}{d - d^p} \theta \sum_{l=1}^{k} \|x_l\|^p$$

*for all* $x_1, \cdots, x_k \in A$.

## 6.5  Stability of Homomorphisms and Derivations in Multi-$C^*$-Ternary Algebras

Using the fixed point method, we prove the Hyers-Ulam stability of homomorphisms and derivations on multi-$C^*$-ternary algebras for the additive functional equation:

$$2f\left( \frac{\sum_{j=1}^{p} x_j}{2} + \sum_{j=1}^{d} y_j \right) = \sum_{j=1}^{p} f(x_j) + 2 \sum_{j=1}^{d} f(y_j).$$

### 6.5.1  Stability of Homomorphisms

Assume that $A$, $B$ are $C^*$-ternary algebras.
For any mapping $f : A \to B$, we define

$$C_\mu f(x_1, \cdots, x_p, y_1, \cdots, y_d)$$

$$:= 2f\left( \frac{\sum_{j=1}^{p} \mu x_j}{2} + \sum_{j=1}^{d} \mu y_j \right) - \sum_{j=1}^{p} \mu f(x_j) - 2 \sum_{j=1}^{d} \mu f(y_j)$$

for all $\mu \in \mathbb{T}^1 := \{ \lambda \in \mathbb{C} : |\lambda| = 1 \}$ and $x_1, \cdots, x_p, y_1, \cdots, y_d \in A$. One can easily show that a mapping $f : A \to B$ satisfies

$$C_\mu f(x_1, \cdots, x_p, y_1, \cdots, y_d) = 0$$

for all $\mu \in \mathbb{T}^1$ and $x_1, \cdots, x_p, y_1, \cdots, y_d \in A$ if and only if

$$f(\mu x + \lambda y) = \mu f(x) + \lambda f(y)$$

for all $\mu, \lambda \in \mathbb{T}^1$ and $x, y \in A$.

Now, we introduce the following lemma for the main results in this section:

**Lemma 6.45.** *Let $\{x_n\}$, $\{y_n\}$ and $\{z_n\}$ be the convergent sequences in A. Then the sequence $\{[x_n, y_n, z_n]\}$ is convergent in A*

*Proof.* Let $x, y, z \in A$ be such that

$$\lim_{n\to\infty} x_n = x, \quad \lim_{n\to\infty} y_n = y, \quad \lim_{n\to\infty} z_n = z.$$

Since

$$[x_n, y_n, z_n] - [x, y, z]$$
$$= [x_n - x, y_n - y, z_n, z] + [x_n, y_n, z] + [x, y_n - y, z_n] + [x_n, y, z_n - z]$$

for all $n \geq 1$, we get

$$\|[x_n, y_n, z_n] - [x, y, z]\| = \|x_n - x\|\|y_n - y\|\|z_n - z\| + \|x_n - x\|\|y_n\|\|z\|$$
$$+ \|x\|\|y_n - y\|\|z_n\| + \|x_n\|\|y\|\|z_n - z\|$$

for all $n \geq 1$ and so

$$\lim_{n\to\infty} [x_n, y_n, z_n] = [x, y, z].$$

This completes the proof. □

Using Theorem 1.3, we prove the Hyers-Ulam stability of homomorphisms in multi-$C^*$–ternary algebras for the following functional equation:

$$C_\mu f(x_1, \cdots, x_m) = 0.$$

**Theorem 6.46.** *Let $((B^k, \|\cdot\|_k) : k \geq 1)$ be a multi-$C^*$-ternary algebra. Let $f : A \to B$ be a mapping for which there exist the functions $\varphi : A^{(p+d)k} \to [0, \infty)$ and $\psi : A^{3k} \to [0, \infty)$ such that*

$$\lim_{n\to\infty} \gamma^{-n} \varphi(\gamma^n x_{11}, \cdots, \gamma^n x_{1p}, \gamma^n y_{11}, \cdots, \gamma^n y_{1p},$$

$$\cdots, \gamma^n x_{k1}, \cdots, \gamma^n x_{kp}, \cdots, \gamma^n y_{k1}, \cdots, \gamma^n y_{kd}) = 0, \tag{6.105}$$

$$\left\|\left(C_\mu f(x_{11}, \cdots, x_{1p}, y_{11},\right.\right.$$

$$\left.\left.\cdots, y_{1d}), \cdots, C_\mu f(x_{k1}, \cdots, x_{kp}, y_{k1}, \cdots, y_{kd})\right)\right\|_k \tag{6.106}$$

$$\leq \varphi(x_{11}, \cdots, x_{1p}, y_{11}, \cdots, y_{1d}, \cdots, x_{k1}, \cdots, x_{kp}, y_{k1}, \cdots, y_{kd}),$$

$$\left\| \left( f([x_1, y_1, z_1]) - [f(x_1), f(y_1), f(z_1)], \right.\right.$$

$$\left.\left. \cdots, f([x_k, y_k, z_k]) - [f(x_k), f(y_k), f(z_k)] \right) \right\|_k$$

$$\leq \psi(x_1, y_1, z_1, \cdots, x_k, y_k, z_k), \tag{6.107}$$

$$\lim_{n\to\infty} \gamma^{-3n} \psi(\gamma^n x_1, \gamma^n y_1, \gamma^n z_1, \cdots, \gamma^n x_k, \gamma^n y_k, \gamma^n z_k) = 0 \tag{6.108}$$

*and*

$$\lim_{n\to\infty} \gamma^{-2n} \psi(\gamma^n x_1, \gamma^n y_1, z_1, \cdots, \gamma^n x_k, \gamma^n y_k, z_k) = 0 \tag{6.109}$$

*for all* $\mu \in \mathbb{T}^1$ *and* $x_{11}, \cdots, x_{1p}, y_{11}, \cdots, y_{1d}, \cdots, x_{k1}, \cdots, x_{kp}, y_{k1}, \cdots, y_{kd}, x_1, \cdots, x_k, y_1, \cdots, y_k, z_1, \cdots, z_k \in A$, *where* $\gamma = \frac{p+2d}{2}$. *If there exists* $L < 1$ *such that*

$$\varphi\left( \overbrace{\gamma x_1, \cdots, \gamma x_1}^{p+d}, \overbrace{\gamma x_2, \cdots, \gamma x_2}^{p+d}, \cdots, \overbrace{\gamma x_k, \cdots, \gamma x_k}^{p+d} \right)$$

$$\leq \gamma L \varphi\left( \overbrace{x_1, \cdots, x_1}^{p+d}, \overbrace{x_2, \cdots, x_2}^{p+d}, \cdots, \overbrace{x_k, \cdots, x_k}^{p+d} \right) \tag{6.110}$$

*for all* $x_1, x_2, \cdots, x_k \in A$, *then there exists a unique multi-$C^*$-ternary algebra homomorphism* $H : A \to B$ *such that*

$$\left\| \left( f(x_1) - H(x_1), \cdots, f(x_k) - H(x_k) \right) \right\|_k$$

$$\leq \frac{1}{(1-L)2\gamma} \varphi\left( \overbrace{x_1, \cdots, x_1}^{p+d}, \overbrace{x_2, \cdots, x_2}^{p+d}, \cdots, \overbrace{x_k, \cdots, x_k}^{p+d} \right) \tag{6.111}$$

*for all* $x_1, \cdots, x_k \in A$.

*Proof.* Let $\mu = 1$ and $x_{ij} = y_{ij} = x_i$ for $1 \leq i \leq k$ in (6.106). Then we have

$$\left\| \left( f(\gamma x_1) - \gamma f(x_1), \cdots, f(\gamma x_k) - \gamma f(x_k) \right) \right\|_k$$

$$\leq \frac{1}{2} \varphi\left( \overbrace{x_1, \cdots, x_1}^{p+d}, \overbrace{x_2, \cdots, x_2}^{p+d}, \cdots, \overbrace{x_k, \cdots, x_k}^{p+d} \right) \tag{6.112}$$

for all $x_1, \cdots, x_k \in A$. Consider the set $E := \{g : A \to B\}$ and introduce the generalized metric on $E$ as follows:

$$d(g,h) = \inf \left\{ C \in \mathbb{R}_+ : \left\| \left( g(x_1) - h(x_1), \cdots, g(x_k) - h(x_k) \right) \right\|_k \right.$$

$$\left. \leq C\varphi \left( \overbrace{x_1, \cdots, x_1}^{p+d}, \overbrace{x_2, \cdots, x_2}^{p+d}, \cdots, \overbrace{x_k, \cdots, x_k}^{p+d} \right), \ \forall x_1, \cdots, x_k \in A \right\}$$

which $(E, d)$ is complete.

Now, we consider the linear mapping $\Lambda : E \to E$ such that

$$\Lambda g(x) := \frac{1}{\gamma} g(\gamma x)$$

for all $x \in A$. Now, we have

$$d(\Lambda g, \Lambda h) \leq L d(g, h)$$

for all $g, h \in E$. Let $g, h \in E$ and $C \in [0, \infty]$ be an arbitrary constant with $d(g, h) \leq C$. From the definition of $d$, we have

$$\left\| \left( g(x_1) - h(x_1), \cdots, g(x_k) - h(x_k) \right) \right\|_k \leq C\varphi \left( \overbrace{x_1, \cdots, x_1}^{p+d}, \cdots, \overbrace{x_k, \cdots, x_k}^{p+d} \right)$$

for all $x_1, \cdots, x_k \in A$. By the assumption and the last inequality, we have

$$\left\| \left( \Lambda g(x_1) - \Lambda h(x_1), \cdots, \Lambda g(x_k) - \Lambda h(x_k) \right) \right\|_k$$

$$= \frac{1}{\gamma} \left\| \left( g(\gamma x_1) - h(\gamma x_1), \cdots, g(\gamma x_k) - h(\gamma x_k) \right) \right\|_k$$

$$\leq \frac{C}{\gamma} \varphi \left( \overbrace{\gamma x_1, \cdots, \gamma x_1}^{p+d}, \cdots, \overbrace{\gamma x_k, \cdots, \gamma x_k}^{p+d} \right)$$

$$\leq C L \varphi \left( \overbrace{x_1, \cdots, x_1}^{p+d}, \cdots, \overbrace{x_k, \cdots, x_k}^{p+d} \right)$$

for all $x_1, \cdots, x_k \in A$ and so

$$\left\| \left( \Lambda f(x_1) - f(x_1), \cdots, \Lambda f(x_k) - f(x_k) \right) \right\|_k$$

$$= \left\| \left( \frac{1}{\gamma} f(\gamma x_1) - f(x_1), \cdots, \frac{1}{\gamma} f(\gamma x_k) - f(x_k) \right) \right\|_k$$

$$= \frac{1}{\gamma} \left\| \left( f(\gamma x_1) - \gamma f(x_1), \cdots, f(\gamma x_k) - \gamma f(x_k) \right) \right\|_k$$

$$\leq \frac{1}{2\gamma} \varphi \left( \overbrace{x_1, \cdots, x_1}^{p+d}, \cdots, \overbrace{x_k, \cdots, x_k}^{p+d} \right)$$

for all $x_1, \cdots, x_k \in A$. Hence $d(\Lambda f, f) \leq \frac{1}{2\gamma}$. By Theorem 1.3, the sequence $\{\Lambda^n f\}$ converges to a fixed point $H$ of $\Lambda$, i.e., $H : A \to B$ is a mapping defined by

$$H(x) = \lim_{n \to \infty} (\Lambda^n f)(x) = \lim_{n \to \infty} \frac{1}{\gamma^n} f(\gamma^n x) \tag{6.113}$$

and $H(\gamma x) = \gamma H(x)$ for all $x \in A$. Also, $H$ is the unique fixed point of $\Lambda$ in the set $E' = \{g \in E : d(f, g) < \infty\}$ and

$$d(H, f) \leq \frac{1}{1-L} d(\Lambda f, f) \leq \frac{1}{(1-L)2\gamma}$$

i.e., the inequality (6.111) holds for all $x_1, \cdots, x_k \in A$. Thus it follows from the definition of $H$, (6.105) and (6.106) that

$$\left\| \left( 2H\left( \frac{\sum_{j=1}^{p} \mu x_{1j}}{2} + \sum_{j=1}^{d} \mu y_{1j} \right) - \sum_{j=1}^{p} \mu H(x_{1j}) - 2 \sum_{j=1}^{d} \mu H(y_{1j}), \right. \right.$$

$$\left. \left. \cdots, 2H\left( \frac{\sum_{j=1}^{p} \mu x_{kj}}{2} + \sum_{j=1}^{d} \mu y_{kj} \right) - \sum_{j=1}^{p} \mu H(x_{kj}) - 2 \sum_{j=1}^{d} \mu H(y_{kj}) \right) \right\|_k$$

$$= \lim_{n \to \infty} \frac{1}{\gamma^n} \left\| \left( 2f\left( \gamma^n \frac{\sum_{j=1}^{p} \mu x_{1j}}{2} + \gamma^n \sum_{j=1}^{d} \mu y_{1j} \right) \right. \right.$$

$$- \sum_{j=1}^{p} \mu f(\gamma^n x_{1j}) - 2 \sum_{j=1}^{d} \mu f(\gamma^n y_{1j}),$$

$$\cdots, 2f\left( \gamma^n \frac{\sum_{j=1}^{p} \mu x_{kj}}{2} + \gamma^n \sum_{j=1}^{d} \mu y_{kj} \right)$$

$$\left. \left. - \sum_{j=1}^{p} \mu f(\gamma^n x_{kj}) - 2 \sum_{j=1}^{d} \mu f(\gamma^n y_{kj}) \right) \right\|_k$$

$$\leq \lim_{n \to \infty} \frac{1}{\gamma^n} \left\| \left( C_\mu f(\gamma^n x_{11}, \cdots, \gamma^n x_{1p}, \gamma^n y_{11}, \cdots, \gamma^n y_{1d}), \right. \right.$$

$$\left. \left. \cdots, C_\mu f(\gamma^n x_{k1}, \cdots, \gamma^n x_{kp}, \gamma^n y_{k1}, \cdots, \gamma^n y_{kd}) \right) \right\|_k$$

$$\leq \lim_{n \to \infty} \frac{1}{\gamma^n} \varphi(\gamma^n x_{11}, \cdots, \gamma^n x_{1p}.\gamma^n y_{11}, \cdots, \gamma^n y_{1d},$$

$$\cdots, \gamma^n x_{k1}, \cdots, \gamma^n x_{kp}, \gamma^n y_{k1}, \cdots, \gamma^n y_{kd}) = 0$$

for all $\mu \in \mathbb{T}^1$ and $x_{11}, \cdots, x_{1p}, y_{11}, \cdots, y_{1d}, \cdots, x_{k1}, \cdots, x_{kp}, y_{k1}, \cdots, y_{kd} \in A$. Hence we have

$$2H\left(\frac{\sum_{j=1}^p \mu x_{ij}}{2} + \sum_{j=1}^d \mu y_{ij}\right) = \sum_{j=1}^p \mu H(x_{ij}) + 2\sum_{j=1}^d \mu H(y_{ij})$$

for all $\mu \in \mathbb{T}^1$, $x_{11}, \cdots, x_{1p}, y_{11}, \cdots, y_{1d}, \cdots, x_{k1}, \cdots, x_{kp}, y_{k1}, \cdots, y_{kd} \in A$ and $1 \le i \le k$ and so

$$H(\lambda x + \mu y) = \lambda H(x) + \mu H(y)$$

for all $\lambda, \mu \in \mathbb{T}^1$ and $x, y \in A$. Therefore, by Lemma 3.12, the mapping $H : A \to B$ is $\mathbb{C}$-linear.

On the other hand, it follows from (6.107) and (6.108) that

$$\left\|\Big(H([x_1, y_1, z_1]) - [H(x_1), H(y_1), H(z_1)],\right.$$

$$\cdots, H([x_k, y_k, z_k]) - [H(x_k), H(y_k), H(z_k)]\Big)\Big\|_k$$

$$= \lim_{n\to\infty} \frac{1}{\gamma^{3n}} \left\|\Big(f\big([\gamma^n x_1, \gamma^n y_1, \gamma^n z_1]\big) - \big[f(\gamma^n x_1), f(\gamma^n y_1), f(\gamma^n z_1)\big],\right.$$

$$\cdots, f\big([\gamma^n x_k, \gamma^n y_k, \gamma^n z_k]\big) - \big[f(\gamma^n x_k), f(\gamma^n y_k), f(\gamma^n z_k)\big]\Big)\Big\|_k$$

$$\le \lim_{n\to\infty} \frac{1}{\gamma^{3n}} \psi(\gamma^n x_1, \gamma^n y_1, \gamma^n z_1, \cdots, \gamma^n x_k, \gamma^n y_k, \gamma^n z_k)$$

$$= 0$$

for all $x_1, y_1, z_1, \cdots, x_k, y_k, z_k \in A$. Thus we have

$$H([x, y, z]) = [H(x), H(y), H(z)]$$

for all $x, y, z \in A$. Thus $H : A \to B$ is a homomorphism satisfying (6.111).

Now, let $T : A \to B$ be another multi-$C^*$-ternary algebras homomorphism satisfying (6.111). Since $d(f, T) \le \frac{1}{(1-L)2\gamma}$ and $T$ is $\mathbb{C}$-linear, we get $T \in E'$ and $(\Lambda T)(x) = \frac{1}{\gamma}(T\gamma x) = T(x)$ for all $x \in A$, i.e., $T$ is a fixed point of $\Lambda$. Since $H$ is the unique fixed point of $\Lambda \in E'$, we have $H = T$. This completes the proof. $\square$

**Theorem 6.47.** *Let $((B^k, \| \cdot \|_k)$ $:$ $k \ge 1)$ be a multi-$C^*$-ternary algebra. Let $f : A \to B$ be a mapping for which there exist the functions $\varphi : A^{(p+d)k} \to [0, \infty)$ and $\psi : A^{3k} \to [0, \infty)$ satisfying the inequalities (6.106) and (6.107) such that*

$$\lim_{n\to\infty} \gamma^n \varphi\Big(\frac{x_{11}}{\gamma^n}, \cdots, \frac{x_{1p}}{\gamma^n}, \frac{y_{11}}{\gamma^n}, \cdots, \frac{y_{1p}}{\gamma^n}, \cdots, \frac{x_{k1}}{\gamma^n},$$

$$\cdots, \frac{x_{kp}}{\gamma^n}, \cdots, \frac{y_{k1}}{\gamma^n}, \cdots, \frac{y_{kd}}{\gamma^n}\Big) = 0, \tag{6.114}$$

$$\lim_{n\to\infty} \gamma^{3n} \psi\Big(\frac{x_1}{\gamma^n}, \frac{y_1}{\gamma^n}, \frac{z_1}{\gamma^n}, \cdots, \frac{x_k}{\gamma^n}, \frac{y_k}{\gamma^n}, \frac{z_k}{\gamma^n}\Big) = 0 \tag{6.115}$$

and

$$\lim_{n\to\infty} \gamma^{2n} \psi\Big(\frac{x_1}{\gamma^n}, \frac{y_1}{\gamma^n}, z_1, \cdots, \frac{x_k}{\gamma^n}, \frac{y_k}{\gamma^n}, z_k\Big) = 0 \tag{6.116}$$

for all $\mu \in \mathbb{T}^1$ and $x_{11}, \cdots, x_{1p}, y_{11}, \cdots, y_{1d}, \cdots, x_{k1}, \cdots, x_{kp}, y_{k1}, \cdots, y_{kd},$ $x_1, \cdots, x_k, y_1, \cdots, y_k, z_1, \cdots, z_k \in A$, where $\gamma = \frac{p+2d}{2}$. If there exists $L < 1$ such that

$$\varphi\Big(\overbrace{\frac{x_1}{\gamma}, \cdots, \frac{x_1}{\gamma}}^{p+d}, \overbrace{\frac{x_2}{\gamma}, \cdots, \frac{x_2}{\gamma}}^{p+d}, \cdots, \overbrace{\frac{x_k}{\gamma}, \cdots, \frac{x_k}{\gamma}}^{p+d}\Big)$$

$$\leq \frac{L}{\gamma}\varphi\Big(\overbrace{x_1, \cdots, x_1}^{p+d}, \overbrace{x_2, \cdots, x_2}^{p+d}, \cdots, \overbrace{x_k, \cdots, x_k}^{p+d}\Big) \tag{6.117}$$

for all $x_1, x_2, \cdots, x_k \in A$, then there exists a unique multi-$C^*$-ternary algebra homomorphism $H : A \to B$ such that

$$\Big\|\big(f(x_1) - H(x_1), \cdots, f(x_k) - H(x_k)\big)\Big\|_k$$

$$\leq \frac{1}{(1-L)2\gamma}\varphi\Big(\overbrace{x_1, \cdots, x_1}^{p+d}, \overbrace{x_2, \cdots, x_2}^{p+d}, \cdots, \overbrace{x_k, \cdots, x_k}^{p+d}\Big) \tag{6.118}$$

for all $x_1, \cdots, x_k \in A$.

*Proof.* If we replace $x_i$ in (6.112) by $\frac{x_i}{\gamma}$ for $1 \leq i \leq k$, then we have

$$\Big\|\big(f(x_1) - \gamma f\big(\frac{1}{x_1}\big), \cdots, f(x_k) - \gamma f\big(\frac{1}{x_k}\big)\big)\Big\|_k$$

$$\leq \frac{1}{2}\varphi\Big(\overbrace{\frac{1}{x_1}, \cdots, \frac{1}{x_1}}^{p+d}, \overbrace{\frac{1}{x_2}, \cdots, \frac{1}{x_2}}^{p+d}, \cdots, \overbrace{\frac{1}{x_k}, \cdots, \frac{1}{x_k}}^{p+d}\Big) \tag{6.119}$$

for all $x_1, \cdots, x_k \in A$. Consider the set $E := \{g : A \to B\}$ and introduce the generalized metric on $E$ as follows:

$$d(g, h) = \inf\left\{C \in \mathbb{R}_+ : \left\|\big(g(x_1) - h(x_1), \cdots, g(x_k) - h(x_k)\big)\right\|_k\right.$$

$$\left. \leq C\varphi\big(\overbrace{x_1, \cdots, x_1}^{p+d}, \overbrace{x_2, \cdots, x_2}^{p+d}, \cdots, \overbrace{x_k, \cdots, x_k}^{p+d}\big), \ \forall x_1, \cdots, x_k \in A\right\},$$

which $(E, d)$ is complete. Now, we consider the linear mapping $\Lambda : E \to E$ such that

$$\Lambda g(x) := \gamma g\Big(\frac{x}{\gamma}\Big)$$

for all $x \in A$. Now, we have

$$d(\Lambda g, \Lambda h) \leq L d(g, h)$$

for all $g, h \in E$. Let $g, h \in E$ and $C \in [0, \infty]$ be an arbitrary constant with $d(g, h) \leq C$. From the definition of $d$, we have

$$\left\|\big(g(x_1) - h(x_1), \cdots, g(x_k) - h(x_k)\big)\right\|_k \leq C\varphi\big(\overbrace{x_1, \cdots, x_1}^{p+d}, \cdots, \overbrace{x_k, \cdots, x_k}^{p+d}\big)$$

for all $x_1, \cdots, x_k \in A$. By the assumption and the last inequality, we have

$$\left\|\big(\Lambda g(x_1) - \Lambda h(x_1), \cdots, \Lambda g(x_k) - \Lambda h(x_k)\big)\right\|_k$$

$$= \gamma \left\|\Big(g\Big(\frac{x_1}{\gamma}\Big) - h\Big(\frac{x_1}{\gamma}\Big), \cdots, g\Big(\frac{x_k}{\gamma}\Big) - h\Big(\frac{x_k}{\gamma}\Big)\Big)\right\|_k$$

$$\leq C\gamma\varphi\big(\overbrace{\frac{x_1}{\gamma}, \cdots, \frac{x_1}{\gamma}}^{p+d}, \cdots, \overbrace{\frac{x_k}{\gamma}, \cdots, \frac{x_k}{\gamma}}^{p+d}\big)$$

$$\leq CL\varphi\big(\overbrace{x_1, \cdots, x_1}^{p+d}, \cdots, \overbrace{x_k, \cdots, x_k}^{p+d}\big)$$

for all $x_1, \cdots, x_k \in A$ and so

$$d(\Lambda g, \Lambda h) \leq L d(g, h)$$

for any $g, h \in E$. It follows from (6.119) that $d(\Lambda f, f) \leq \frac{1}{2\gamma}$. Therefore, according to Theorem 1.3, the sequence $\{\Lambda^n f\}$ converges to a fixed point $H$ of $\Lambda$, i.e., $H : A \to B$ is a mapping defined by

$$H(x) = \lim_{n \to \infty} (\Lambda^n f)(x) = \lim_{n \to \infty} \gamma^n f\left(\frac{x}{\gamma^n}\right) \tag{6.120}$$

for all $x \in A$.

The rest of the proof is similar to the proof of Theorem 6.46 and so we omit it. This completes the proof.                    □

**Theorem 6.48.** *Let $r$ and $\theta$ be non-negative real numbers such that $r \notin [1, 3]$ and $((B^k, \| \cdot \|_k) : k \geq 1)$ be a multi-$C^*$-ternary algebra. Let $f : A \to B$ be a mapping such that*

$$\left\|\Big(C_\mu f(x_{11}, \cdots, x_{1p}, y_{11}, \cdots, y_{1d}),\right.$$

$$\left. \cdots, C_\mu f(x_{k1}, \cdots, x_{kp}, y_{k1}, \cdots, y_{kd})\Big)\right\|_k \tag{6.121}$$

$$\leq \theta\left(\sum_{j=1}^{p} \|x_{1j}\|_A^r + \sum_{j=1}^{d} \|y_{1j}\|_A^r + \cdots + \sum_{j=1}^{p} \|x_{kj}\|_A^r + \sum_{j=1}^{d} \|y_{kj}\|_A^r\right)$$

*and*

$$\left\|\Big(f([x_1, y_1, z_1]) - [f(x_1), f(y_1), f(z_1)],\right.$$

$$\left. \cdots, f([x_k, y_k, z_k]) - [f(x_k), f(y_k), f(z_k)]\Big)\right\|_k \tag{6.122}$$

$$\leq \theta(\|x_1\|_A^r \cdot \|y_1\|_A^r \cdot \|z_1\|_A^r + \cdots + \|x_k\|_A^r \cdot \|y_k\|_A^r \cdot \|z_k\|_A^r)$$

*for all $\mu \in \mathbb{T}^1$ and $x_{11}, \cdots, x_{1p}, y_{11}, \cdots, y_{1d}, \cdots, x_{k1}, \cdots, x_{kp}, y_{k1}, \cdots, y_{kd}$, $x_1, \cdots, x_k, y_1, \cdots, y_k, z_1, \cdots, z_k \in A$. Then there exists a unique $C^*$-ternary algebra homomorphism $H : A \to B$ such that*

$$\left\|\Big(f(x_1) - H(x_1), \cdots, f(x_k) - H(x_k)\Big)\right\|_B$$

$$\leq \frac{2^r(p+d)\theta}{|2(p+2d)^r - (p+2d)2^r|}(\|x_1\|_A^r + \cdots + \|x_k\|_A^r) \tag{6.123}$$

*for all $x_1, \cdots, x_k \in A$.*

*Proof.* The proof follows from Theorem 6.46 by taking

$$\varphi(x_{11}, \cdots, x_{1p}, y_{11}, \cdots, y_{1d}, \cdots, x_{k1}, \cdots, x_{kp}, y_{k1}, \cdots, y_{kd})$$

$$:= \theta\left(\sum_{j=1}^{p} \|x_{ij}\|_A^r + \sum_{j=1}^{d} \|y_{ij}\|_A^r + \cdots + \sum_{j=1}^{p} \|x_{kj}\|_A^r + \sum_{j=1}^{d} \|y_{kj}\|_A^r\right),$$

$$\psi(x_1, y_1, z_1, \cdots, x_k, y_k, z_k)$$

$$:= \theta(\|x_1\|_A^r \cdot \|y_1\|_A^r \cdot \|z_1\|_A^r + \cdots + \|x_k\|_A^r \cdot \|y_k\|_A^r \cdot \|z_k\|_A^r)$$

for all $\mu \in \mathbb{T}^1$ and $x_{11}, \cdots, x_{1p}, y_{11}, \cdots, y_{1d}, \cdots, x_{k1}, \cdots, x_{kp}, y_{k1}, \cdots, y_{kd},$
$x_1, \cdots, x_k, y_1, \cdots, y_k, z_1, \cdots, z_k \in A$, $L = 2^{1-r}(p + 2d)^{r-1}$, when $0 < r < 1$, and
$L = 2 - 2^{1-r}(p + 2d)^{r-1}$, when $r > 3$. This completes the proof.          $\square$

**Theorem 6.49.** *Let* $((B^k, \|\cdot\|_k) : k \geq 1)$ *be a multi-$C^*$-ternary algebra. Let*
$f : A \to B$ *be a mapping for which there exist the functions* $\varphi : A^{(p+d)k} \to [0, \infty)$
*and* $\psi : A^{3k} \to [0, \infty)$ *such that*

$$\lim_{n\to\infty} d^{-n}\varphi(d^n x_{11}, \cdots, d^n x_{1p}, d^n y_{11}, \cdots, d^n y_{1p},$$

$$\cdots, d^n x_{k1}, \cdots, d^n x_{kp}, \cdots, d^n y_{k1}, \cdots, d^n y_{kd}) = 0, \qquad (6.124)$$

$$\|(C_\mu f(x_{11}, \cdots, x_{1p}, y_{11}, \cdots, y_{1d}),$$

$$\cdots, C_\mu f(x_{k1}, \cdots, x_{kp}, y_{k1}, \cdots, y_{kd}))\|_k \qquad (6.125)$$

$$\leq \varphi(x_{11}, \cdots, x_{1p}, y_{11}, \cdots, y_{1d}, \cdots, x_{k1}, \cdots, x_{kp}, y_{k1}, \cdots, y_{kd}),$$

$$\|(f([x_1, y_1, z_1]) - [f(x_1), f(y_1), f(z_1)],$$

$$\cdots, f([x_k, y_k, z_k]) - [f(x_k), f(y_k), f(z_k)])\|_k$$

$$\leq \psi(x_1, y_1, z_1, \cdots, x_k, y_k, z_k), \qquad (6.126)$$

$$\lim_{n\to\infty} d^{-3n}\psi(d^n x_1, d^n y_1, d^n z_1, \cdots, d^n x_k, d^n y_k, d^n z_k) = 0 \qquad (6.127)$$

*and*

$$\lim_{n\to\infty} d^{-2n}\psi(d^n x_1, d^n y_1, z_1, \cdots, d^n x_k, d^n y_k, z_k) = 0 \qquad (6.128)$$

*for all* $\mu \in \mathbb{T}^1$ *and* $x_{11}, \cdots, x_{1p}, y_{11}, \cdots, y_{1d}, \cdots, x_{k1}, \cdots, x_{kp}, y_{k1}, \cdots, y_{kd},$
$x_1, \cdots, x_k, y_1, \cdots, y_k, z_1, \cdots, z_k \in A$, *where* $\gamma = \frac{p+2d}{2}$. *If there exists* $L < 1$ *such*
*that*

$$\varphi\left(\overbrace{dx_1, \cdots, dx_1}^{p+d}, \overbrace{dx_2, \cdots, dx_2}^{p+d}, \cdots, \overbrace{dx_k, \cdots, dx_k}^{p+d}\right) \qquad (6.129)$$

$$\leq dL\varphi\left(\overbrace{0, \cdots, 0}^{p}, \overbrace{x_1, \cdots, x_1}^{d}, \overbrace{0, \cdots, 0}^{p}, \overbrace{x_2, \cdots, x_2}^{d}, \cdots, \overbrace{0, \cdots, 0}^{p}, \overbrace{x_k, \cdots, x_k}^{d}\right)$$

*for all $x_1, x_2, \cdots, x_k \in A$, then there exists a unique homomorphism $H : A \to B$ such that*

$$\|(f(x_1) - H(x_1), \cdots, f(x_k) - H(x_k))\|_k$$

$$\leq \frac{1}{(1-L)2d} \varphi\Big( \overbrace{0, \cdots, 0}^{p}, \overbrace{x_1, \cdots, x_1}^{d}, \overbrace{0, \cdots, 0}^{p}, \qquad\qquad (6.130)$$

$$\overbrace{x_2, \cdots, x_2}^{d}, \cdots, \overbrace{0, \cdots, 0}^{p}, \overbrace{x_k, \cdots, x_k}^{d} \Big)$$

*for all $x_1, \cdots, x_k \in A$.*

*Proof.* Let $\mu = 1$ and $x_{ij} = 0, \quad y_{ij} = x_i$ for $1 \leq i \leq k$ in (6.125). Then we get

$$\|(f(dx_1) - df(x_1), \cdots, f(dx_k) - df(x_k))\|_k \qquad\qquad (6.131)$$

$$\leq \frac{1}{2}\varphi\Big( \overbrace{0, \cdots, 0}^{p}, \overbrace{x_1, \cdots, x_1}^{d}, \overbrace{0, \cdots, 0}^{p}, \overbrace{x_2, \cdots, x_2}^{d}, \cdots, \overbrace{0, \cdots, 0}^{p}, \overbrace{x_k, \cdots, x_k}^{d} \Big)$$

for all $x_1, \cdots, x_k \in A$. Consider the set $E := \{g : A \to B\}$ and introduce the generalized metric on $E$ as follows:

$$d(g, h)$$

$$= \inf\{C \in \mathbb{R}_+ : \|(g(x_1) - h(x_1), \cdots, g(x_k) - h(x_k))\|_k, \ \forall x_1, \cdots, x_k \in A\}$$

$$\leq C\varphi\Big( \overbrace{0, \cdots, 0}^{p}, \overbrace{x_1, \cdots, x_1}^{d}, \overbrace{0, \cdots, 0}^{p}, \overbrace{x_2, \cdots, x_2}^{d}, \cdots, \overbrace{0, \cdots, 0}^{p}, \overbrace{x_k, \cdots, x_k}^{d} \Big),$$

which $(E, d)$ is complete.

Now, we consider the linear mapping $\Lambda : E \to E$ defined by

$$\Lambda g(x) := \frac{1}{d} g(dx)$$

for all $x \in A$. Now, we have

$$d(\Lambda g, \Lambda h) \leq L d(g, h)$$

for all $g, h \in E$. Let $g, h \in E$ and let $C \in [0, \infty]$ be an arbitrary constant with $d(g, h) \leq C$. From the definition of $d$, we have

$$\|(g(x_1) - h(x_1), \cdots, g(x_k) - h(x_k))\|_k$$

$$\leq C\varphi\Big( \overbrace{0, \cdots, 0}^{p}, \overbrace{x_1, \cdots, x_1}^{d}, \cdots, \overbrace{0, \cdots, 0}^{p}, \overbrace{x_k, \cdots, x_k}^{d} \Big)$$

for all $x_1, \cdots, x_k \in A$. By the assumption and the last inequality, we have

$$\|(\Lambda g(x_1) - \Lambda h(x_1), \cdots, \Lambda g(x_k) - \Lambda h(x_k))\|_k$$
$$= \frac{1}{d}\|(g(dx_1) - h(dx_1), \cdots, (g(dx_k) - h(dx_k)\|_k$$
$$\leq \frac{C}{d}\varphi\Big(\overbrace{0,\cdots,0}^{p}, \overbrace{dx_1, \cdots, dx_1}^{d}, \cdots, \overbrace{0,\cdots,0}^{p}, \overbrace{dx_k, \cdots, dx_k}^{d}\Big)$$
$$\leq CL\varphi\Big(\overbrace{0,\cdots,0}^{p}, \overbrace{x_1, \cdots, x_1}^{d}, \cdots, \overbrace{0,\cdots,0}^{p}, \overbrace{x_k, \cdots, x_k}^{d}\Big)$$

for all $x_1, \cdots, x_k \in A$. Thus we have

$$\|(\Lambda f(x_1) - f(x_1), \cdots, \Lambda f(x_k) - f(x_k))\|_k$$
$$= \Big\|\big(\frac{1}{d}f(dx_1) - f(x_1), \cdots, \frac{1}{d}f(dx_k) - f(x_k)\big)\Big\|_k$$
$$= \frac{1}{d}\|(f(dx_1) - df(x_1), \cdots, f(dx_k) - df(x_k))\|_k$$
$$\leq \frac{1}{2d}\varphi\Big(\overbrace{0,\cdots,0}^{p}, \overbrace{x_1, \cdots, x_1}^{d}, \cdots, \overbrace{0,\cdots,0}^{p}, \overbrace{x_k, \cdots, x_k}^{d}\Big)$$

for all $x_1, \cdots, x_k \in A$. Hence $d(\Lambda f, f) \leq \frac{1}{2d}$. By Theorem 1.3, the sequence $\{\Lambda^n f\}$ converges to a fixed point $H$ of $\Lambda$, i.e., $H : A \to B$ is a mapping defined by

$$H(x) = \lim_{n\to\infty} (\Lambda^n f)(x) = \lim_{n\to\infty} \frac{1}{d^n} f(d^n x) \tag{6.132}$$

and $H(dx) = dH(x)$ for all $x \in A$. Also, $H$ is the unique fixed point of $\Lambda$ in the set $E' = \{g \in E : d(f, g) < \infty\}$ and

$$d(H, f) \leq \frac{1}{1-L}d(\Lambda f, f) \leq \frac{1}{(1-L)2d},$$

i.e., the inequality (6.130) holds for all $x_1, \cdots, x_k \in A$. It follows from the definition of $H$, (6.124) and (6.125) that

$$\Big\|\Big(2H\Big(\frac{\sum_{j=1}^{p}\mu x_{1j}}{2} + \sum_{j=1}^{d}\mu y_{1j}\Big) - \sum_{j=1}^{p}\mu H(x_{1j}) - 2\sum_{j=1}^{d}\mu H(y_{1j}),$$
$$\cdots, 2H\Big(\frac{\sum_{j=1}^{p}\mu x_{kj}}{2} + \sum_{j=1}^{d}\mu y_{kj}\Big) - \sum_{j=1}^{p}\mu H(x_{kj}) - 2\sum_{j=1}^{d}\mu H(y_{kj})\Big)\Big\|_k$$

$$= \lim_{n\to\infty} \frac{1}{d^n} \left\| \left( 2f\left(d^n \frac{\sum_{j=1}^p \mu x_{1j}}{2} + d^n \sum_{j=1}^d \mu y_{1j}\right) \right. \right.$$

$$- \sum_{j=1}^p \mu f(d^n x_{1j}) - 2 \sum_{j=1}^d \mu f(d^n y_{1j}),$$

$$\cdots, 2f\left(d^n \frac{\sum_{j=1}^p \mu x_{kj}}{2} + d^n \sum_{j=1}^d \mu y_{kj}\right)$$

$$\left. \left. - \sum_{j=1}^p \mu f(d^n x_{kj}) - 2 \sum_{j=1}^d \mu f(d^n y_{kj})\right) \right\|_k$$

$$\leq \lim_{n\to\infty} \frac{1}{d^n} \left\| \left( C_\mu f(d^n x_{11}, \cdots, d^n x_{1p}, d^n y_{11}, \cdots, d^n y_{1d}), \right. \right.$$

$$\left. \left. \cdots, C_\mu f(d^n x_{k1}, \cdots, d^n x_{kp}, d^n y_{k1}, \cdots, d^n y_{kd})\right) \right\|_k$$

$$+ \lim_{n\to\infty} \frac{1}{d^n} \varphi(d^n x_{11}, \cdots, d^n x_{1p}, d^n y_{11}, \cdots, d^n y_{1d},$$

$$\cdots, d^n x_{k1}, \cdots, d^n x_{kp}, d^n y_{k1}, \cdots, d^n y_{kd}) = 0$$

for all $\mu \in \mathbb{T}^1$ and $x_{11}, \cdots, x_{1p}, y_{11}, \cdots, y_{1d}, \cdots, x_{k1}, \cdots, x_{kp}, y_{k1}, \cdots, y_{kd} \in A$. Hence we have

$$2H\left(\frac{\sum_{j=1}^p \mu x_{ij}}{2} + \sum_{j=1}^d \mu y_{ij}\right) = \sum_{j=1}^p \mu H(x_{ij}) + 2 \sum_{j=1}^d \mu H(y_{ij})$$

for all $\mu \in \mathbb{T}^1$, $x_{11}, \cdots, x_{1p}, y_{11}, \cdots, y_{1d}, \cdots, x_{k1}, \cdots, x_{kp}, y_{k1}, \cdots, y_{kd} \in A$ and $1 \leq i \leq k$ and so

$$H(\lambda x + \mu y) = \lambda H(x) + \mu H(y)$$

for all $\lambda, \mu \in \mathbb{T}^1$ and $x, y \in A$. Therefore, by Lemma 3.12, the mapping $H : A \to B$ is $\mathbb{C}$-linear.

On the other hand, it follows from (6.126) and (6.127) that

$$\left\| H([x_1, y_1, z_1]) - [H(x_1), H(y_1), H(z_1)], \right.$$

$$\left. \cdots, H([x_k, y_k, z_k]) - [H(x_k), H(y_k), H(z_k)] \right\|_k$$

$$= \lim_{n \to \infty} \frac{1}{d^{3n}} \left\| f\left( [d^n x_1, d^n y_1, d^n z_1] \right) \right.$$

$$- \left[ f(d^n x_1), f(d^n y_1), f(d^n z_1) \right],$$

$$\cdots, f\left( [d^n x_k, d^n y_k, d^n z_k] \right)$$

$$- \left. \left[ f(d^n x_k), f(d^n y_k), f(d^n z_k) \right] \right\|_k$$

$$\leq \lim_{n \to \infty} \frac{1}{d^{3n}} \psi(d^n x_1, d^n y_1, d^n z_1, \cdots, d^n x_k, d^n y_k, d^n z_k)$$

$$= 0$$

for all $x_1, y_1, z_1, \cdots, x_k, y_k, z_k \in A$. Thus

$$H([x, y, z]) = [H(x), H(y), H(z)]$$

for all $x, y, z \in A$. Thus $H : A \to B$ is a multi-$C^*$-ternary algebra homomorphism satisfying (6.129).

Now, let $T : A \to B$ be another $C^*$-ternary algebras homomorphism satisfying (6.130). Since $d(f, T) \leq \frac{1}{(1-L)2d}$ and $T$ is $\mathbb{C}$-linear, we have $T \in E'$ and $(\Lambda T)(x) = \frac{1}{d}(T\gamma x) = T(x)$ for all $x \in A$, i.e., $T$ is a fixed point of $\Lambda$. Since $H$ is the unique fixed point of $\Lambda \in E'$, we have $H = T$. This completes the proof. $\square$

**Theorem 6.50.** *Let $r, s, \theta$ be nonnegative real numbers such that $0 < r \neq 1$, $0 < s \neq 3$ and let $d \geq 2$. Suppose that $f : A \to B$ is a mapping with $f(0) = 0$ satisfying (6.121) and*

$$\left\| \left( f([x_1, y_1, z_1]) - [f(x_1), f(y_1), f(z_1)], \right. \right.$$

$$\cdots, f([x_k, y_k, z_k]) - [f(x_k), f(y_k), f(z_k)] \right) \Big\|_k \tag{6.133}$$

$$\leq \theta \left( \|x_1\|_A^s \cdot \|y_1\|_A^s \cdot \|z_1\|_A^s + \cdots + \|x_k\|_A^s \cdot \|y_k\|_A^s \cdot \|z_k\|_A^s \right)$$

*for all $\mu \in \mathbb{T}^1$ and $x_1, \cdots, x_k, y_1, \cdots, y_k, z_1, \cdots, z_k \in A$. Then there exists a unique $C^*$–ternary algebra homomorphism $H : A \to B$ such that*

$$\left\| \left( f(x_1) - H(x_1), \cdots, f(x_k) - H(x_k) \right) \right\|_K$$

$$\leq \frac{d\theta}{2|d - d^r|} \left( \|x_1\|_A^r + \cdots + \|x_K\|_A^r \right) \tag{6.134}$$

*for all $x_1, \cdots, x_k \in A$*

*Proof.* Let $0 < r < 1$ and $0 < s < 3$. Similarly, one can prove the theorem for other cases. The proof follows from Theorem 6.49 by taking

$$\varphi(x_{11}, \cdots, x_{1p}, y_{11}, \cdots, y_{1d}, \cdots, x_{k1}, \cdots, x_{kp}, y_{k1}, \cdots, y_{kd})$$

$$:= \theta\left( \sum_{j=1}^{p} \|x_{1j}\|_A^r + \sum_{j=1}^{d} \|y_{1j}\|_A^r + \cdots + \sum_{j=1}^{p} \|x_{kj}\|_A^r + \sum_{j=1}^{d} \|y_{kj}\|_A^r \right),$$

$$\psi(x_1, y_1, z_1, \cdots, x_k, y_k, z_k)$$

$$:= \theta(\|x_1\|_A^s \cdot \|y_1\|_A^s \cdot \|z_1\|_A^s + \cdots + \|x_k\|_A^s \cdot \|y_k\|_A^s \cdot \|z_k\|_A^s)$$

for all $\mu \in \mathbb{T}^1$ and $x_{11}, \cdots, x_{1p}, y_{11}, \cdots, y_{1d}, \cdots, x_{k1}, \cdots, x_{kp}, y_{k1}, \cdots, y_{kd}, x_1,$ $\cdots, x_k, y_1, \cdots, y_k, z_1, \cdots, z_k \in A$ and $L = d^{r-1}$, when $0 < r < 1$ and $0 < s < 3$ and $L = 2 - d^{r-1}$, when $r > 1$ and $s > 3$. This completes the proof. $\qquad\square$

Now, assume that $A$ is a unital multi-$C^*$-ternary algebra with the norm $\| \cdot \|$ and the unit $e$ and $B$ is a unital $C^*$-ternary algebra with the norm $\| \cdot \|$ and the unit $e'$.

We investigate homomorphisms in multi-$C^*$-ternary algebras associated with the following functional equation:

$$C_\mu f(x_1, \cdots, x_p, y_1, \cdots, y_d) = 0.$$

**Theorem 6.51.** *Let $r < 1$, $\theta$ be nonnegative real numbers and $f : A \to B$ be a mapping satisfying (6.121) and (6.122). If there exist a real number $\lambda > 1$ (resp., $0 < \lambda < 1$) and an element $x_0 \in A$ such that $\lim_{n\to\infty} \frac{1}{\lambda^n} f(\lambda^n x_0) = e'$ (resp., $\lim_{n\to\infty} \lambda^n f(\frac{x_0}{\lambda^n}) = e'$), then the mapping $f : A \to B$ is a multi-$C^*$-ternary algebra homomorphism.*

*Proof.* By using the proof of Theorem 6.48, there exists a unique multi-$C^*$-ternary algebra homomorphism $H : A \to B$ satisfying (6.123). It follows from (6.123) that

$$H(x) = \lim_{n\to\infty} \frac{1}{\lambda^n} f(\lambda^n x) \quad \left( \text{resp., } H(x) = \lim_{n\to\infty} \lambda^n f\left(\frac{x}{\lambda^n}\right) \right)$$

for all $x \in A$ and $\lambda > 1$ ($0 < \lambda < 1$). Therefore, by the assumption, we get that $H(x_0) = e'$. Let $\lambda > 1$ and $\lim_{n\to\infty} \frac{1}{\lambda^n} f(\lambda^n x_0) = e'$. It follows from (6.122) that

$$\left\| \Big( [H(x_1), H(y_1), H(z_1)] - [H(x_1), H(y_1), f(z_1)], \right.$$

$$\cdots, [H(x_k), H(y_k), H(z_k)] - [H(x_k), H(y_k), f(z_k)] \Big) \right\|$$

$$= \left\| \Big( H[x_1, y_1, z_1] - [H(x_1), H(y_1), f(z_1)], \right.$$

$$\cdots, H[x_k, y_k, z_k] - [H(x_k), H(y_k), f(z_k)] \Big) \right\|$$

$$= \lim_{n\to\infty} \frac{1}{\lambda^{2n}} \left\| \left( f([\lambda^n x_1, \lambda^n y_1, z_1]) - [f(\lambda^n x_1), f(\lambda^n y_1), f(z_1)], \right. \right.$$

$$\left. \left. \cdots, f([\lambda^n x_k, \lambda^n y_k, z_k]) - [f(\lambda^n x_k), f(\lambda^n y_k), f(z_k)] \right) \right\|$$

$$\leq \lim_{n\to\infty} \frac{\lambda^{rn}}{\lambda^{3n}} \theta(\|x_1\|_A^r \cdot \|y_1\|_A^r \cdot \|z_1\|_A^r + \cdots + \|x_k\|_A^r \cdot \|y_k\|_A^r \cdot \|z_k\|_A^r)$$

$$= 0$$

for all $x_1, \cdots, x_k \in A$ and so

$$[H(x), H(y), H(z)] = [H(x), H(y), f(z)]$$

for all $x, y, z \in A$. Letting $x = y = x_0$ in the last equality, we get $f(z) = H(z)$ for all $z \in A$. Similarly, one can show that $H(x) = f(x)$ for all $x \in A$ when $0 < \lambda < 1$ and $\lim_{n\to\infty} \lambda^n f(\frac{x_0}{\lambda^n}) = e'$.

Similarly, one can show the theorem for the case $\lambda > 1$. Therefore, the mapping $f : A \to B$ is a multi-$C^*$-ternary algebra homomorphism. This completes the proof.                                                                                    □

*Remark 6.52.* Let $r > 1$, $\theta$ be nonnegative real numbers and $f : A \to B$ be a mapping satisfying (6.121) and (6.122). If there exist a real number $\lambda > 1$ (resp., $0 < \lambda < 1$) and an element $x_0 \in A$ such that $\lim_{n\to\infty} \frac{1}{\lambda^n} f(\lambda^n x_0) = e'$ (resp., $\lim_{n\to\infty} \lambda^n f(\frac{x_0}{\lambda^n}) = e'$), then the mapping $f : A \to B$ is a multi-$C^*$-ternary algebra homomorphism.

## 6.5.2   Stability of Derivations in Multi-$C^*$-Ternary Algebras

Assume that $A$ is a $C^*$-ternary algebra with the norm $\| \cdot \|$.

Park [231] proved the Hyers-Ulam stability of derivations on $C^*$-ternary algebras for the following functional equation:

$$C_\mu f(x_1, \cdots, x_p, y_1, \cdots, y_d) = 0.$$

For any mapping $f : A \to A$, let

$$\mathbf{D}f(x, y, z) = f([x, y, z]) - [f(x), y, z] - [x, f(y), z] - [x, y, f(z)]$$

for all $x, y, z \in A$.

**Theorem 6.53.** *Let $((A^k, \| \cdot \|_k) : k \in \mathbb{N})$ be a multi-$C^*$-ternary algebra. Let $f : A \to A$ be a mapping for which there exist the functions $\varphi : A^{(p+d)k} \to [0, \infty)$ and $\psi : A^{3k} \to [0, \infty)$ satisfying the inequalities (6.105), (6.106) and (6.108) such that*

$$\left\|\left(\mathbf{D}f(x_1, y_1, z_1), \cdots, \mathbf{D}f(x_k, y_k, z_k)\right)\right\|$$

$$\leq \psi(x_1, y_1, z_1, \cdots, x_k, y_k, z_k) \qquad (6.135)$$

for all $\mu \in \mathbb{T}^1$ and $x_{11}, \cdots, x_{1p}, y_{11}, \cdots, y_{1d}, \cdots, x_{k1}, \cdots, x_{kp}, y_{k1}, \cdots, y_{kd},$ $x_1, \cdots, x_k, y_1, \cdots, y_k, z_1, \cdots, z_k \in A$, where $\gamma = \frac{p+2d}{2}$. If there exists $L < 1$ such that

$$\varphi\left(\overbrace{\gamma x_1, \cdots, \gamma x_1}^{p+d}, \overbrace{\gamma x_2, \cdots, \gamma x_2}^{p+d}, \cdots, \overbrace{\gamma x_k, \cdots, \gamma x_k}^{p+d}\right)$$

$$\leq \gamma L\varphi\left(\overbrace{x_1, \cdots, x_1}^{p+d}, \overbrace{x_2, \cdots, x_2}^{p+d}, \cdots, \overbrace{x_k, \cdots, x_k}^{p+d}\right) \qquad (6.136)$$

for all $x_1, x_2, \cdots, x_k \in A$, then there exists a unique multi-$C^*$-ternary derivation $\delta : A \to B$ such that

$$\left\|\left(f(x_1) - \delta(x_1), \cdots, f(x_k) - \delta(x_k)\right)\right\|_k$$

$$\leq \frac{1}{(1-L)2\gamma}\varphi\left(\overbrace{x_1, \cdots, x_1}^{p+d}, \overbrace{x_2, \cdots, x_2}^{p+d}, \cdots, \overbrace{x_k, \cdots, x_k}^{p+d}\right) \qquad (6.137)$$

for all $x_1, \cdots, x_k \in A$.

*Proof.* By the same reasoning as in the proof of Theorem 6.46, there exists a unique $\mathbb{C}$-linear mapping $\delta : A \to A$ satisfying (6.135). The mapping $\delta : A \to A$ is given by

$$\delta(x) = \lim_{n\to\infty} (\Lambda^n f)(x) = \lim_{n\to\infty} \frac{1}{\gamma^n} f(\gamma^n x) \qquad (6.138)$$

and $\delta(\gamma x) = \gamma \delta(x)$ for all $x \in A$. Also, $H$ is the unique fixed point of $\Lambda$ in the set $E' = \{g \in E : d(f, g) < \infty\}$ and

$$d(\delta, f) \leq \frac{1}{1-L}d(\Lambda f, f) \leq \frac{1}{(1-L)2\gamma},$$

i.e., the inequality (6.110) holds for all $x_1, \cdots, x_k \in A$. It follows from the definition of $\delta$, (6.105), (6.106) and (6.138) that

$$\left\|\left(C_\mu \delta(x_{11}, \cdots, x_{1p}y_{11}, \cdots y_{1d}), \cdots, C_\mu \delta(x_{k1}, \cdots, x_{kp}y_{k1}, \cdots y_{kd})\right)\right\|_k$$

$$= \lim_{n\to\infty} \frac{1}{\gamma^n}\left\|\left(C_\mu f(\gamma^n x_{11}, \cdots, \gamma^n x_{1p}, \gamma^n y_{11}, \cdots, \gamma^n y_{1d}),\right.\right.$$

$$\cdots, C_\mu f(\gamma^n x_{k1}, \cdots, \gamma^n x_{kp}, \gamma^n y_{k1}, \cdots, \gamma^n y_{kd})\Big)\Big\|_k$$

$$\leq \lim_{n \to \infty} \frac{1}{\gamma^n} \varphi(\gamma^n x_{11}, \cdots, \gamma^n x_{1p} \cdot \gamma^n y_{11}, \cdots, \gamma^n y_{1d},$$

$$\cdots, \gamma^n x_{k1}, \cdots, \gamma^n x_{kp}, \gamma^n y_{k1}, \cdots, \gamma^n y_{kd})$$

$$= 0$$

for all $\mu \in \mathbb{T}^1$ and $x_{11}, \cdots, x_{1p}, y_{11}, \cdots, y_{1d}, \cdots, x_{k1}, \cdots, x_{kp}, y_{k1}, \cdots, y_{kd} \in A$. Hence we have

$$2\delta\Big(\frac{\sum_{j=1}^p \mu x_{ij}}{2} + \sum_{j=1}^d \mu y_{ij}\Big) = \sum_{j=1}^p \mu \delta(x_{ij}) + 2 \sum_{j=1}^d \mu \delta(y_{ij})$$

for all $\mu \in \mathbb{T}^1$, $x_{11}, \cdots, x_{1p}, y_{11}, \cdots, y_{1d}, \cdots, x_{k1}, \cdots, x_{kp}, y_{k1}, \cdots, y_{kd} \in A$ and $1 \leq i \leq k$ and so

$$\delta(\lambda x + \mu y) = \lambda \delta(x) + \mu \delta(y)$$

for all $\lambda, \mu \in \mathbb{T}^1$ and $x, y \in A$. Therefore, by Lemma 3.12, the mapping $\delta : A \to B$ is $\mathbb{C}$-linear.

On the other hand, it follows from (6.108) and (6.135) that

$$\Big\|\Big(D\delta(x_1, y_1, z_1), \cdots, D\delta(x_k, y_k, z_k)\Big)\Big\|_k$$

$$= \lim_{n \to \infty} \frac{1}{\gamma^{3n}} \Big\| f\Big(Df(\gamma^n x_1, \gamma^n y_1, \gamma^n z_1), \cdots, f(\gamma^n x_k, \gamma^n y_k, \gamma^n z_k)\Big)\Big\|$$

$$\leq \lim_{n \to \infty} \frac{1}{\gamma^{3n}} \psi(\gamma^n x_1, \gamma^n y_1, \gamma^n z_1, \cdots, \gamma^n x_k, \gamma^n y_k, \gamma^n z_k)$$

$$= 0$$

for all $x_1, y_1, z_1, \cdots, x_k, y_k, z_k \in A$ and so

$$\begin{aligned}
(\delta([x_1, y_1, z_1]), &\cdots, \delta([x_k, y_k, z_k])) \\
&= ([\delta(x_1), (y_1), (z_1)] + [x_1, \delta(y_1), z_1] + [x_1, y_1, \delta(z_1)], \quad (6.139) \\
&\quad \cdots, [\delta(x_k), (y_k), (z_k)] + [x_k, \delta(y_k), z_k] + [x_k, y_k, \delta(z_k)])
\end{aligned}$$

for all $x, y, z \in A$ and so the mapping $\delta : A \to A$ is a $C^*$-ternary derivation. It follows from (6.135) and (6.108) that

$$\left\| \Big( \delta[x_1, y_1, z_1] - [\delta(x_1), y_1, z_1] - [x_1, \delta(y_1), z_1] - [x, y, f(z_1)], \right.$$

$$\left. \cdots, \delta[x_k, y_k, z_k] - [\delta(x_k), y_k, z_k] - [x_k, \delta(y_k), z_k] - [x, y, f(z_k)] \Big) \right\|$$

$$= \lim_{n \to \infty} \frac{1}{\gamma^{2n}} \left\| \Big( f[\gamma^n x_1, \gamma^n y_1, z_1] - [f(\gamma^n x_1), \gamma^n y_1, z_1] \right.$$

$$- [\gamma^n x_1, f(\gamma^n y_1), z_1] - [\gamma^n x_1, \gamma^n y_1, f(z_1)],$$

$$\cdots, f[\gamma^n x_k, \gamma^n y_k, z_k] - [f(\gamma^n x_k), \gamma^n y_k, z_k]$$

$$\left. - [\gamma^n x_k, f(\gamma^n y_k), z_k] - [\gamma^n x_k, \gamma^n y_k, f(z_k)] \Big) \right\|$$

$$\leq \lim_{n \to \infty} \frac{1}{\gamma^{2n}} \psi(\gamma^n x_1, \gamma^n y_1, z_1, \cdots, \gamma^n x_k, \gamma^n y_k, z_k)$$

$$= 0$$

for all $x_1, y_1, z_1, \cdots, x_k, y_k, z_k \in A$ and so

$$(\delta[x, y, z]) = [\delta(x), y, z] + [x, \delta(y), z] + [x, y, f(z)] \tag{6.140}$$

for all $x, y, z \in A$. Hence it follows from (6.139) and (6.140) that

$$[x, y, \delta(z)] = [x, y, f(z)] \tag{6.141}$$

for all $x, y, z \in A$. Letting $x = y = f(z) - \delta(z)$ in (6.141), we have

$$\|f(z) - \delta(z)\|^3 = \left\| \Big[ f(z) - \delta(z), f(z) - \delta(z), f(z) - \delta(z) \Big] \right\| = 0 \tag{6.142}$$

for all $z_1, \cdots, z_k \in A$ and hence $f(z) = \delta(z)$ for all $z \in A$. Therefore, the mapping $f : A \to A$ is a multi-$C^*$–ternary derivation. This completes the proof. $\qquad\square$

**Corollary 6.54.** *Let $r < 1$, $s < 2$, $\theta$ be non-negative real numbers and $f : A \to A$ be a mapping satisfying (6.121) and*

$$\left\| \Big( \mathbf{D}f(x_1, y_1, z_1), \cdots, \mathbf{D}f(x_k, y_k, z_k) \Big) \right\|$$

$$\leq \theta(\|x_1\|_A^s \cdot \|y_1\|_A^s \cdot \|z_1\|_A^s + \cdots + \|x_k\|_A^s \cdot \|y_k\|_A^s \cdot \|z_k\|_A^s)$$

*for all $x_1, y_1, z_1, \cdots, x_k, y_k, z_k \in A$. Then the mapping $f : A \to A$ is a multi-$C^*$–ternary derivation.*

*Proof.* Define

$$\varphi(x_{11}, \cdots, x_{1p}, y_{11}, \cdots, y_{1d}, \cdots, x_{k1}, \cdots, x_{kp}, y_{k1}, \cdots, y_{kd})$$

$$= \theta \Big( \sum_{j=1}^{p} \|x_{1j}\|_A^r + \sum_{j=1}^{d} \|y_{1j}\|_A^r, \cdots, \sum_{j=1}^{p} \|x_{kj}\|_A^r + \sum_{j=1}^{d} \|y_{kj}\|_A^r \Big)$$

and

$$\psi(x_1, y_1, z_1, \cdots, x_k, y_k, z_k)$$

$$= \theta\left(\|x_1\|_A^s \cdot \|y_1\|_A^s \cdot \|z_1\|_A^s + \cdots + \|x_k\|_A^s \cdot \|y_k\|_A^s \cdot \|z_k\|_A^s\right)$$

for all $x_1, y_1, z_1, \cdots, x_k, y_k, z_k, x_{11}, \cdots, x_{1p}, y_{11}, \cdots, y_{1d}, \cdots, x_{k1}, \cdots, x_{kp},$ $y_{k1}, \cdots, y_{kd} \in A$ and applying Theorem 6.53. Then we get the desired result. $\square$

**Theorem 6.55.** *Let $((A^k, \| \cdot \|_k) : k \geq 1)$ be a multi-$C^*$-ternary algebra. Let $f : A \rightarrow A$ be a mapping for which there exist the functions $\varphi : A^{(p+d)k} \rightarrow [0, \infty)$ and $\psi : A^{3k} \rightarrow [0, \infty)$ satisfying (6.106), (6.114), (6.115) and (6.135) for all $\mu \in \mathbb{T}^1$ and $x_{11}, \cdots, x_{1p}, y_{11}, \cdots, y_{1d}, \cdots, x_{k1}, \cdots, x_{kp}, y_{k1}, \cdots, y_{kd}, x_1, \cdots, x_k,$ $y_1, \cdots, y_k, z_1, \cdots, z_k \in A$, where $\gamma = \frac{p+2d}{2}$. If there exists $L < 1$ such that*

$$\varphi\left(\overbrace{\frac{x_1}{\gamma}, \cdots, \frac{x_1}{\gamma}}^{p+d}, \overbrace{\frac{x_2}{\gamma}, \cdots, \frac{x_2}{\gamma}}^{p+d}, \cdots, \overbrace{\frac{x_k}{\gamma}, \cdots, \frac{x_k}{\gamma}}^{p+d}\right) \qquad (6.143)$$

$$\leq \frac{L}{\gamma}\varphi\left(\overbrace{x_1, \cdots, x_1}^{p+d}, \overbrace{x_2, \cdots, x_2}^{p+d}, \cdots, \overbrace{x_k, \cdots, x_k}^{p+d}\right) \qquad (6.144)$$

*for all $x_1, x_2, \cdots, x_k \in A$, then there exists a unique multi-$C^*$-ternary algebra homomorphism $\delta : A \rightarrow A$ such that*

$$\|(f(x_1) - \delta(x_1), \cdots, f(x_k) - \delta(x_k)\|_k$$

$$\leq \frac{1}{(1-L)2\gamma}\varphi\left(\overbrace{x_1, \cdots, x_1}^{p+d}, \overbrace{x_2, \cdots, x_2}^{p+d}, \cdots, \overbrace{x_k, \cdots, x_k}^{p+d}\right) \qquad (6.145)$$

*for all $x_1, \cdots, x_k \in A$.*

*Proof.* By the same reasoning as in the proof of Theorem 6.47, there exists a unique $\mathbb{C}$-linear mapping $\delta : A \rightarrow A$ satisfying (6.135). The rest of the proof is similar to the proof of Theorem 6.53 and so we omit it. $\square$

# Chapter 7
# Stability of Functional Equations in Non-Archimedean Banach Algebras

In [203], Moslehian and Rassias proved the Hyers-Ulam stability of the Cauchy and quadratic functional equations in non-Archimedean normed spaces. After their results, some papers (see, for instance, [71, 83, 103]) on the stability of other equations in such spaces have been published.

Next, Eshaghi-Gordji et al. and Cho et al. applied the direct method or the fixed point method to prove the stability of some functional equations in non-Archimedean Banach algebras. In this chapter, we study the directions mentioned above and apply the fixed point method to show the Hyers-Ulam stability of some wide classes of functional equations in non-Archimedean Banach algebras and non-Archimedean $C^*$-algebras.

In Sect. 7.1, we extend the results presented in Chap. 3 to the setting of non-Archimedean $C^*$-algebras.

In Sect. 7.2, we extend the results presented in Chap. 2 to the setting of non-Archimedean $C^*$-algebras. In fact, by using the fixed point method, we prove the Hyers-Ulam stability of homomorphisms and derivations on non-Archimedean $C^*$-algebras and non-Archimedean Lie $C^*$-algebras for the following additive functional equation:

$$\sum_{i=1}^{m} f\left(mx_i + \sum_{j=1, j\neq i}^{m} x_j\right) + f\left(\sum_{i=1}^{m} x_i\right) = 2f\left(\sum_{i=1}^{m} mx_i\right)$$

for each $m \geq 2$.

© Springer International Publishing Switzerland 2015
Y.J. Cho et al., *Stability of Functional Equations in Banach Algebras*, DOI 10.1007/978-3-319-18708-2_7

## 7.1   Stability of Jensen Type Functional Equations: The Fixed Point Approach

Using the fixed point method, we prove the Hyers-Ulam stability of homomorphisms in non-Archimedean $C^*$-algebras and non-Archimedean Lie $C^*$-algebras and derivations on non-Archimedean $C^*$-algebras and non-Archimedean Lie $C^*$-algebras for the following Jensen type functional equation:

$$f\left(\frac{x+y}{2}\right) + f\left(\frac{x-y}{2}\right) = f(x).$$

### 7.1.1   Stability of Homomorphisms in Non-Archimedean C*-Algebras

Throughout this section, assume that $A$ is a non-Archimedean $C^*$-algebra with the norm $\|\cdot\|_A$ and $B$ is a non-Archimedean $C^*$-algebra with the norm $\|\cdot\|_B$.

For any mapping $f : A \rightarrow B$, we define

$$D_\mu f(x, y) := \mu f\left(\frac{x+y}{2}\right) + \mu f\left(\frac{x-y}{2}\right) - f(\mu x) \tag{7.1}$$

for all $\mu \in \mathbb{T}^1 := \{v \in \mathbb{C} : |v| = 1\}$ and $x, y \in A$.

Now, we prove the Hyers-Ulam stability of homomorphisms in non-Archimedean $C^*$-algebras for the functional equation $D_\mu f(x, y) = 0$.

**Theorem 7.1.** *Let $f : A \rightarrow B$ be a mapping for which there exist the functions $\varphi, \psi : A^2 \rightarrow [0, \infty)$ and $\eta : A \rightarrow [0, \infty)$ such that*

$$\|D_\mu f(x, y)\|_B \leq \varphi(x, y), \tag{7.2}$$

$$\|f(xy) - f(x)f(y)\|_B \leq \psi(x, y) \tag{7.3}$$

*and*

$$\|f(x^*) - f(x)^*\|_B \leq \eta(x) \tag{7.4}$$

*for all $\mu \in \mathbb{T}^1$ and $x, y \in A$. If there exists $L < 1$ such that $|2| < 1$ and*

$$\varphi(2x, 2y) \leq |2|L\varphi(x, y), \tag{7.5}$$

$$\psi(2x, 2y) \leq |4|L\psi(x, y) \tag{7.6}$$

*and*

$$\eta(2x) \leq |2|L\eta(x) \tag{7.7}$$

*for all* $x, y \in A$, *then there exists a unique non-Archimedean* $C^*$*-algebra homomorphism* $H : A \to B$ *such that*

$$\|f(x) - H(x)\|_B \le \frac{L}{1 - L} \varphi(x, 0) \tag{7.8}$$

*for all* $x \in A$.

*Proof.* It follows from (7.5), (7.6), (7.7) and $L < 1$ that

$$\lim_{n \to \infty} \frac{1}{|2|^n} \varphi(2^n x, 2^n y) = 0, \tag{7.9}$$

$$\lim_{n \to \infty} \frac{1}{|2|^{2n}} \psi(2^n x, 2^n y) = 0 \tag{7.10}$$

and

$$\lim_{n \to \infty} \frac{1}{|2|^n} \eta(2^n x) = 0 \tag{7.11}$$

for all $x, y \in A$. Consider the set $X := \{g : A \to B\}$ and introduce the generalized metric on $X$ as follows:

$$d(g, h) = \inf\{C \in \mathbb{R}_+ : \|g(x) - h(x)\|_B \le C\varphi(x, 0), \ \forall x \in A\},$$

which $(X, d)$ is complete.

Now, we consider the linear mapping $J : X \to X$ defined by

$$Jg(x) := \frac{1}{2} g(2x)$$

for all $x \in A$. Now, we have

$$d(Jg, Jh) \le Ld(g, h)$$

for all $g, h \in X$. Letting $\mu = 1$ and $y = 0$ in (7.2), we have

$$\left\| 2f\left(\frac{x}{2}\right) - f(x) \right\|_B \le \varphi(x, 0) \tag{7.12}$$

for all $x \in A$ and so

$$\left\| f(x) - \frac{1}{2} f(2x) \right\|_B \le \frac{1}{|2|} \varphi(2x, 0) \le L\varphi(x, 0)$$

for all $x \in A$. Hence $d(f, Jf) \le L$. By Theorem 1.3, there exists a mapping $H : A \to B$ such that

(1) $H$ is a fixed point of $J$, i.e.,

$$H(2x) = 2H(x) \tag{7.13}$$

for all $x \in A$. The mapping $H$ is a unique fixed point of $J$ in the set

$$Y = \{g \in X : d(f, g) < \infty\}.$$

This implies that $H$ is a unique mapping satisfying (7.13) such that there exists $C \in (0, \infty)$ such that

$$\|H(x) - f(x)\|_B \leq C\varphi(x, 0)$$

for all $x \in A$;

(2) $d(J^n f, H) \to 0$ as $n \to \infty$. This implies the equality

$$\lim_{n \to \infty} \frac{f(2^n x)}{2^n} = H(x) \tag{7.14}$$

for all $x \in A$;

(3) $d(f, H) \leq \frac{1}{1-L} d(f, Jf)$, which implies the inequality

$$d(f, H) \leq \frac{L}{1 - L}.$$

This implies that the inequality (7.8) holds.

It follows from (7.5) and (7.14) that

$$\left\| H\left(\frac{x+y}{2}\right) + H\left(\frac{x-y}{2}\right) - H(x) \right\|_B$$

$$= \lim_{n \to \infty} \frac{1}{|2|^n} \|f(2^{n-1}(x+y)) + f(2^{n-1}(x-y)) - f(2^n x)\|_B$$

$$\leq \lim_{n \to \infty} \frac{1}{|2|^n} \varphi(2^n x, 2^n y)$$

$$= 0$$

for all $x, y \in A$. Then we have

$$H\left(\frac{x+y}{2}\right) + H\left(\frac{x-y}{2}\right) = H(x) \tag{7.15}$$

for all $x, y \in A$. Letting $z = \frac{x+y}{2}$ and $w = \frac{x-y}{2}$ in (7.15), we have

$$H(z) + H(w) = H(z + w)$$

for all $z, w \in A$ and so the mapping $H : A \to B$ is Cauchy additive, i.e.,

$$H(z + w) = H(z) + H(w)$$

for all $z, w \in A$. Letting $y = x$ in (7.2), we get

$$\mu f(x) = f(\mu x)$$

for all $\mu \in \mathbb{T}^1$ and all $x \in A$. By the similar method as in above, we have

$$\mu H(x) = H(\mu x)$$

f or all $\mu \in \mathbb{T}^1$ and $x \in A$. Thus one can show that the mapping $H : A \to B$ is $\mathbb{C}$-linear. It follows from (7.6) that

$$
\begin{aligned}
\|H(xy) - H(x)H(y)\|_B &= \lim_{n \to \infty} \frac{1}{|4|^n} \|f(4^n xy) - f(2^n x)f(2^n y)\|_B \\
&\leq \lim_{n \to \infty} \frac{1}{|4|^n} \psi(2^n x, 2^n y) \\
&= 0
\end{aligned}
$$

for all $x, y \in A$ and so

$$H(xy) = H(x)H(y)$$

for all $x, y \in A$. It follows from (7.7) that

$$
\begin{aligned}
\|H(x^*) - H(x)^*\|_B &= \lim_{n \to \infty} \frac{1}{|2|^n} \|f(2^n x^*) - f(2^n x)^*\|_B \\
&\leq \lim_{n \to \infty} \frac{1}{|2|^n} \eta(2^n x) \\
&= 0
\end{aligned}
$$

for all $x \in A$ and so

$$H(x^*) = H(x)^*$$

for all $x \in A$. Thus $H : A \to B$ is a non-Archimedean $C^*$-algebra homomorphism satisfying (7.8). This completes the proof.                                                              $\square$

**Corollary 7.2.** *Let* $r < 1$, $\theta$ *be nonnegative real numbers and* $f : A \rightarrow B$ *be a mapping such that*

$$\|D_\mu f(x,y)\|_B \leq \theta(\|x\|_A^r + \|y\|_A^r), \tag{7.16}$$

$$\|f(xy) - f(x)f(y)\|_B \leq \theta(\|x\|_A^r + \|y\|_A^r) \tag{7.17}$$

*and*

$$\|f(x^*) - f(x)^*\|_B \leq \theta\|x\|_A^r \tag{7.18}$$

*for all* $\mu \in \mathbb{T}^1$ *and* $x, y \in A$. *Then there exists a unique non-Archimedean* $C^*$-*algebra homomorphism* $H : A \rightarrow B$ *such that*

$$\|f(x) - H(x)\|_B \leq \frac{|2|^r \theta}{|2| - |2|^r}\|x\|_A^r \tag{7.19}$$

*for all* $x \in A$.

*Proof.* The proof follows from Theorem 7.1 by taking

$$\varphi(x,y) = \psi(x,y) := \theta(\|x\|_A^r + \|y\|_A^r),$$

$$\eta(x) := \theta(\|x\|_A^r)$$

for all $x, y \in A$ and $L = |2|^{r-1}$.    □

**Theorem 7.3.** *Let* $f : A \rightarrow B$ *be a mapping for which there exist the functions* $\varphi, \psi : A^2 \rightarrow [0, \infty)$ *and* $\eta : A \rightarrow [0, \infty)$ *satisfying* (7.2), (7.3) *and* (7.4). *If there exists* $L < 1$ *such that* $|2| < 1$ *and*

$$|2|\varphi\left(\frac{x}{2}, \frac{y}{2}\right) \leq L\varphi(x,y), \tag{7.20}$$

$$|4|\psi\left(\frac{x}{2}, \frac{y}{2}\right) \leq L\psi(x,y) \tag{7.21}$$

*and*

$$|2|\eta\left(\frac{x}{2}\right) \leq L\eta(x) \tag{7.22}$$

*for all* $x, y \in A$, *then there exists a unique non-Archimedean* $C^*$-*algebra homomorphism* $H : A \rightarrow B$ *such that*

$$\|f(x) - H(x)\|_B \leq \frac{L}{|2| - |2|L}\varphi(x, 0) \tag{7.23}$$

*for all* $x \in A$.

*Proof.* It follows from (7.20), (7.21), (7.22) and $L < 1$ that

$$\lim_{n\to\infty} |2|^n \varphi\left(\frac{x}{2^n}, \frac{y}{2^n}\right) = 0,$$

$$\lim_{n\to\infty} |2|^{2n} \psi\left(\frac{x}{2^n}, \frac{y}{2^n}\right) = 0$$

and

$$\lim_{n\to\infty} |2|^n \eta\left(\frac{x}{2^n}\right) = 0$$

for all $x, y \in A$. We consider the linear mapping $J : X \to X$ defined by

$$Jg(x) := 2g\left(\frac{x}{2}\right)$$

for all $x \in A$. It follows from (7.12) that

$$\left\|f(x) - 2f\left(\frac{x}{2}\right)\right\|_B \leq \varphi\left(\frac{x}{2}, 0\right) \leq \frac{L}{|2|}\varphi(x, 0)$$

for all $x \in A$. Hence $d(f, Jf) \leq \frac{L}{|2|}$. By Theorem 1.3, there exists a mapping $H : A \to B$ such that

(1) $H$ is a fixed point of $J$, i.e.,

$$H(2x) = 2H(x) \tag{7.24}$$

for all $x \in A$. The mapping $H$ is a unique fixed point of $J$ in the set

$$Y = \{g \in X : d(f, g) < \infty\}.$$

This implies that $H$ is a unique mapping satisfying (7.24) such that there exists $C \in (0, \infty)$ such that

$$\|H(x) - f(x)\|_B \leq C\varphi(x, 0)$$

for all $x \in A$;

(2) $d(J^n f, H) \to 0$ as $n \to \infty$. This implies the equality

$$\lim_{n\to\infty} 2^n f\left(\frac{x}{2^n}\right) = H(x)$$

for all $x \in A$;

(3) $d(f, H) \leq \frac{1}{1-L}d(f, Jf)$, which implies the inequality

$$d(f, H) \leq \frac{L}{|2| - |2|L},$$

which implies that the inequality (7.23) holds.

The rest of the proof is similar to the proof of Theorem 7.1. This completes the proof. □

**Corollary 7.4.** *Let $r > 2$, $\theta$ be nonnegative real numbers and $f : A \rightarrow B$ be a mapping satisfying (7.16), (7.17) and (7.18). Then there exists a unique non-Archimedean $C^*$-algebra homomorphism $H : A \rightarrow B$ such that*

$$\|f(x) - H(x)\|_B \leq \frac{\theta}{|2|^r - |2|} \|x\|_A^r \tag{7.25}$$

*for all $x \in A$.*

*Proof.* The proof follows from Theorem 7.3 by taking

$$\varphi(x, y) = \psi(x, y) := \theta(\|x\|_A^r + \|y\|_A^r), \quad \eta(x) := \theta \|x\|_A^r$$

for all $x, y \in A$ and $L = |2|^{1-r}$. □

**Theorem 7.5.** *Let $f : A \rightarrow B$ be an odd mapping for which there exist the functions $\varphi, \psi : A^2 \rightarrow [0, \infty)$ and $\eta : A \rightarrow [0, \infty)$ satisfying (7.2), (7.3) and (7.4). If there exists $L < 1$ such that*

$$\varphi(x, 3x) \leq |2|L\varphi\left(\frac{x}{2}, \frac{3x}{2}\right)$$

*for all $x \in A$ and (7.5), (7.6) and (7.7) hold then there exists a unique non-Archimedean $C^*$-algebra homomorphism $H : A \rightarrow B$ such that*

$$\|f(x) - H(x)\|_B \leq \frac{1}{|2| - |2|L}\varphi(x, 3x) \tag{7.26}$$

*for all $x \in A$.*

*Proof.* Consider the set $X := \{g : A \rightarrow B\}$ and introduce the generalized metric on $X$ as follows:

$$d(g, h) = \inf\{C \in \mathbb{R}_+ : \|g(x) - h(x)\|_B \leq C\varphi(x, 3x), \ \forall x \in A\},$$

which $(X, d)$ is complete. Now, we consider the linear mapping $J : X \rightarrow X$ defined by

$$Jg(x) := \frac{1}{2}g(2x)$$

for all $x \in A$. Now, we have

$$d(Jg, Jh) \le Ld(g, h)$$

for all $g, h \in X$. Letting $\mu = 1$ and replacing $y$ by $3x$ in (7.2), we have

$$\|f(2x) - 2f(x)\|_B \le \varphi(x, 3x) \tag{7.27}$$

for all $x \in A$ and so

$$\left\| f(x) - \frac{1}{2} f(2x) \right\|_B \le \frac{1}{|2|} \varphi(x, 3x)$$

for all $x \in A$. Hence $d(f, Jf) \le \frac{1}{|2|}$. By Theorem 1.3, there exists a mapping $H : A \to B$ such that

(1) $H$ is a fixed point of $J$, i.e.,

$$H(2x) = 2H(x) \tag{7.28}$$

for all $x \in A$. The mapping $H$ is a unique fixed point of $J$ in the set

$$Y = \{g \in X : d(f, g) < \infty\}.$$

This implies that $H$ is a unique mapping satisfying (7.28) such that there exists $C \in (0, \infty)$ such that

$$\|H(x) - f(x)\|_B \le C\varphi(x, 3x)$$

for all $x \in A$;

(2) $d(J^n f, H) \to 0$ as $n \to \infty$. This implies the equality

$$\lim_{n \to \infty} \frac{f(2^n x)}{2^n} = H(x)$$

for all $x \in A$;

(3) $d(f, H) \le \frac{1}{1-L} d(f, Jf)$, which implies the inequality

$$d(f, H) \le \frac{1}{|2| - |2|L}.$$

This implies that the inequality (7.26) holds.

The rest of the proof is similar to the proof of Theorem 7.1. This completes the proof. $\qquad\square$

**Corollary 7.6.** *Let* $r < \frac{1}{2}$, $\theta$ *be nonnegative real numbers and* $f : A \to B$ *be an odd mapping such that*

$$\|D_\mu f(x, y)\|_B \leq \theta \cdot \|x\|_A^r \cdot \|y\|_A^r, \tag{7.29}$$

$$\|f(xy) - f(x)f(y)\|_B \leq \theta \cdot \|x\|_A^r \cdot \|y\|_A^r \tag{7.30}$$

*and*

$$\|f(x^*) - f(x)^*\|_B \leq \theta \|x\|_A^{2r} \tag{7.31}$$

*for all* $\mu \in \mathbb{T}^1$ *and* $x, y \in A$. *Then there exists a unique non-Archimedean* $C^*$-*algebra homomorphism* $H : A \to B$ *such that*

$$\|f(x) - H(x)\|_B \leq \frac{3^r\theta}{|2| - |2|^{2r}} \|x\|_A^{2r} \tag{7.32}$$

*for all* $x \in A$.

*Proof.* The proof follows from Theorem 7.5 by taking

$$\varphi(x, y) = \psi(x, y) := \theta \cdot \|x\|_A^r \cdot \|y\|_A^r, \quad \eta(x) := \theta \cdot \|x\|_A^r$$

for all $x, y \in A$ and $L = |2|^{2r-1}$.                              □

**Theorem 7.7.** *Let* $f : A \to B$ *be an odd mapping for which there exist the functions* $\varphi, \psi : A^2 \to [0, \infty)$ *and* $\eta : A \to [0, \infty)$ *satisfying* (7.2), (7.3), *and* (7.4). *If there exists* $L < 1$ *such that*

$$\varphi(x, 3x) \leq \frac{1}{|2|} L\varphi(2x, 6x)$$

*for all* $x \in A$, *and also* (7.20), (7.21) *and* (7.22) *hold, then there exists a unique non-Archimedean* $C^*$-*algebra homomorphism* $H : A \to B$ *such that*

$$\|f(x) - H(x)\|_B \leq \frac{L}{|2| - |2|L} \varphi(x, 3x) \tag{7.33}$$

*for all* $x \in A$.

*Proof.* We consider the linear mapping $J : X \to X$ defined by

$$Jg(x) := 2g\left(\frac{x}{2}\right)$$

for all $x \in A$. It follows from (7.27) that

$$\left\|f(x) - 2f(\frac{x}{2})\right\|_B \leq \varphi\left(\frac{x}{2}, \frac{3x}{2}\right) \leq \frac{L}{|2|}\varphi(x, 3x)$$

for all $x \in A$. Hence $d(f, Jf) \le \frac{L}{2}$. By Theorem 1.3, there exists a mapping $H : A \to B$ such that

(1) $H$ is a fixed point of $J$, i.e.,

$$H(2x) = 2H(x) \tag{7.34}$$

for all $x \in A$. The mapping $H$ is a unique fixed point of $J$ in the set

$$Y = \{g \in X : d(f, g) < \infty\}.$$

This implies that $H$ is a unique mapping satisfying (7.34) such that there exists $C \in (0, \infty)$ such that

$$\|H(x) - f(x)\|_B \le C\varphi(x, 3x)$$

for all $x \in A$;

(2) $d(J^n f, H) \to 0$ as $n \to \infty$. This implies the equality

$$\lim_{n \to \infty} 2^n f\left(\frac{x}{2^n}\right) = H(x)$$

for all $x \in A$;

(3) $d(f, H) \le \frac{1}{1-L} d(f, Jf)$, which implies the inequality

$$d(f, H) \le \frac{L}{2 - 2L},$$

which implies that the inequality (7.33) holds.

The rest of the proof is similar to the proof of Theorem 7.1. $\square$

**Corollary 7.8.** *Let $r > 1$, $\theta$ be nonnegative real numbers and $f : A \to B$ be an odd mapping satisfying (7.29), (7.30) and (7.31). Then there exists a unique non-Archimedean $C^*$-algebra homomorphism $H : A \to B$ such that*

$$\|f(x) - H(x)\|_B \le \frac{\theta}{|2|^{2r} - |2|} \|x\|_A^{2r} \tag{7.35}$$

*for all $x \in A$.*

*Proof.* The proof follows from Theorem 7.7 by taking

$$\varphi(x, y) = \psi(x, y) := \theta \cdot \|x\|_A^r \cdot \|y\|_A^r, \quad \eta(x) := \theta \cdot \|x\|_A^r$$

for all $x, y \in A$ and $L = |2|^{1-2r}$. $\square$

### 7.1.2  Stability of Derivations in Non-Archimedean
####      $C^*$-Algebras

Assume that $A$ is a non-Archimedean $C^*$-algebra with the norm $\| \cdot \|_A$.

Now, we prove the Hyers-Ulam stability of derivations in non-Archimedean $C^*$-algebras for the functional equation $D_\mu f(x, y) = 0$.

**Theorem 7.9.** *Let $f : A \to A$ be a mapping for which there exist the functions $\varphi, \psi : A^2 \to [0, \infty)$ such that*

$$\|D_\mu f(x, y)\|_A \le \varphi(x, y) \tag{7.36}$$

*and*

$$\|f(xy) - f(x)y - xf(y)\|_A \le \psi(x, y) \tag{7.37}$$

*for all $\mu \in \mathbb{T}^1$ and all $x, y \in A$. If there exists $L < 1$ such that*

$$\varphi(x, 0) \le |2|L\varphi\left(\frac{x}{2}, 0\right)$$

*for all $x \in A$, (7.5) and (7.6) hold. Then there exists a unique non-Archimedean derivation $\delta : A \to A$ such that*

$$\|f(x) - \delta(x)\|_A \le \frac{L}{1 - L}\varphi(x, 0) \tag{7.38}$$

*for all $x \in A$.*

*Proof.* By the same reasoning as in the proof of Theorem 7.1, there exists a unique involution $\mathbb{C}$-linear mapping $\delta : A \to A$ satisfying (7.38). Also, the mapping $\delta : A \to A$ is given by

$$\delta(x) = \lim_{n \to \infty} \frac{f(2^n x)}{2^n}$$

for all $x \in A$. It follows from (7.37) that

$$\begin{aligned}
\|\delta(xy) &- \delta(x)y - x\delta(y)\|_A \\
&= \lim_{n \to \infty} \frac{1}{|4|^n}\|f(4^n xy) - f(2^n x) \cdot 2^n y - 2^n xf(2^n y)\|_A \\
&\le \lim_{n \to \infty} \frac{1}{|4|^n}\psi(2^n x, 2^n y) \\
&= 0
\end{aligned}$$

for all $x, y \in A$. Then we have

$$\delta(xy) = \delta(x)y + x\delta(y)$$

for all $x, y \in A$. Thus $\delta : A \to A$ is a derivation satisfying (7.38). This completes the proof.    $\square$

**Corollary 7.10.** *Let $r < 1$, $\theta$ be nonnegative real numbers and $f : A \to A$ be a mapping such that*

$$\|D_\mu f(x, y)\|_A \leq \theta(\|x\|_A^r + \|y\|_A^r) \tag{7.39}$$

*and*

$$\|f(xy) - f(x)y - xf(y)\|_A \leq \theta(\|x\|_A^r + \|y\|_A^r) \tag{7.40}$$

*for all $\mu \in \mathbb{T}^1$ and $x, y \in A$. Then there exists a unique derivation $\delta : A \to A$ such that*

$$\|f(x) - \delta(x)\|_A \leq \frac{|2|^r \theta}{|2| - |2|^r} \|x\|_A^r \tag{7.41}$$

*for all $x \in A$.*

*Proof.* The proof follows from Theorem 7.9 by taking

$$\varphi(x, y) = \psi(x, y) := \theta(\|x\|_A^r + \|y\|_A^r)$$

for all $x, y \in A$ and $L = |2|^{r-1}$.                                       $\square$

*Remark 7.11.* Let $f : A \to A$ be a mapping for which there exist the functions $\varphi, \psi : A^2 \to [0, \infty)$ satisfying (7.36) and (7.37). If there exists $L < 1$ such that

$$\varphi(x, 0) \leq \frac{1}{|2|} L \varphi(2x, 0)$$

for all $x \in A$, (7.20) and (7.21) hold, then there exists a unique derivation $\delta : A \to A$ such that

$$\|f(x) - \delta(x)\|_A \leq \frac{L}{|2| - |2|L} \varphi(x, 0) \tag{7.42}$$

for all $x \in A$.

**Corollary 7.12.** *Let $r > 2$, $\theta$ be nonnegative real numbers and $f : A \to A$ be a mapping satisfying (7.39) and (7.40). Then there exists a unique derivation $\delta : A \to A$ such that*

$$\|f(x) - \delta(x)\|_A \leq \frac{\theta}{|2|^r - |2|} \|x\|_A^r \tag{7.43}$$

*for all $x \in A$.*

*Proof.* The proof follows from Remark 7.11 by taking

$$\varphi(x, y) = \psi(x, y) := \theta(\|x\|_A^r + \|y\|_A^r)$$

for all $x, y \in A$ and $L = |2|^{1-r}$.                                                        □

**Remark 7.13.** For the inequalities controlled by the product of powers of norms, one can obtain similar results to Theorems 7.5, 7.7 and Corollaries 7.6, 7.8.

### 7.1.3  Stability of Homomorphisms in Non-Archimedean Lie $C^*$-Algebras

A non-Archimedean $C^*$-algebra $\mathcal{C}$ endowed with the Lie product

$$[x, y] := \frac{xy - yx}{2}$$

on $\mathcal{C}$ is called a *non-Archimedean Lie $C^*$-algebra* (see [224, 225, 227]).

**Definition 7.14.** Let $A$ and $B$ be non-Archimedean Lie $C^*$-algebras. A $\mathbb{C}$-linear mapping $H : A \to B$ is called a *non-Archimedean Lie $C^*$-algebra homomorphism* if

$$H([x, y]) = [H(x), H(y)]$$

for all $x, y \in A$.

Throughout this section, assume that $A$ is a non-Archimedean Lie $C^*$-algebra with the norm $\| \cdot \|_A$ and $B$ is a non-Archimedean Lie $C^*$-algebra with the norm $\| \cdot \|_B$.

Now, we prove the Hyers-Ulam stability of homomorphisms in non-Archimedean Lie $C^*$-algebras for the functional equation $D_\mu f(x, y) = 0$.

**Theorem 7.15.** *Let $f : A \to B$ be a mapping for which there are functions $\varphi, \psi : A^2 \to [0, \infty)$ satisfying (7.2) such that*

$$\|f([x, y]) - [f(x), f(y)]\|_B \leq \psi(x, y) \tag{7.44}$$

*for all $x, y \in A$. If there exists $L < 1$ such that*

$$\varphi(x, 0) \leq |2|L\varphi\left(\frac{x}{2}, 0\right)$$

*for all $x \in A$, and also (7.5) and (7.6) hold, then there exists a unique non-Archimedean Lie $C^*$-algebra homomorphism $H : A \to B$ satisfying (7.8).*

*Proof.* By the same reasoning as in the proof of Theorem 7.1, there exists a unique $\mathbb{C}$-linear mapping $H : A \to B$ satisfying (7.8). Also, the mapping $H : A \to B$ is given by

$$H(x) = \lim_{n\to\infty} \frac{f(2^n x)}{2^n}$$

for all $x \in A$. It follows from (7.44) that

$$\|H([x, y]) - [H(x), H(y)]\|_B$$

$$= \lim_{n\to\infty} \frac{1}{|4|^n} \|f(4^n[x, y]) - [f(2^n x), f(2^n y)]\|_B$$

$$\leq \lim_{n\to\infty} \frac{1}{|4|^n} \psi(2^n x, 2^n y)$$

$$= 0$$

for all $x, y \in A$. Then we have

$$H([x, y]) = [H(x), H(y)]$$

for all $x, y \in A$. Thus $H : A \to B$ is a non-Archimedean Lie $C^*$-algebra homomorphism satisfying (7.8). This completes the proof. $\qquad\square$

**Corollary 7.16.** *Let $r < 1$, $\theta$ be nonnegative real numbers and $f : A \to B$ be a mapping satisfying (7.16) such that*

$$\|f([x, y]) - [f(x), f(y)]\|_B \leq \theta(\|x\|_A^r + \|y\|_A^r) \tag{7.45}$$

*for all $x, y \in A$. Then there exists a unique non-Archimedean Lie $C^*$-algebra homomorphism $H : A \to B$ satisfying (7.19).*

*Proof.* The proof follows from Theorem 7.15 by taking

$$\varphi(x, y) = \psi(x, y) := \theta(\|x\|_A^r + \|y\|_A^r)$$

for all $x, y \in A$ and $L = |2|^{r-1}$. $\qquad\square$

*Remark 7.17.* Let $f : A \to B$ be a mapping for which there exist the functions $\varphi, \psi : A^2 \to [0, \infty)$ and $\eta : A \to [0, \infty)$ satisfying (7.2), (7.5), (7.6) and (7.44). If there exists $L < 1$ such that

$$\varphi(x, 0) \leq \frac{1}{|2|} L \varphi(2x, 0)$$

for all $x \in A$, then there exists a unique non-Archimedean Lie $C^*$-algebra homomorphism $H : A \to B$ satisfying (7.23).

**Corollary 7.18.** *Let $r > 2$, $\theta$ be nonnegative real numbers and $f : A \to B$ be a mapping satisfying (7.16) and (7.45). Then there exists a unique non-Archimedean Lie $C^*$-algebra homomorphism $H : A \to B$ satisfying (7.25).*

*Proof.* The proof follows from Remark 7.17 by taking

$$\varphi(x, y) = \psi(x, y) := \theta(\|x\|_A^r + \|y\|_A^r)$$

for all $x, y \in A$ and $L = |2|^{1-r}$.                                              □

*Remark 7.19.* For the inequalities controlled by the product of powers of norms, one can obtain similar results to Theorems 7.5, 7.7 and their corollaries.

### 7.1.4  Stability of Non-Archimedean Lie Derivations in $C^*$-Algebras

First, we give the following definition on the non-Archimedean Lie derivation in a non-Archimedean Lie $C^*$-algebra.

**Definition 7.20.** Let $A$ be a non-Archimedean Lie $C^*$-algebra. A $\mathbb{C}$-linear mapping $\delta : A \to A$ is called a *non-Archimedean Lie derivation* if

$$\delta([x, y]) = [\delta(x), y] + [x, \delta(y)]$$

for all $x, y \in A$.

Assume that $A$ is a non-Archimedean Lie $C^*$-algebra with the norm $\| \cdot \|_A$.

Now, we prove the Hyers-Ulam stability of non-Archimedean Lie derivations on non-Archimedean Lie $C^*$-algebras for the functional equation

$$D_\mu f(x, y) = 0.$$

**Theorem 7.21.** *Let $f : A \to A$ be a mapping for which there exists a function $\varphi, \psi : A^2 \to [0, \infty)$ satisfying (7.5), (7.6) and (7.36) such that*

$$\|f([x, y]) - [f(x), y] - [x, f(y)]\|_A \leq \psi(x, y) \tag{7.46}$$

*for all $x, y \in A$. If there exists $L < 1$ such that*

$$\varphi(x, 0) \leq |2|L\varphi\left(\frac{x}{2}, 0\right)$$

*for all $x \in A$. Then there exists a unique non-Archimedean Lie derivation $\delta : A \to A$ satisfying (7.38).*

*Proof.* By the same reasoning as in the proof of Theorem 7.1, there exists a unique involution $\mathbb{C}$-linear mapping $\delta : A \to A$ satisfying (7.38). Also, the mapping $\delta : A \to A$ is given by

$$\delta(x) = \lim_{n \to \infty} \frac{f(2^n x)}{2^n}$$

for all $x \in A$. It follows from (7.44) that

$$
\begin{aligned}
\|\delta([x, y]) - [\delta(x), y] - [x, \delta(y)]\|_A \\
= \lim_{n \to \infty} \frac{1}{|4|^n} \|f(4^n[x, y]) - [f(2^n x), 2^n y] - [2^n x, f(2^n y)]\|_A \\
\leq \lim_{n \to \infty} \frac{1}{|4|^n} \psi(2^n x, 2^n y) \\
= 0
\end{aligned}
$$

for all $x, y \in A$. Then we have

$$\delta([x, y]) = [\delta(x), y] + [x, \delta(y)]$$

for all $x, y \in A$. Thus $\delta : A \to A$ is a non-Archimedean Lie derivation satisfying (7.38). This completes the proof. $\qquad\square$

**Corollary 7.22.** *Let $r < 1$, $\theta$ be nonnegative real numbers and $f : A \to A$ be a mapping satisfying (7.39) such that*

$$\|f([x, y]) - [f(x), y] - [x, f(y)]\|_A \leq \theta(\|x\|_A^r + \|y\|_A^r) \qquad (7.47)$$

*for all $x, y \in A$. Then there exists a unique non-Archimedean Lie derivation $\delta : A \to A$ satisfying (7.41).*

*Proof.* The proof follows from Theorem 7.15 by taking

$$\varphi(x, y) = \psi(x, y) := \theta(\|x\|_A^r + \|y\|_A^r), \qquad \eta(x) := \theta\|x\|_A^r$$

for all $x, y \in A$ and $L = |2|^{r-1}$. $\qquad\square$

*Remark 7.23.* Let $f : A \to A$ be a mapping for which there exist functions $\varphi, \psi : A^2 \to [0, \infty)$ and $\eta : A \to [0, \infty)$ satisfying (7.20), (7.21), (7.22), (7.36) and (7.46). If there exists $L < 1$ such that

$$\varphi(x, 0) \leq \frac{1}{|2|} L\varphi(2x, 0)$$

for all $x \in A$, then there exists a unique non-Archimedean Lie derivation $\delta : A \to A$ satisfying (7.42).

**Corollary 7.24.** *Let $r > 2$, $\theta$ be nonnegative real numbers and $f : A \to A$ be a mapping satisfying (7.39) and (7.47). Then there exists a unique non-Archimedean Lie derivation $\delta : A \to A$ satisfying (7.43).*

*Proof.* The proof follows from Remark 7.23 by taking

$$\varphi(x, y) = \psi(x, y) := \theta(\|x\|_A^r + \|y\|_A^r)$$

for all $x, y \in A$ and $L = |2|^{1-r}$. $\qquad\qquad\qquad\qquad\qquad\qquad\qquad\qquad\Box$

*Remark 7.25.* For the inequalities controlled by the product of powers of norms, one can obtain similar results to Theorems 7.5, 7.7 and their corollaries.

## 7.2   Stability for $m$-Variable Additive Functional Equations

Using the fixed point method, we prove the Hyers-Ulam stability of homomorphisms and derivations on non-Archimedean $C^*$-algebras and non-Archimedean Lie $C^*$-algebras for the following additive functional equation:

$$\sum_{i=1}^{m} f\left(mx_i + \sum_{j=1, j\neq i}^{m} x_j\right) + f\left(\sum_{i=1}^{m} x_i\right) = 2f\left(\sum_{i=1}^{m} mx_i\right) \qquad (7.48)$$

for each $m \geq 2$. For any mapping $f : A \to B$, we define

$$D_\mu f(x_1, \cdots, x_m)$$

$$:= \sum_{i=1}^{m} \mu f\left(mx_i + \sum_{j=1, j\neq i}^{m} x_j\right) + f\left(\mu \sum_{i=1}^{m} x_i\right) - 2f\left(\mu \sum_{i=1}^{m} mx_i\right)$$

for all $\mu \in \mathbb{T}^1 := \{v \in \mathbb{C} : |v| = 1\}$ and $x_1, \cdots, x_m \in A$.

### 7.2.1   Stability of Homomorphisms and Derivations in Non-Archimedean $C^*$-Algebras

Now, we prove the Hyers–Ulam stability of homomorphisms in non-Archimedean $C^*$-algebras for the functional equation $D_\mu f(x_1, \cdots, x_m) = 0$.

**Theorem 7.26.** *Let $f : A \to B$ be a mapping for which there exist the functions $\varphi : A^m \to [0, \infty)$, $\psi : A^2 \to [0, \infty)$ and $\eta : A \to [0, \infty)$ such that $|m| < 1$ is far from zero and*

$$\|D_\mu f(x_1, \cdots, x_m)\|_B \leq \varphi(x_1, \cdots, x_m), \qquad\qquad (7.49)$$

$$\|f(xy) - f(x)f(y)\|_B \leq \psi(x, y) \qquad\qquad\qquad (7.50)$$

*and*

$$\|f(x^*) - f(x)^*\|_B \leq \eta(x) \tag{7.51}$$

*for all $\mu \in \mathbb{T}^1$ and $x_1, \cdots, x_m, x, y \in A$. If there exists $L < 1$ such that*

$$\varphi(mx_1, \cdots, mx_m) \leq |m|L\varphi(x_1, \cdots, x_m), \tag{7.52}$$

$$\psi(mx, my) \leq |m|^2 L\psi(x, y) \tag{7.53}$$

*and*

$$\eta(mx) \leq |m|L\eta(x) \tag{7.54}$$

*for all $x, y, x_1, \cdots, x_m \in A$, then there exists a unique non-Archimedean $C^*$-algebra homomorphism $H : A \to B$ such that*

$$\|f(x) - H(x)\|_B \leq \frac{1}{|m| - |m|L}\varphi(x, 0, \cdots, 0) \tag{7.55}$$

*for all $x \in A$.*

*Proof.* It follows from (7.52), (7.53), (7.54) and $L < 1$ that

$$\lim_{n\to\infty} \frac{1}{|m|^n}\varphi(m^n x_1, \cdots, m^n x_m) = 0, \tag{7.56}$$

$$\lim_{n\to\infty} \frac{1}{|m|^{2n}}\psi(m^n x, m^n y) = 0 \tag{7.57}$$

and

$$\lim_{n\to\infty} \frac{1}{|m|^n}\eta(m^n x) = 0 \tag{7.58}$$

for all $x, y, x_1, \cdots, x_m \in A$. Let us define $\Omega$ to be the set of all mappings $g : A \to B$ and introduce a generalized metric on $\Omega$ as follows:

$$d(g, h) \tag{7.59}$$
$$= \inf\{k \in (0, \infty) : \|g(x) - h(x)\|_B < k\phi(x, 0, \cdots, 0), \ \forall x \in A\},$$

which $(\Omega, d)$ is a generalized complete metric space. Now, we consider the function $J : \Omega \to \Omega$ defined by $Jg(x) = \frac{1}{m}g(mx)$ for all $x \in A$ and $g \in \Omega$. Note that, for all $g, h \in \Omega$,

$$d(g, h) < k \implies \|g(x) - h(x)\|_B < k\phi(x, 0, \cdots, 0)$$

$$\implies \left\| \frac{1}{m} g(mx) - \frac{1}{m} h(mx) \right\|_B < \frac{k}{|m|} \phi(mx, 0, \cdots, 0)$$

$$\implies \left\| \frac{1}{m} g(mx) - \frac{1}{m} h(mx) \right\|_B < kL\phi(mx, 0, \cdots, 0)$$

$$\implies d(Jg, Jh) < kL. \tag{7.60}$$

From this, it is easy to see that

$$d(Jg, Jk) \leq Ld(g, h)$$

for all $g, h \in \Omega$, that is, $J$ is a self-function of $\Omega$ with the Lipschitz constant $L$. Putting $\mu = 1, x = x_1$ and $x_2 = x_3 = \cdots = x_m = 0$ in (7.49) we have

$$\|f(mx) - mf(x)\|_B \leq \phi(x, 0, \cdots, 0) \tag{7.61}$$

for all $x \in A$. Then

$$\left\| f(x) - \frac{1}{m} f(mx) \right\|_B \leq \frac{1}{|m|} \phi(x, 0, \cdots, 0) \tag{7.62}$$

for all $x \in A$, that is,

$$d(Jf, f) \leq \frac{1}{|m|} < \infty.$$

Now, from the fixed point method, it follows that there exists a fixed point $H$ of $J$ in $\Omega$ such that

$$H(x) = \lim_{n \to \infty} \frac{1}{m^n} f(m^n x) \tag{7.63}$$

for all $x \in A$ since $\lim_{n \to \infty} d(J^n f, H) = 0$.

On the other hand, it follows from (7.49), (7.56) and (7.63) that

$$\|D_\mu H(x_1, \cdots, x_m)\|_B = \lim_{n \to \infty} \left\| \frac{1}{m^n} Df(m^n x_1, \cdots, m^n x_m) \right\|_B$$

$$\leq \lim_{n \to \infty} \frac{1}{|m|^n} \phi(m^n x_1, \cdots, m^n x_m) \tag{7.64}$$

$$= 0.$$

By the similar method as in above, we have

$$\mu H(mx) = H(m\mu x)$$

for all $\mu \in \mathbb{T}^1$ and $x \in A$. Thus one can show that the mapping $H : A \to B$ is
$\mathbb{C}$-linear. It follows from (7.50), (7.57) and (7.63) that

$$
\begin{aligned}
\|H(xy) &- H(x)H(y)\|_B \\
&= \lim_{n\to\infty} \frac{1}{|m|^{2n}} \|f(m^{2n}xy) - f(m^n x)f(m^n y)\|_B \\
&\leq \lim_{n\to\infty} \frac{1}{|m|^{2n}} \psi(m^n x, m^n y) \\
&= 0
\end{aligned}
\tag{7.65}
$$

for all $x, y \in A$ and so $H(xy) = H(x)H(y)$ for all $x, y \in A$. Thus $H : \mathcal{A} \to \mathcal{B}$ is a
homomorphism satisfying (7.55).

Also, by (7.51), (7.58), (7.63) and the similar method, we have $H(x^*) = H(x)^*$.
This completes the proof.                                                                 □

**Corollary 7.27.** *Let $r > 1$, $\theta$ be nonnegative real numbers and $f : A \to B$ be a
mapping such that*

$$
\|D_\mu f(x_1, \cdots, x_m)\|_B \leq \theta \cdot (\|x_1\|_A^r + \|x_2\|_A^r + \cdots + \|x_m\|_A^r),
$$

$$
\|f(xy) - f(x)f(y)\|_B \leq \theta \cdot (\|x\|_A^r \cdot \|y\|_A^r)
\tag{7.66}
$$

*and*

$$
\|f(x^*) - f(x)^*\|_B \leq \theta \cdot \|x\|_A^r
$$

*for all $\mu \in \mathbb{T}^1$ and $x_1, \cdots, x_m, x, y \in A$. Then there exists a unique non-Archimedean
$C^*$-algebra homomorphism $H : A \to B$ such that*

$$
\|f(x) - H(x)\|_B \leq \frac{\theta}{|m| - |m|^r} \|x\|_A^r
\tag{7.67}
$$

*for all $x \in A$.*

*Proof.* The proof follows from Theorem 7.26 by taking

$$
\varphi(x_1, \cdots, x_m) := \theta \cdot (\|x_1\|_A^r + \|x_2\|_A^r + \cdots + \|x_m\|_A^r),
$$

$$
\psi(x, y) := \theta \cdot (\|x\|_A^r \cdot \|y\|_A^r),
$$

$$
\eta(x) := \theta \cdot \|x\|_A^r
\tag{7.68}
$$

for all $x_1, \cdots, x_m, x, y \in A$ and $L = |m|^{r-1}$.                                □

Now, we prove the Hyers-Ulam stability of derivations on non-Archimedean $C^*$-algebras for the functional equation

$$D_\mu f(x_1, \cdots, x_m) = 0.$$

*Remark 7.28.* Let $f : A \to A$ be a mapping for which there exist the functions $\varphi : A^m \to [0, \infty)$, $\psi : A^2 \to [0, \infty)$ and $\eta : A \to [0, \infty)$ such that $|m| < 1$ is far from zero and

$$\|D_\mu f(x_1, \cdots, x_m)\|_A \le \varphi(x_1, \cdots, x_m),$$

$$\|f(xy) - f(x)y - xf(y)\|_A \le \psi(x, y) \tag{7.69}$$

and

$$\|f(x^*) - f(x)^*\|_A \le \eta(x)$$

for all $\mu \in \mathbb{T}^1$ and $x_1, \cdots, x_m, x, y \in A$. If there exists $L < 1$ such that (7.52), (7.53) and (7.54) hold, then there exists a unique non-Archimedean $C^*$-algebra derivation $\delta : A \to A$ such that

$$\|f(x) - \delta(x)\|_A \le \frac{1}{|m| - |m|L} \varphi(x, 0, \cdots, 0)$$

for all $x \in A$.

## 7.2.2 Stability of Homomorphisms and Derivations in Non-Archimedean Lie $C^*$-Algebras

Now, we prove the Hyers-Ulam stability of homomorphisms in non-Archimedean Lie $C^*$-algebras for the functional equation $D_\mu f(x_1, \cdots, x_m) = 0$.

**Theorem 7.29.** *Let $f : A \to B$ be a mapping for which there exist the functions $\varphi : A^m \to [0, \infty)$ and $\psi : A^2 \to [0, \infty)$ such that (7.49) and (7.51) hold and*

$$\|f([x, y]) - [f(x), f(y)]\|_B \le \psi(x, y) \tag{7.70}$$

*for all $\mu \in \mathbb{T}^1$ and $x, y \in A$. If there exists $L < 1$ and (7.52) and (7.53) hold, then there exists a unique non-Archimedean Lie $C^*$-algebra homomorphism $H : A \to B$ such that (7.55) hold.*

*Proof.* By the same reasoning as in the proof of Theorem 7.26, we can find the mapping $H : A \to B$ given by

$$H(x) = \lim_{n \to \infty} \frac{f(m^n x)}{m^n} \tag{7.71}$$

for all $x \in A$. It follows from (7.53) and (7.71) that

$$
\begin{aligned}
&\|H([x,y]) - [H(x), H(y)]\|_B \\
&\quad = \lim_{n \to \infty} \frac{1}{|m|^{2n}} \|f(m^{2n}[x,y]) - [f(m^n x), f(m^n y)]\|_B \\
&\quad \leq \lim_{n \to \infty} \frac{1}{|m|^{2n}} \psi(m^n x, m^n y) \tag{7.72} \\
&\quad = 0
\end{aligned}
$$

for all $x, y \in A$ and so

$$H([x,y]) = [H(x), H(y)] \tag{7.73}$$

for all $x, y \in A$. Thus $H : A \to B$ is a non-Archimedean Lie $C^*$-algebra homomorphism satisfying (7.55). This completes the proof. $\qquad\square$

**Corollary 7.30.** *Let $r > 1$, $\theta$ be nonnegative real numbers and $f : A \to B$ be a mapping such that*

$$\|D_\mu f(x_1, \cdots, x_m)\|_B \leq \theta(\|x_1\|_A^r + \|x_2\|_A^r + \cdots + \|x_m\|_A^r),$$

$$\|f([x,y]) - [f(x), f(y)]\|_B \leq \theta \cdot \|x\|_A^r \cdot \|y\|_A^r \tag{7.74}$$

*and*

$$\|f(x^*) - f(x)^*\|_B \leq \theta \cdot \|x\|_A^r$$

*for all $\mu \in \mathbb{T}^1$ and $x_1, \cdots, x_m, x, y \in A$. Then there exists a unique non-Archimedean Lie $C^*$-algebra homomorphism $H : A \to B$ such that*

$$\|f(x) - H(x)\|_B \leq \frac{\theta}{|m| - |m|^r} \|x\|_A^r \tag{7.75}$$

*for all $x \in A$.*

*Proof.* The proof follows from Theorem 7.29 and the method similar to Corollary 7.27. $\qquad\square$

**Definition 7.31.** Let $A$ be a non-Archimedean Lie $C^*$-algebra. A $\mathbb{C}$-linear mapping $\delta : A \to A$ is called a *non-Archimedean Lie derivation* if

$$\delta([x, y]) = [\delta(x), y] + [x, \delta(y)]$$

for all $x, y \in A$.

Now, we prove the Hyers-Ulam stability of derivations on non-Archimedean Lie $C^*$-algebras for the functional equation

$$D_\mu f(x_1, \cdots, x_m) = 0.$$

**Theorem 7.32.** *Let $f : A \to A$ be a mapping for which there exist the functions $\varphi : A^m \to [0, \infty)$ and $\psi : A^2 \to [0, \infty)$ such that (7.49) and (7.51) hold and*

$$\|f([x, y]) - [f(x), y] - [x, f(y)]\|_A \leq \psi(x, y) \tag{7.76}$$

*for all $x, y \in A$. If there exists $L < 1$ and (7.52) and (7.53) hold, then there exists a unique non-Archimedean Lie derivation $\delta : A \to A$ such that such that (7.55) holds.*

*Proof.* It is straight forward to show, there exists a unique $\mathbb{C}$-linear mapping $\delta : A \to A$ satisfying (7.55) and the mapping $\delta : A \to A$ is given by

$$\delta(x) = \lim_{n \to \infty} \frac{f(m^n x)}{m^n} \tag{7.77}$$

for all $x \in A$. It follows from (7.53) and (7.75) that

$$\|\delta([x, y]) - [\delta(x), y] - [x, \delta(y)]\|_A$$

$$= \lim_{n \to \infty} \frac{1}{|m|^{2n}} \|f(m^{2n}[x, y]) - [f(m^n x), \cdot m^n y] - [m^n x, f(m^n y)]\|_A$$

$$\leq \lim_{n \to \infty} \frac{1}{|m|^{2n}} \psi(m^n x, m^n y) \tag{7.78}$$

$$= 0$$

for all $x, y \in A$ and so

$$\delta([x, y]) = [\delta(x), y] + [x, \delta(y)]$$

for all $x, y \in A$. Thus $\delta : A \to A$ is a non-Archimedean Lie derivation satisfying (7.55). This completes the proof. $\square$

# References

1. Abellanas, L., Alonso, L.: A general setting for Casimir invariants. J. Math. Phys. **16**, 1580–1584 (1975)
2. Abramov, V., Kerner, R., Le Roy, B.: Hypersymmetry: a $\mathbb{Z}_3$-graded generalization of supersymmetry. J. Math. Phys. **38**, 1650–1669 (1997)
3. Aczél, J.: The general solution of two functional equations by reduction to functions additive in two variables and with aid of Hamel-bases. Glasnik Mat.-Fiz. Astronom. Drustvo Mat. Fiz. Hrvatske **20**, 65–73 (1965)
4. Aczél, J.: Lectures on Functional equations and Their Applications. Academic, New York/London (1966)
5. Aczél, J.: A Short Course on Functional Equations. D. Reidel Publishing Company, Dordrecht (1987)
6. Aczél, J., Dhombres, J.: Functional Equations in Several Variables. Cambridge University Press, Cambridge (1989)
7. Agarwal, R.P., Cho, Y.J., Park, C., Saadati, R.: Approximate homomorphisms and derivation in multi-Banach algebras. Comment. Math. **51**, 23–38 (2011)
8. Agarwal, R.P., Cho, Y.J., Saadati, R., Wang, S.: Nonlinear $L$-fuzzy stability of cubic functional equations. J. Inequal. Appl. **2012**, 19 pp. (2012)
9. Alli, G., Sewell, G.L.: New methods and structures in the theory of the multi-mode Dicke laser model. J. Math. Phys. **36**, 5598–5626 (1995)
10. Almira, J.M., Luther, U.: Inverse closedness of approximation algebras. J. Math. Anal. Appl. **314**, 30–44 (2006)
11. Amyari, M., Moslehian, M.S.: Approximately ternary semigroup homomorphisms. Lett. Math. Phys. **77**, 1–9 (2006)
12. J. An: On an additive functional inequality in normed modules over a $C^*$-algebra. J. Korea Soc. Math. Educ. Ser. B. Pure Appl. Math. **15**, 393–400 (2008)
13. An, J., Cui, J., Park, C.: Jordan $*$-derivations on $C^*$-algebras and $JC^*$-algebras. Abstr. Appl. Anal. **2008**, Article ID 410437 (2008)
14. Andrews, L.C.: Special Functions for Engineers and Applied Mathematicians. MacMilan, New York (1985)
15. Antoine, J.P., Inoue, A., Trapani, C.: $O^*$-dynamical systems and $*$-derivations of unbounded operator algebras. Math. Nachr. **204**, 5–28 (1999)
16. Antoine, J.P., Inoue, A., Trapani, C.: Partial $*$-Algebras and Their Operator Realizations. Kluwer Academic Publishers, Dordrecht (2002)
17. Aoki, T.: On the stability of the linear transformation in Banach spaces. J. Math. Soc. Jpn. **2**, 64–66 (1950)

18. Ara, P., Mathieu, M.: Local Multipliers of $C^*$-Algebras. Springer, London (2003)
19. Baak, C.: Cauchy-Rassias stability of Cauchy-Jensen additive mappings in Banach spaces. Acta Math. Sin. **22**, 1789–1796 (2006)
20. Baak, C., Boo, D., Rassias, Th.M.: Generalized additive mapping in Banach modules and isomorphisms between $C^*$-algebras. J. Math. Anal. Appl. **314**, 150–156 (2006)
21. Bae, J., Park, W.: Approximate bi-homomorphisms and bi-derivations in $C^*$-ternary algebras. Bull. Korean Math. Soc. **47**, 195–209 (2010)
22. Bae, J., Park, W.: Generalized Ulam-Hyers stability of $C^*$-ternary algebra 3-homomorphisms for a functional equation. J. Chungcheong Math. Soc. **24**, 147–162 (2011)
23. Bagarello, F.: Applications of topological ∗-algebras of unbounded operators. J. Math. Phys. **39**, 6091–6105 (1998)
24. Bagarello, F.: Fixed point results in topological ∗-algebras of unbounded operators. Publ. RIMS Kyoto Univ. **37**, 397–418 (2001)
25. Bagarello, F.: Applications of topological ∗-algebras of unbounded operators to modified quons. Nuovo Cimento B **117**, 593–611 (2002)
26. Bagarello, F., Karwowski, W.: Partial ∗-algebras of closed linear operators in Hilbert space. Publ. RIMS Kyoto Univ. **21**, 205–236 (1985); **22**, 507–511 (1986)
27. Bagarello, F., Morchio, G.: Dynamics of mean-field spin models from basic results in abstract differential equations. J. Stat. Phys. **66**, 849–866 (1992)
28. Bagarello, F., Sewell, G.L.: New structures in the theory of the laser model *II*: microscopic dynamics and a non-equilibrium entropy principle. J. Math. Phys. **39**, 2730–2747 (1998)
29. Bagarello, F., Trapani, C.: Almost mean field Ising model: an algebraic approach. J. Stat. Phys. **65**, 469–482 (1991)
30. Bagarello, F., Trapani, C.: A note on the algebraic approach to the "almost" mean field Heisenberg model. Nuovo Cimento B **108**, 779–784 (1993)
31. Bagarello, F., Trapani, C.: States and representations of $CQ^*$-algebras. Ann. Inst. H. Poincaré **61**, 103–133 (1994)
32. Bagarello, F., Trapani, C.: The Heisenberg dynamics of spin systems: a quasi-∗-algebras approach. J. Math. Phys. **37**, 4219–4234 (1996)
33. Bagarello, F., Trapani, C.: $CQ^*$-algebras: structure properties. Publ. RIMS Kyoto Univ. **32**, 85–116 (1996)
34. Bagarello, F., Trapani, C.: Morphisms of certain Banach $C^*$-modules. Publ. RIMS Kyoto Univ. **36**, 681–705 (2000)
35. Bagarello, F., Trapani, C.: Algebraic dynamics in $O^*$-algebras: a perturbative approach. J. Math. Phys. **43**, 3280–3292 (2002)
36. Bagarello, F., Inoue, A., Trapani, C.: Some classes of topological quasi ∗-algebras. Proc. Am. Math. Soc. **129**, 2973–2980 (2001)
37. Bagarello, F., Inoue, A., Trapani, C.: ∗-derivations of quasi-∗-algebras. Int. J. Math. Math. Sci. **21**, 1077–1096 (2004)
38. Bagarello, F., Inoue, A., Trapani, C.: Exponentiating derivations of quasi-∗-algebras: possible approaches and applications. Int. J. Math. Math. Sci. **2005**, 2805–2820 (2005)
39. Bagarello, F., Trapani, C., Triolo, S.: Quasi-∗-algebras of measurable operators. Studia Math. **172**, 289–305 (2006)
40. Bahyrycz, A., Piszczek, M.: Hyperstability of the Jensen functional equation. Acta Math. Hung. **142**, 353–365 (2014)
41. Bahyrycz, A., Brzdęk, J., Piszczek, M., Sikorska, J.: Hyperstability of the Fréchet equation and a characterization of inner product spaces. J. Funct. Spaces Appl. **2013**, Article ID 496361, 6 pp. (2013)
42. Bahyrycz, A., Brzdęk, J., Leśniak, Z.: On approximate solutions of the generalized Volterra integral equation. Nonlinear Anal. Real World Appl. **20**, 59–66 (2014)
43. Baker, J.A.: The stability of certain functional equations. Proc. Am. Math. Soc. **112**, 729–732 (1991)

44. Baktash, E., Cho, Y.J., Jalili, M., Saadati, R., Vaezpour, S.M.: On the stability of cubic mappings and quadratic mappings in random normed spaces. J. Inequal. Appl. **2008**, Article ID 902187, 11 pp. (2008)

45. Banach, S.: Sur les operations dans les ensembles abstraits et leur application aux euations integrales. Fundam. Math. **3**, 133–181 (1922)

46. Bazunova, N., Borowiec, A., Kerner, R.: Universal differential calculus on ternary algebras. Lett. Math. Phys. **67**, 195–206 (2004)

47. Belaid, B., Elqorachi, E., Rassias, Th.M.: On the Hyers–Ulam stability of approximately Pexider mappings. Math. Inequal. Appl. **11**, 805–818 (2008)

48. Benyamini, Y., Lindenstrauss, J.: Geometric Nonlinear Functional Analysis, Vol. 1. Colloquium Publications, vol. 48. American Mathematical Society, Providence (2000)

49. Birkhoff, G.: Orthogonality in linear metric spaces. Duke Math. J. **1**, 169–172 (1935)

50. Bodaghi, A., Park, C.: Jordan $*$-derivations and quadratic Jordan $*$-derivations on real $C^*$-algebras and real $JC^*$-algebras. Int. J. Geom. Methods Mod. Phys. **10**(10), 1350051, 16 pp. (2013)

51. Bourgin, D.G.: Classes of transformations and bordering transformations. Bull. Am. Math. Soc. **57**, 223–237 (1951)

52. Brešar, M., Zalar, B.: On the structure of Jordan $*$-derivations. Colloq. Math. **LXIII**, 163–171 (1992)

53. Brillouët-Belluot, N., Brzdek, J., Ciepliński, K.: On some developments in Ulam's type stability. Abstr. Appl. Anal. **2012**, Article ID 716936, 41 pp. (2012)

54. Brown, L., Pedersen, G.: $C^*$-algebras of real rank zero. J. Funct. Anal. **99**, 131–149 (1991)

55. Brzdęk, J.: Remarks on hyperstability of the Cauchy functional equation. Aequ. Math. **86**, 255–267 (2013)

56. Brzdęk, J., Ciepliński, K.: A fixed point approach to the stability of functional equations in non–Archimedean metric spaces. Nonlinear Anal. **74**, 6861–6867 (2011)

57. Brzdęk, J., Sikorska, J.: A conditional exponential functional equation and its stability. Nonlinear Anal. **72**, 2923–2934 (2010)

58. Brzdęk, J., Chudziak, J., Páles, Zs.: A fixed point approach to stability of functional equations. Nonlinear Anal. **74**, 6728–6732 (2011)

59. Brzdęk, J., Ciepliński, K.: Hyperstability and superstability. Abstr. Appl. Anal. **2013**, Article ID 401756, 13 pp. (2013)

60. Brzdęk, J., Cădariu, L., Ciepliński, K.: Fixed point theory and the Ulam stability. J. Funct. Spaces **2014**, Article ID 829419, 16 pp. (2014)

61. Brzdęk, J., Ciepliński, K., Leśniak, Z.: On Ulam's type stability of the linear equation and related issues. Discret. Dyn. Nat. Soc. **2014**, Article ID 536791, 14 pp. (2014)

62. Cădariu, L., Radu, V.: Fixed points and the stability of Jensen's functional equation. J. Inequal. Pure Appl. Math. **4**(1), Article 4 (2003)

63. Cădariu, L., Radu, V.: On the stability of the Cauchy functional equation: a fixed point approach. In: Iteration Theory (ECIT '02). Grazer Mathematische Berichte, vol. 346, pp. 43–52. Karl-Franzens-Universität, Graz (2004)

64. Cădariu, L., Radu, V.: Fixed point methods for the generalized stability of functional equations in a single variable. Fixed Point Theory Appl. **2008**, Article ID 749392, 15 pp. (2008)

65. Carlsson, S.O.: Orthogonality in normed linear spaces. Ark. Mat. **4**, 297–318 (1962)

66. Castro, L.P., Ramos, A.: Hyers-Ulam-Rassias stability for a class of nonlinear Volterra integral equations. Banach J. Math. Anal. **3**, 47–56 (2009)

67. Cauchy, A.L.: Cours d'analyse de l'Ecole Polytechnique. Vol. 1, Analyse algebrique. Debure, Paris (1821)

68. Cayley, A.: On the 34 concomitants of the ternary cubic. Am. J. Math. **4**, 1–15 (1881)

69. Chahbi, A., Bounader, N.: On the generalized stability of d'Alembert functional equation. J. Nonlinear Sci. Appl. **6**, 198–204 (2013)

70. Cho, Y.J., Saadati, R.: Lattictic non-Archimedean random stability of ACQ functional equation. Adv. Differ. Equ. **2011**, 12 pp. (2011)

71. Cho, Y.J., Park, C., Saadati, R.: Functional inequalities in non-Archimedean Banach spaces. Appl. Math. Lett. **60**, 1994–2002 (2010)
72. Cho, Y.J., Eshaghi Gordji, M., Zolfaghari, S.: Solutions and stability of generalized mixed type QC functional equations in random normed spaces. J. Inequal. Appl. **2010**, Article ID 403101, 16 pp. (2010)
73. Cho, Y.J., Park, C., Rassias, Th.M., Saadati, R.: Inner product spaces and functional equations. J. Comput. Anal. Appl. **13**, 296–304 (2011)
74. Cho, Y.J., Saadati, R., Shabanian, S., Vaezpour, S.M.: On solution and stability of a two-variable functional equations. Discret. Dyn. Nat. Soc. **2011**, Article ID 527574, 18 pp. (2011)
75. Cho, Y.J., Kang, J.I., Saadati, R.: Fixed points and stability of additive functional equations on the Banach algebras. J. Comput. Anal. Appl. **14**, 1103–1111 (2012)
76. Cho, Y.J., Kang, S.M., Sadaati, R.: Nonlinear random stability via fixed-point method. J. Appl. Math. **2012**, Article ID 902931, 44 pp. (2012)
77. Cho, Y.J., Saadati, R., Vahidi, J.: Approximation of homomorphisms and derivations on non-Archimedean Lie $C^*$-algebras via fixed point method. Discret. Dyn. Nat. Soc. **2012**, Article ID 373904, 9 pp. (2012)
78. Cho, Y.J., Rassias, Th.M., Saadati, R.: Stability of Functional Equations in Random Normed Spaces. Springer, New York (2013)
79. Cho, Y.J., Saadati, R., Yang, Y.O.: Random-ternary algebras and application. J. Inequal. Appl. **2015**, 26 (2015)
80. Cholewa, P.W.: Remarks on the stability of functional equations. Aequ. Math. **27**, 76–86 (1984)
81. Chu, H.Y., Kang, D.S., Rassias, Th.M.: On the stability of a mixed $n$-dimensional quadratic functional equation. Bull. Belg. Math. Soc.–Simon Stevin **15**, 9–24 (2008)
82. Ciepliński, K.: Generalized stability of multi-additive mappings. Appl. Math. Lett. **23**, 1291–1294 (2010)
83. Cieplinski, K.: Stability of multi-additive mappings in non-Archimedean normed spaces. J. Math. Anal. Appl. **373**, 376–383 (2011)
84. Ciepliński, K.: On the generalized Hyers-Ulam stability of multi-quadratic mappings. Comput. Math. Appl. **62**, 3418–3426 (2011)
85. Ciepliński, K.: Stability of multi-additive mappings in $\beta$-Banach spaces. Nonlinear Anal. **75**, 4205–4212 (2012)
86. Ciepliński, K.: Applications of fixed point theorems to the Hyers-Ulam stability of functional equations-a Survey. Ann. Funct. Anal. **3**, 151–164 (2012)
87. Czerwik, S.: On the stability of the quadratic mapping in normed spaces. Abh. Math. Sem. Univ. Hambg. **62**, 59–64 (1992)
88. Czerwik, P.: Functional Equations and Inequalities in Several Variables. World Scientific Publishing Company, River Edge/Hong Kong/Singapore/London (2002)
89. Czerwik, S.: Stability of Functional Equations of Ulam-Hyers-Rassias Type. Hadronic Press, Palm Harbor (2003)
90. Dales, H.G.: Banach Algebras and Automatic Continuity. London Mathematical Society Monographs, New Series, vol. 24, Oxford University Press, Oxford (2000)
91. Dales, H.G., Moslehian, M.S.: Stability of mappings on multi-normed spaces. Glasg. Math. J. **49**, 321–332 (2007)
92. Dales, H.G., Polyakov, M.E.: Multi-normed spaces. Dissertationes Math. (Rozprawy Mat.) **488**, 165 pp. (2012)
93. Daletskii, Y.L., Takhtajan, L.: Leibniz and Lie algebra structures for Nambu algebras. Lett. Math. Phys. **39**, 127–141 (1997)
94. Dehghanian, M., Modarres Mosadegh, S.M.S., Park, C., Shin, D.Y.: $C^*$-ternary 3-derivations on $C^*$-ternary algebras. J. Inequal. Appl. **2013**, 124 (2013)
95. Diaz, J., Margolis, B.: A fixed point theorem of the alternative for contractions on a generalized complete metric space. Bull. Am. Math. Soc. **74**, 305–309 (1968)
96. Dixmier, J.: $C^*$-Algebras. North-Holland Publishing Company, Amsterdam/New York/Oxford (1977)

97. Ebadian, A., Ghobadipour, N., Baghban, H.: Stability of bi-$\theta$-derivations on $JB^*$-triples. Int. J. Geom. Methods Mod. Phys. **9**, Article ID 1250051, 12 pp. (2012)

98. Ekhaguere, G.O.S.: Partial $W^*$-dynamical systems. In: Current Topics in Operator Algebras, Proceedings of the Satellite Conference of ICM-90, pp. 202–217, World Scientific, Singapore (1991)

99. Epifanio, G., Trapani, C.: Quasi-∗-algebras valued quantized fields. Ann. Inst. H. Poincaré **46**, 175–185 (1987)

100. Eshaghi Gordji, M., Ghobadipour, N.: Stability of $(\alpha, \beta, \gamma)$-derivations on Lie $C^*$-algebras. Int. J. Geom. Methods Mod. Phys. **7**, 1093–1102 (2010)

101. Eshaghi Gordji, M., Khodaei, H.: Stability of Functional Equations. Lap Lambert Academic Publishing, Saarbrücken (2010)

102. Eshaghi Gordji, M., Khodaei, H.: A fixed point technique for investigating the stability of $(\alpha, \beta, \gamma)$-derivations on Lie $C^*$-algebras. Nonlinear Anal. **76**, 52–57 (2013)

103. Eshaghi Gordji, M., Savadkouhi, M.B.: Stability of cubic and quartic functional equations in non-Archimedean spaces. Acta Appl. Math. **110**, 1321–1329 (2010)

104. Eshaghi Gordji, M., Cho, Y.J., Ghaemi, M.B., Majani, H.: Approximately quintic and sextic mappings form $r$-divisible groups into Šerstnev probabilistic Banach spaces: fixed point method. Discret. Dyn. Nat. Soc. **2011**, Article ID 572062, 16 pp. (2011)

105. Eshaghi Gordji, M., Ghaemi, M.B., Cho, Y.J., Majani, M.: A general system of Euler-Lagrange-type quadratic functional equations in Menger probabilistic non-Archimedean 2-normed spaces. Abstr. Appl. Anal. Vol. **2011**, Article ID 208163, 21 pp. (2011)

106. Eshaghi Gordji, M., Ghaemi, B., Rassias, J.M., Alizadeh, B.: Nearly ternary quadratic higher derivations on non-Archimedean ternary Banach algebras: a fixed point approach. Abstr. Appl. Anal. **2011**, Article ID 417187, 18 pp. (2011)

107. Eshaghi Gordji, M., Ramezani, M., Cho, Y.J., Baghani, H.: Approximate Lie brackets: a fixed point approach. J. Inequal. Appl. **2012**, 125 (2012)

108. Eskandani, G.Z.: On the Hyers-Ulam-Rassias stability of an additive functional equation in quasi-Banach spaces. J. Math. Anal. Appl. **345**, 405–409 (2008)

109. Eskandani, G.Z., Gavruta, P.: Hyers-Ulam-Rassias stability of pexiderized Cauchy functional equation in 2-Banach spaces. J. Nonlinear Sci. Appl. **5**(Special issue), 459–465 (2012)

110. Eskandani, G.Z., Rassias, Th.M.: Hyers-Ulam-Rassias stability of derivations in proper $JCQ^*$-triples. Mediterr. J. Math. **10**, 1391–1400 (2013)

111. Faiziev, V.A., Rassias, Th.M.: The space of pseudocharacters on semigroups. Nonlinear Funct. Anal. Appl. **5**, 107–126 (2000)

112. Faiziev, V.A., Rassias, Th.M., Sahoo, P.K.: The space of $(\psi, \gamma)$–additive mappings on semigroups. Trans. Am. Math. Soc. **354**, 4455–4472 (2002)

113. Fechner, W.: Stability of a functional inequalities associated with the Jordan-von Neumann functional equation. Aequ. Math. **71**, 149–161 (2006)

114. Fleming, R.J., Jamison, J.E.: Isometries on Banach Spaces: Function Spaces. Monographs and Surveys in Pure and Applied Mathematics, vol. 129. Chapman & Hall/CRC, Boca Raton/London/New York/Washington, DC (2003)

115. Forti, G.L.: An existence and stability theorem for a class of functional equations. Stochastica **4**, 22–30 (1980)

116. Forti, G.L.: Comments on the core of the direct method for proving Hyers-Ulam stability of functional equations. J. Math. Anal. Appl. **295**, 127–133 (2004)

117. Forti, G.L.: Elementary remarks on Ulam-Hyers stability of linear functional equations. J. Math. Anal. Appl. **328**, 109–118 (2007)

118. Fredenhagen, K., Hertel, J.: Local algebras of observables and pointlike localized fields. Commun. Math. Phys. **80**, 555–561 (1981)

119. Gajda, Z.: On stability of additive mappings. Int. J. Math. Math. Sci. **14**, 431–434 (1991)

120. Gao, Z.X., Cao, H.X., Zheng, W.T., Xu, L.: Generalized Hyers-Ulam-Rassias stability of functional inequalities and functional equations. J. Math. Inequal. **3**, 63–77 (2009)

121. Gauss, C.F.: Theoria moyus corporum caelestium. Perthes–Besser, Hamburg (1809)

122. Găvruta, P.: On the stability of some functional equations. In: Stability of Mappings of Hyers-Ulam Type, pp. 93–98. Hadronic Press. Palm Harbor (1994)
123. Găvruta, P.: A generalization of the Hyers-Ulam-Rassias stability of approximately additive mappings. J. Math. Anal. Appl. **184**, 431–436 (1994)
124. Găvruta, P.: On a problem of G. Isac and Th.M. Rassias concerning the stability of mappings. J. Math. Anal. Appl. **261**, 543–553 (2001)
125. Găvruta, P.: On the Hyers-Ulam-Rassias stability of the quadratic mappings. Nonlinear Funct. Anal. Appl. **9**, 415–428 (2004)
126. Gilányi, A.: Eine zur Parallelogrammgleichung äquivalente Ungleichung. Aequ. Math. **62**, 303–309 (2001)
127. Gilányi, A.: On a problem by K. Nikodem. Math. Inequal. Appl. **5**, 707–710 (2002)
128. Goodearl, K.R.: Notes on Real and Complex $C^*$-Algebras. Shiva Mathematics Series IV. Shiva Publishing, Cheshire (1982)
129. Grabiec, A.: The generalized Hyers-Ulam stability of a class of functional equations. Publ. Math. Debr. **48**, 217–235 (1996)
130. Gudder, S., Strawther, D.: Orthogonally additive and orthogonally increasing functions on vector spaces. Pac. J. Math. **58**, 427–436 (1975)
131. Haag, R., Kastler, D.: An algebraic approach to quantum field theory. J. Math. Phys. **5**, 848–861 (1964)
132. Hardy, G.H., Littlewood, J.E., Polya, G.: Inequalities, 2nd edn. Cambridge University Press, Cambridge (1952)
133. Hyers, D.H.: On the stability of the linear functional equation. Proc. Natl. Acad. Sci. USA **27**, 222–224 (1941)
134. Hyers, D.H.: The stability of homomorphisms and related topics. In: Rassias, Th.M. (ed.), Global Analysis Analysis on Manifolds. Teubner-Texte zur Mathematik, pp. 140–153. B.G. Teubner, Leipzig (1983)
135. Hyers, D.H., Rassias, Th.M.: Approximate homomorphisms. Aequ. Math. **44**, 125–153 (1992)
136. Hyers, D.H., Isac, G., Rassias, Th.M.: Stability of Functional Equations in Several Variables. Birkhäuser, Basel (1998)
137. Hyers, D.H., Isac, G., Rassias, Th.M.: On the asymptoticity aspect of Hyers-Ulam stability of mappings. Proc. Am. Math. Soc. **126**, 425–430 (1998)
138. Isac, G., Rassias, Th.M.: On the Hyers-Ulam stability of additive mappings. J. Approx. Theory **72**, 131–137 (1993)
139. Isac, G., Rassias, Th.M.: Stability of $\psi$-additive mappings: applications to nonlinear analysis. Int. J. Math. Math. Sci. **19**, 219–228 (1996)
140. Jacobson, N.: Lie Algebras. Dover, New York (1979)
141. James, R.C.: Orthogonality in normed linear spaces. Duke Math. J. **12**, 291–302 (1945)
142. Jesen, B., Karpf, J., Thorup, A.: Some functional equations in groups and rings. Math. Scand. **22**, 257–265 (1968)
143. Johnson, B.E.: Approximately multiplicative maps between Banach algebras. J. Lond. Math. Soc. **37**, 294–316 (1988)
144. Jun, K., Kim, H.: On the stability of Appolonius' equation. Bull. Belg. Math. Soc. Simon Stevin **11**, 615–624 (2004)
145. Jun, K., Lee, Y.: A generalization of the Hyers-Ulam-Rassias stability of Jensen's equation. J. Math. Anal. Appl. **238**, 305–315 (1999)
146. Jun, K., Kim, B., Shin, D.: On Hyers-Ulam-Rassias stability of the Pexider equation. J. Math. Anal. Appl. **239**, 20–29 (1999)
147. Jun, K.W., Kim, H.M., Rassias, J.M.: Extended Hyers-Ulam stability for Cauchy-Jensen mappings. J. Differ. Equ. Appl. **13**, 1139–1153 (2007)
148. Jung, S.-M.: On the Hyers-Ulam-Rassias stability of approximately additive mappings. J. Math. Anal. Appl. **204**, 221–226 (1996)
149. Jung, S.-M.: Hyers-Ulam-Rassias Stability of Functional Equations in Mathematical Analysis. Hadronic Press, Palm Harbor (2001)

150. Jung, S.-M.: Hyers-Ulam stability of a system of first order linear differential equations with constant coefficients. J. Math. Anal. Appl. **320**, 549–561 (2006)
151. Jung, S.-M.: A fixed point approach to the stability of an equation of the square spiral. Banach J. Math. Anal. **1**, 148–153 (2007)
152. Jung, S.-M.: A fixed point approach to the stability of a Volterra integral equation. Fixed Point Theory Appl. **2007**, Article ID 57064, 9 pp. (2007)
153. Jung, S.-M.: A fixed point approach to the stability of isometries. J. Math. Anal. Appl. **329**, 879–890 (2007)
154. Jung, S.-M.: A fixed point approach to the stability of a Volterra integral equation. Fixed Point theory Appl. **2007**, Article ID 57064, 9 pp. (2007)
155. Jung, Y.S., Chang, I.S.: The stability of a cubic type functional equation with the fixed point alternative. J. Math. Anal. Appl. **306**, 752–760 (2005)
156. Jung, S.-M., Kim, T.S.: A fixed point approach to the stability of the cubic functional equation. Bol. Soc. Mat. Mexicana **12**, 51–57 (2006)
157. Jung, S.-M., Lee, Z.H.: A fixed point approach to the stability of quadratic functional equation with involution. Fixed Point Theory Appl. **2008**, Article ID 732086, 11 pp. (2008)
158. Jung, S.-M., Rassias, Th.M.: Ulam's problem for approximate homomorphisms in connection with Bernoulli's differential equation. Appl. Math. Comput. **187**, 223–227 (2007)
159. Jung, S.-M.: Hyers-Ulam stability of linear differential equations of first order. Appl. Math. Lett. **22**, 70–74 (2009)
160. Jung, S.-M.: Hyers–Ulam–Rassias Stability of Functional Equations in Nonlinear Analysis. Springer Optimization and Its Applications, vol. 48. Springer, New York/Dordrecht/Heidelberg/London (2011)
161. Jung, S.-M., Rassias, Th.M.: Generalized Hyers–Ulam stability of Riccati differential equation. Math. Inequal. Appl. **11**, 777–782 (2008)
162. Jung, S.-M., Rassias, J.M.: A fixed point approach to the stability of a functional equation of the spiral of Theodorus. Fixed Point Theory Appl. **2008**, Article ID 945010, 7 pp. (2008)
163. Jung, S.-M., Rassias, M.Th.: A linear functional equation of third order associated to the Fibonacci numbers. Abstr. Appl. Anal. **2014**, Article ID 137468, 7 pp. (2014)
164. Jung, S.-M., Kim, T.S., Lee, K.S.: A fixed point approach to the stability of quadratic functional equation. Bull. Korean Math. Soc. **43**, 531–541 (2006)
165. Jung, S.-M., Popa, D., Rassias, M.Th.: On the stability of the linear functional equation in a single variable on complete metric groups. J. Glob. Optim. **59**, 165–171 (2014)
166. Jung, S.-M., Rassias, M.Th., Mortici, C.: On a functional equation of trigonometric type. Appl. Math. Comput. **252**, 294–303 (2015)
167. Kadison, R.V., Pedersen, G.: Means and convex combinations of unitary operators. Math. Scand. **57**, 249–266 (1985)
168. Kadison, R.V., Ringrose, J.R.: Fundamentals of the Theory of Operator Algebras. Academic, New York (1983)
169. Kapranov, M., Gelfand, I.M., Zelevinskii, A.: Discriminants, Resultants and Multidimensional Determinants. Birkhäuser, Berlin (1994)
170. Kenari, H.M., Saadati, R., Park, C.: Homomorphisms and derivations in $C^*$-ternary algebras via fixed point method. Adv. Differ. Equ. **2012**, 13 pp. (2012)
171. Kerner, R.: The cubic chessboard. Geometry and physics. Class. Quantum Gravity **14**(1A), A203–A225 (1997)
172. Kerner, R.: Ternary algebraic structures and their applications in physics. Univ. P. and M. Curie, Paris (2000, preprint)
173. Khodaei, H., Eshaghi Gordji, M., Kim, S.S., Cho, Y.J.: Approximation of radical functional equations related to quadratic and quartic mappings. J. Math. Anal. Appl. **397**, 284–297 (2012)
174. Khrennikov, A.: Non-Archimedean Analysis: Quantum Paradoxes, Dynamical Systems and Biological Models. Mathematics and Its Applications. vol. 427, Kluwer Academic Publishers, Dordrecht (1997)

175. Kim, G.H., Dragomir, S.S.: The stability of the generalized d'Alembert and Jensen functional equations II. Int. J. Math. Math. Sci. **2006**, Article ID 43185, 12 pp. (2006)
176. Kochanek, T., Lewicki, M.: Stability problem for number-theoretically multiplicative functions. Proc. Am. Math. Soc. **135**, 2591–2597 (2007)
177. Kuczma, M.: Functional equations on restricted domains. Aequ. Math. **18**, 1–34 (1978)
178. Kuczma, M.: An Introduction to the Theory of Functional Equations and Inequalities. Uniwersytet Slaski, Warszawa/Krakow/Katowice (1985)
179. Lassner, G.: Algebras of unbounded operators and quantum dynamics. Physica **124 A**, 471–480 (1984)
180. Lee, S.H., Park, C.: Hyers-Ulam-Rassias stability of isometric homomorphisms in quasi-Banach algebras. J. Comput. Anal. Appl. **10**, 39–51 (2008)
181. Lee, J.R., Park, C., Shin, D.Y.: On the stability of generalized additive functional inequalities in Banach spaces. J. Inequal. Appl. **2008**, Article ID 210626, 13 pp. (2008)
182. Lee, J.R., Park, C., Shin, D.Y.: Stability of an additive functional inequality in proper $CQ^*$-algebras. Bull. Korean Math. Soc. **48**, 853–871 (2011)
183. Lee, Y.H., Jung, S.-M., Rassias, M.Th.: On an n-dimensional mixed type additive and quadratic functional equation. Appl. Math. Comput. **228**, 13–16 (2014)
184. Legendre, A.M.: Elements de geometrie, Note II. Didot, Paris (1791)
185. Li, Y., Shen, Y.: Hyers-Ulam stability of nonhomogeneous linear differential equations of second order. Int. J. Math. Math. Sci. **2009**, Article ID 576852, 7 pp. (2009)
186. Lu, G., Park, C.: Additive functional inequalities in Banach spaces. J. Inequal. Appl. **2012**, 294 (2012)
187. Lungu, N., Popa, D.: Hyers-Ulam stability of a first order partial differential equation. J. Math. Anal. Appl. **385**, 86–91 (2012)
188. Luxemburg, W.A.J.: On the convergence of successive approximations in the theory of ordinary differential equations, II. Koninkl, Nederl. Akademie van Wetenschappen, Amsterdam, Proc. Ser. A (5) 61; Indag. Math. **20**, 540–546 (1958)
189. Miheţ, D.: The fixed point method for fuzzy stability of the Jensen functional equation. Fuzzy Sets Syst. **160**, 1663–1667 (2009)
190. Miheţ, D., Radu, V.: On the stability of the additive Cauchy functional equation in random normed spaces. J. Math. Anal. Appl. **343**, 567–572 (2008)
191. Mirzavaziri, M., Moslehian, M.S.: A fixed point approach to stability of a quadratic equation. Bull. Braz. Math. Soc. (N.S.) **37**, 361–376 (2006)
192. Miura, T., Miyajima, S., Takahasi, S.E.: Hyers-Ulam stability of linear differential operator with constant coefficients. Math. Nachr. **258**, 90–96 (2003)
193. Miura, T., Miyajima, S., Takahasi, S.E.: A characterization of Hyers-Ulam stability of first order linear differential operators. J. Math. Anal. Appl. **286**, 136–146 (2003)
194. Miura, T., Jung, S.-M., Takahasi, S.E.: Hyers-Ulam-Rassias stability of the Banach space valued linear differential equations $y' = \lambda y$. J. Korean Math. Soc. **41**, 995–1005 (2004)
195. Mlesnite, O.: Existence and Ulam-Hyers stability results for coincidence problems. J. Nonlinear Sci. Appl. **6**, 108–116 (2013)
196. Mohammadi, M., Cho, Y.J., Park, C., Vetro, P., Saadati, R.: Random stability of an additive-quadratic-quartic functional equation. J. Inequal. Appl. **2010**, Article ID 754210, 18 pp. (2010)
197. Montigny, M., Patera, J.: Discrete and continuous graded contractions of Lie algebras and superalgebras. J. Phys. A **24**, 525–547 (1991)
198. Morchio, G., Strocchi, F.: Mathematical structures for long range dynamics and symmetry breaking. J. Math. Phys. **28**, 622–635 (1987)
199. Mortici, C., Rassias, M.Th., Jung, S.-M.: On the stability of a functional equation associated with the Fibonacci numbers. Abstr. Appl. Anal. **2014**, Article ID 546046, 6 pp. (2014)
200. Moslehian, M.S.: On the orthogonal stability of the pexiderized quadratic equation. J. Differ. Equ. Appl. **11**, 999–1004 (2005)
201. Moslehian, M.S.: Almost derivations on $C^*$-ternary rings. Bull. Belg. Math. Soc.-Simon Stevin **14**, 135–142 (2007)

202. Moslehian, M.S.: Superstability of higher derivations in multi-Banach algebras. Tamsui Oxf. J. Math. Sci. **24**, 417–427 (2008)
203. Moslehian, M.S., Rassias, Th.M.: Stability of functional equations in non-Archimedean spaces. Appl. Anal. Discret. Math. **1**, 325–334 (2007)
204. Moslehian, M.S., Rassias, Th.M.: Orthogonal stability of additive type equations. Aequ. Math. **73**, 249–259 (2007)
205. Moslehian, M.S., Rassias, Th.M.: Generalized Hyers-Ulam stability of mappings on normed Lie triple systems. Math. Inequal. Appl. **11**, 371–380 (2008)
206. Moslehian, M.S., Nikodem, K., Popa, D.: Asymptotic aspect of the quadratic functional equation in multi-normed spaces. J. Math. Anal. Appl. **355**, 717–724 (2009)
207. Moszner, Z.: On the stability of functional equations. Aequ. Math. **77**, 33–38 (2009)
208. Najati, A., Cho, Y.J.: Generalized Hyers-Ulam stability of the pexiderized Cauchy functional equation in non-Archimedean spaces. Fixed Point Theory Appl. **2011**, Article ID 309026, 11 pp. (2011)
209. Najati, A., Park, C.: The pexiderized Apollonius-Jensen type additive mapping and isomorphisms between $C^*$-algebras. J. Differ. Equ. Appl. **14**, 459–479 (2008)
210. Najati, A., Park, C.: Stability of a generalized Euler-Lagrange type additive mapping and homomorphisms in $C^*$-algebras *II*. J. Nonlinear Sci. Appl. **3**, 123–143 (2010)
211. Najati, A., Rahimi, A.: A fixed point approach to the stability of a generalized Cauchy functional equation. Banach J. Math. Anal. **2**, 105–112 (2008)
212. Najati, A., Ranjbari, A.: Stability of homomorphisms for a 3D Cauchy-Jensen type functional equation on $C^*$-ternary algebras. J. Math. Anal. Appl. **341**, 62–79 (2008)
213. Najati, A., Rassias, Th.M.: Stability of homomorphisms and $(\theta, \phi)$-derivations. Appl. Anal. Discret. Math. **3**, 264–281 (2009)
214. Najati, A., Kang, J.I., Cho, Y.J.: Local stability of the pexiderized Cauchy and Jensen's equations in fuzzy spaces. J. Inequal. Appl. **2011**, 78 (2011)
215. Najatim, A., Rassias, Th.M.: Stability of a mixed functional equation in several variables on Banach modules. Nonlinear Anal. **72**, 1755–1767 (2010)
216. Nikoufar, E., Rassias, Th.M.: $\theta$-centralizers on semiprime Banach *-algebras. Ukr. Math. J. **66**, 300–310 (2014)
217. Novotný, P., Hrivnák, J.: On $(\alpha, \beta, \gamma)$-derivations of Lie algebras and corresponding invariant functions. J. Geom. Phys. **58**, 208–217 (2008)
218. O'Regan, D., Rassias, J.M., Saadati, R.: Approximations of ternary Jordan homomorphisms and derivations in multi-C* ternary algebras. Acta Math. Hung. **134**, 99–114 (2012)
219. Pallu de la Barriére, R.: Algèbres unitaires et espaces d'Ambrose. Ann. Ecole Norm. Sup. **70**, 381–401 (1953)
220. Park, C.: On the stability of the linear mapping in Banach modules. J. Math. Anal. Appl. **275**, 711–720 (2002)
221. Park, C.: Linear functional equations in Banach modules over a $C^*$-algebra. Acta Appl. Math. **77**, 125–161 (2003)
222. Park, C.: Modified Trif's functional equations in Banach modules over a $C^*$-algebra and approximate algebra homomorphisms. J. Math. Anal. Appl. **278**, 93–108 (2003)
223. Park, C.: On an approximate automorphism on a $C^*$-algebra. Proc. Am. Math. Soc. **132**, 1739–1745 (2004)
224. Park, C.: Lie *-homomorphisms between Lie $C^*$-algebras and Lie *-derivations on Lie $C^*$-algebras. J. Math. Anal. Appl. **293**, 419–434 (2004)
225. Park, C.: Approximate homomorphisms on $JB^*$-triples. J. Math. Anal. Appl. **306**, 375–381 (2005)
226. Park, C.: Homomorphisms between Lie $JC^*$-algebras and Cauchy-Rassias stability of Lie $JC^*$-algebra derivations. J. Lie Theory **15**, 393–414 (2005)
227. Park, C.: Homomorphisms between Poisson $JC^*$-algebras. Bull. Braz. Math. Soc. **36**, 79–97 (2005)
228. Park, C.: Isomorphisms between unital $C^*$-algebras. J. Math. Anal. Appl. **307**, 753–762 (2005)

229. Park, C.: Hyers-Ulam-Rassias stability of a generalized Euler-Lagrange type additive mapping and isomorphisms between $C^*$-algebras. Bull. Belg. Math. Soc.–Simon Stevin **13**, 619–631 (2006)

230. Park, C.: Isomorphisms between $C^*$-ternary algebras. J. Math. Phys. **47**(10), 103512 (2006)

231. Park, C.: Isomorphisms between $C^*$-ternary algebras. J. Math. Anal. Appl. **327**, 101–115 (2007)

232. Park, C.: Fixed points and Hyers-Ulam-Rassias stability of Cauchy-Jensen functional equations in Banach algebras. Fixed Point Theory Appl. **2007**, Article ID 50175 (2007)

233. Park, C.: Hyers-Ulam-Rassias stability of homomorphisms in quasi-Banach algebras. Bull. Sci. Math. **132**, 87–96 (2008)

234. Park, C.: Generalized Hyers-Ulam stability of functional equations: a fixed point approach. Taiwan. J. Math. **14**, 1591–1608 (2010)

235. Park, C.: Square root and 3rd root functional equations in $C^*$-algebras: an fixed point approach. J. Nonlinear Anal. Optim. **2**, 27–34 (2011)

236. Park, C.: Orthogonal stability of a cubic-quartic functional equation. J. Nonlinear Sci. Appl. **5**(Special issue), 28–36 (2012)

237. Park, C.: Stability of bi-$\theta$-derivations on $JB^*$-triples: revisited. Int. J. Geom. Methods Mod. Phys. **11**(3), 1450015, 10 pp. (2014)

238. Park, C., An, J.S.: Isomorphisms in quasi-Banach algebras. Bull. Korean Math. Soc. **45**, 111–118 (2008)

239. Park, C., Cui, J.: Generalized stability of $C^*$-ternary quadratic mappings. Abstr. Appl. Anal. **2007**, Article ID 23282 (2007)

240. Park, C., Hou, J.: Homomorphisms between $C^*$-algebras associated with the Trif functional equation and linear derivations on $C^*$-algebras. J. Korean Math. Soc. **41**, 461–477 (2004)

241. Park, C., Najati, A.: Homomorphisms and derivations in $C^*$-algebras. Abstr. Appl. Anal. **2007**, Article ID 80630 (2007)

242. Park, C., Park, W.: On the Jensen's equation in Banach modules. Taiwan. J. Math. **6**, 523–531 (2002)

243. Park, C., Park, J.: Generalized Hyers-Ulam stability of an Euler-Lagrange type additive mapping. J. Differ. Equ. Appl. **12**, 1277–1288 (2006)

244. Park, C., Rassias, Th.M.: Hyers-Ulam stability of a generalized Apollonius type quadratic mapping. J. Math. Anal. Appl. **322**, 371–381 (2006)

245. Park, C., Rassias, Th.M.: Fixed points and generalized Hyers-Ulam stability of quadratic functional equations. J. Math. Inequal. **1**, 515–528 (2007)

246. Park, C., Rassias, Th.M.: Homomorphisms between JC*-algebras. Studia Univ. "Babes–Bolyai", Math. **53**, 43–55 (2008)

247. Park, C., Rassias, Th.M.: Homomorphisms in $C^*$-ternary algebras and $JB^*$-triples. J. Math. Anal. Appl. **337**, 13–20 (2008)

248. Park, C., Rassias, Th.M.: Homomorphisms and derivations in proper $JCQ^*$-triples. J. Math. Anal. Appl. **337**, 1404–1414 (2008)

249. Park, C., Rassias, Th.M.: On the stability of orthogonal functional equations. Tamsui Oxf. J. Math. Sci. **24**, 355–365 (2008)

250. Park, C., Rassias, J.M.: Stability of the Jensen-type functional equation in $C^*$-algebras: a fixed point approach. Abstr. Appl. Anal.**2009**, Article ID 360432, 17 pp. (2009)

251. Park, C., Rassias, Th.M.: Fixed points and stability of functional equations. In: Pardalos, P.M., Rassias, Th.M., Khan, A.A. (eds.), Nonlinear Analysis and Variational Problems. Springer Optimization and Its Applications, vol. 35, pp. 125–134. Springer, New York (2010)

252. Park, C., Saadati, R.: Approximation of a generalized additive mapping in multi-Banach modules and isomorphisms in multi-$C^*$-algebras: a fixed-point approach. Adv. Differ. Equ. **2012**, 162, 14 pp. (2012)

253. Park, C., Cho, Y., Han, M.: Stability of functional inequalities associated with Jordan–von Neumann type additive functional equations. J. Inequal. Appl. **2007**, Article ID 41820 (2007)

254. Park, C., An, J.S., Moradlou, F.: Additive functional inequalities in Banach modules. J. Inequal. Appl. **2008**, Article ID 592504, 10 pp. (2008)

255. Park, C., Boo, D.H., Rassias, Th.M.: Approximately additive mappings over $p$-adic fields. J. Chungcheong Math. Soc. **21**, 1–14 (2008)
256. Park, C., Lee, J.R., Shin, D.Y.: Fixed points and stability of functional equations associated with inner product spaces. In: Difference Equations and Applications, pp. 235–242, Ugur-Bahesehir University Publishing Company, Istanbul (2009)
257. Park, C., Lee, J.R., Rassias, Th.M., Saadati, R.: Fuzzy ∗-homomorphisms and fuzzy ∗-derivations in induced fuzzy $C^*$-algebras. Math. Comput. Model. **54**, 2027–2039 (2011)
258. Park, C., Kenary, H.A., Kim, S.O.: Positive-additive functional equations in $C^*$-algebras. Fixed Point Theory **13**, 613–622 (2012)
259. Popa, D., Rasa, I.: On the Hyers-Ulam stability of the linear differential equation. J. Math. Anal. Appl. **381**, 530–537 (2011)
260. Popa, D., Rasa, I.: Hyers-Ulam stability of the linear differential operator with nonconstant coefficients. Appl. Math. Comput. **291**, 1562–1568 (2012)
261. Popa, D., Rasa, I.: The Fréchet functional equation with application to the stability of certain operators. J. Approx. Theory **164**, 138–144 (2012)
262. Popovych, R., Boyko, V., Nesterenko, M., Lutfullin, M.: Realizations of real low–dimensional Lie algebras. J. Phys. A **36**, 7337–7360 (2003)
263. Pourpasha, M.M., Rassias, Th.M., Saadati, R., Vaezpour, S.M.: The stability of some differential equations. Math. Probl. Eng. **2011**, Article ID 128479, 15 pp. (2011)
264. Prastaro, A., Rassias, Th.M.: Ulam stability in geometry of PDE's. Nonlinear Funct. Anal. Appl. **8**, 259–278 (2003)
265. Radu, V.: The fixed point alternative and the stability of functional equations. Fixed Point Theory **4**, 91–96 (2003)
266. Rand, D., Winternitz, P., Zassenhaus, H.: On the identification of Lie algebra given by its structure constants I. Direct decompositions, Levi decompositions and nil radicals. Linear Algebra Appl. **109**, 197–246 (1988)
267. Rassias, Th.M.: On the stability of the linear mapping in Banach spaces. Proc. Am. Math. Soc. **72**, 297–300 (1978)
268. Rassias, J.M.: On approximation of approximately linear mappings by linear mappings. J. Funct. Anal. **46**, 126–130 (1982)
269. Rassias, J.M.: On approximation of approximately linear mappings by linear mappings. Bull. Sci. Math. **108**, 445–446 (1984)
270. Rassias, Th.M.: New characterizations of inner product spaces. Bulletin des Sciences Mathematiques, 2ed serie, Paris **108**, 95–99 (1984)
271. Rassias, Th.M.: On the stability of mappings. Rendiconti del Seminario Matematico e Fisico di Milano **58**, 91–99 (1988)
272. Rassias, J.M.: Solution of a problem of Ulam. J. Approx. Theory **57**, 268–273 (1989)
273. Rassias, Th.M.: Topics in Mathematical Analysis. A Volume Dedicated to the Memory of A. L. Cauchy. World Scientific Publishing Company, Singapore/Teaneck/London (1989)
274. Rassias, Th.M.: Problem 16: 2. Report of the 27th international symposium on functional equations. Aequationes Mathematicae, Bielsko-Bia la, Poland **39**, 292–293 (1990)
275. Rassias, Th.M.: On a modified Hyers-Ulam sequence. J. Math. Anal. Appl. **158**, 106–113 (1991)
276. Rassias, Th.M.: Inner Product Spaces and Applications. Pitman Research Notes in Mathematics Series, No. 376. Addison Wesley Longman, Essex (1997)
277. Rassias, Th.M.: Nonlinear Mathematical Analysis and Applications. Hadronic Press, Palm Harbor (1998)
278. Rassias, Th.M.: Approximation Theory and Applications. Hadronic Press, Palm Harbor (1998)
279. Rassias, Th.M.: On the stability of the quadratic functional equation and its applications. Studia Univ. Babes-Bolyai **XLIII**, 89–124 (1998)
280. Rassias, Th.M.: New Approaches in Nonlinear Analysis. Hadronic Press, Palm Harbor (1999)
281. Rassias, Th.M.: Functional Equations and Inequalities. Kluwer Academic Publishers, Dordrecht/Boston/London (2000)

282. Rassias, Th.M.: Survey on Classical Inequalities. Kluwer Academic Publishers, Dordrecht/Boston/London (2000)
283. Rassias, Th.M.: The problem of S. M. Ulam for approximately multiplicative mappings. J. Math. Anal. Appl. **246**, 352–378 (2000)
284. Rassias, Th.M.: On the stability of functional equations in Banach spaces. J. Math. Anal. Appl. **251**, 264–284 (2000)
285. Rassias, Th.M.: On the stability of functional equations and a problem of Ulam. Acta Appl. Math. **62**, 23–130 (2000)
286. Rassias, Th.M.: On the stability of functional equations originated by a problem of Ulam. Mathematica **44**, 39–75 (2002)
287. Rassias, Th.M.: On the stability of minimum points. Mathematica **45**, 93–104 (2003)
288. Rassias, Th.M.: Functional Equations, Inequalities and Applications. Kluwer Academic Publishers, Dordrecht/Boston/London (2003)
289. Rassias, Th.M.: Handbook of Functional Equations – Functional Inequalities. Springer, New York (2014)
290. Rassias, Th.M.: Handbook of Functional Equations – Stability Theory. Springer, New York (2014)
291. Rassias, Th.M., Brzdek, J.: Functional Equations in Mathematical Analysis – In Honor of S. M. Ulam. Springer, New York (2012)
292. Rassias, Th.M., Šemrl, P.: On the behavior of mappings which do not satisfy Hyers–Ulam stability. Proc. Am. Math. Soc. **114**, 989–993 (1992)
293. Rassias, Th.M., Šemrl, P.: On the Hyers-Ulam stability of linear mappings. J. Math. Anal. Appl. **173**, 325–338 (1993)
294. Rassias, Th.M., Shibata, K.: Variational problem of some quadratic functionals in complex analysis. J. Math. Anal. Appl. **228**, 234–253 (1998)
295. Rassias, Th.M., Tabor, J.: What is left of Hyers–Ulam stability? J. Nat. Geom. **1**, 65–69 (1992)
296. Rassias, Th.M., Tabor, J.: Stability of Mappings of Hyers–Ulam Type. Hadronic Press, Palm Harbor (1994)
297. Rassias, Th.M., Toth, L.: Topics in Mathematical Analysis and Applications. Springer, New York (2014)
298. Rätz, J.: On approximately additive mappings. In: Beckenbach, E.F. (ed.), General Inequalities 2. International Series of Numerical Mathematics, vol. 47, pp. 233–251. Birkhäuser, Basel (1980)
299. Ravi, K., Thandapani, E., Senthil Kumar, B.V.: Solution and stability of a reciprocal type functional equation in several variables. J. Nonlinear Sci. Appl. **7**, 18–27 (2014)
300. Rolewicz, S.: Metric Linear Spaces. PWN-Polish Scientific Publishers/Reidel, Dordrecht (1984)
301. Saadati, R., Sadeghi, Gh.: Approximate homomorphisms and derivations in proper JCQ*-triples via a fixed point method. Expo. Math. **31**, 87–97 (2013)
302. Saadati, R., Sadeghi, Gh., Rassias, Th.M.: Approximate generalized additive mappings in proper multi–$CQ^*$–algebras. Filomat **28** , 677–694 (2014)
303. Saadati, R., Cho, Y.J., Vahidi, J.: The stability of the quartic functional equation in various spaces. Comput. Math. Appl. **60**, 1994–2002 (2010)
304. Saadati, R., Rassias, Th.M., Cho, Y.J.: Approximate $(\alpha, \beta, \gamma)$-derivation on random Lie $C^*$-algebras. RACSAM **109**, 1–10 (2015)
305. Sahoo, P.K., Riedel, T.: Mean Value Theorem and Functional Equations. World Scientific Publishing Company, Singapore/River Edge/London/Hong Kong (1998)
306. Schwaiger, J.: Remark 12. Report on the 25th international symposium on functional equations. Aequationes Mathematicae, Hamburg-Rissen, Germany 35, 120–121 (1985)
307. Šemrl, P.: On Jordan *-derivations and an application. Colloq. Math. **59**, 241–251 (1990)
308. Šemrl, P.: Quadratic functionals and Jordan *-derivations. Stuidia Math. **97**, 157–163 (1991)
309. Sewell, G.L.: Quantum Mechanics and Its Emergent Macrophysics. Princeton University Press, Princeton/Oxford (2002)

310. Shilkret, N.: Non-Archimedian Banach algebras. Ph.D. thesis, Polytechnic University, ProQuest LLC (1968)
311. Shulman, E.V.: Group representations and stability of functional equations. J. Lond. Math. Soc. **54**, 111–120 (1996)
312. Skof, F.: Proprietà locali e approssimazione di operatori. Rend. Sem. Mat. Fis. Milano **53**, 113–129 (1983)
313. Smital, J.: On Functions and Functional Equations. Adam Hilger, Bristol/Philadelphia (1988)
314. Streater, R.F., Wightman, A.S.: PCT, Spin and Statistics and All That. Benjamin, New York (1964)
315. Sundaresan, K.: Orthogonality and nonlinear functionals on Banach spaces. Proc. Am. Math. Soc. **34**, 187–190 (1972)
316. Takhtajan, L.: On foundation of the generalized Nambu mechanics. Commun. Math. Phys. **160**, 295–315 (1994)
317. Thirring, W., Wehrl, A.: On the mathematical structure of the B.C.S.-model. Commun. Math. Phys. **4**, 303–314 (1967)
318. Trapani, C.: Quasi-∗-algebras of operators and their applications. Rev. Math. Phys. **7**, 1303–1332 (1995)
319. Trapani, C.: Some seminorms on quasi-∗-algebras. Studia Math. **158**, 99–115 (2003)
320. Trapani, C.: Bounded elements and spectrum in Banach quasi ∗-algebras. Studia Math. **172**, 249–273 (2006)
321. Ulam, S.M.: A Collection of Mathematical Problems. Interscience Publishers, New York (1960)
322. Ulam, S.M.: Problems in Modern Mathematics. Wiley, New York (1960)
323. Upmeier, H.: Jordan Algebras in Analysis, Operator Theory, and Quantum Mechanics. Regional Conference Series in Mathematics No. 67. American Mathematical Society, Providence (1987)
324. Urs, C.: Ulam-Hyers stability for coupled fixed points of contractive type operators. J. Nonlinear Sci. Appl. **6**, 124–136 (2013)
325. Vainerman, L., Kerner, R.: On special classes of $n$-algebras. J. Math. Phys. **37**, 2553–2565 (1996)
326. Villena, A.R.: Derivations on Jordan–Banach algebras. Studia Math. **118**, 205–229 (1996)
327. Wang, G., Zhou, M., Sun, L.: Hyers-Ulam stability of linear differential equations of first order. Appl. Math. Lett. **21**, 1024–1028 (2008)
328. Wang, Z., Li, X., Rassias, Th.M.: Stability of an additive-cubic-quartic functional equation in multi-Banach spaces. Abstr. Appl. Anal. **2011**, Article ID 536520, 11 pp. (2011)
329. Wyrobek, W.: Orthogonally additive functions modulo a discrete subgroup. Aequ. Math. **78**, 63–69 (2009). Springer, New York (2009)
330. Zariski, O., Samuel, P.: Commutative Algebra. Van Nostrand, Princeton (1958)
331. Zeidler, E.: Nonlinear Functional Analysis and Its Applications I: Fixed-Point Theorems. Springer, New York (1986)
332. Zettl, H.: A characterization of ternary rings of operators. Adv. Math. **48**, 117–143 (1983)

# Index

© Springer International Publishing Switzerland 2015
Y.J. Cho et al., *Stability of Functional Equations in Banach Algebras*, DOI 10.1007/978-3-319-18708-2

Printed in the United States
By Bookmasters